ΔΕΔΟΜΕΝΑ

Euclid's *Data*

or

The Importance of Being Given

To Ulvemor (Mother Wolf)
sine qua non

ACTA HISTORICA SCIENTIARUM NATURALIUM ET MEDICINALIUM
Edidit Biblioteca Universitatis Hauniensis • Vol. 45

ΔΕΔΟΜΕΝΑ

Euclid's *Data*

or

The Importance of Being Given

The Greek Text
translated and explained

by

CHRISTIAN MARINUS TAISBAK

Museum Tusculanum Press
University of Copenhagen
2003

© 2003 Museum Tusculanum Press & Chr. Marinus Taisbak
Cover design and graphics by Henrik Maribo
Lay-out and typing by C.M. Taisbak
Printed in Denmark by Narayana Press, Gylling
ISBN 87 7289 815 1
ISSN 0065 1311

Cover illustration:
Data 56 & similia

Published with financial support from
The Danish Research Council for the Humanities
Velux Fonden

Museum Tusculanum Press
Njalsgade 92
DK-2300 Copenhagen S
Denmark
www.mtp.dk

CONTENTS

Preface 7

Notation and Abbreviations 10

Introduction 13

Chapter 1. Definitions 17
    Definitions 1–4: The Foundation 17
        The Plane 19
        Three questions and one 19
        Six parallel statements 21
        Latent Co-actors 24
        Mathematicians Unemployed or Admitted Tools ? 27
        The Helping Hand 28
        By geometrical methods 29
        Magnitude 30
        Starting Points 30
    Definitions 5–8: Circles 33
    Definitions 9–12: Magnitudes by a Given Greater. 34
    Definitions 13–15: Lines and Direction 35

Chapter 2. Magnitudes and Ratio I. Dt 1–9 37
    Dt 1–4. Four Axiomatic (?) Fundamentals 37
    Dt 5–9. Metamorphoses of Ratios 44

Chapter 3. By a Given Greater than in Ratio. Dt 10–21 57
    Definition 11 57
    Synopsis of Dt 10–21 59

Chapter 4. Magnitudes and Ratio II. Dt 22–24 85

Chapter 5. Position. Distance, Direction, Parallels. Dt 25–38 93

Chapter 6. Form. Triangles and Polygons. Dt 39–55 115
    Definition 3 115
    Dt 55 The Main Theorem? 116
    Triangles 118

Chapter 7. Equiangular parallelograms I. Reciprocal proportion. Dt 56 147

Chapter 8. Application of areas I. Dt 57–62 151

Chapter 9. Ratio and Angles. Dt 63–67 165

Chapter 10. Equiangular parallelograms II. Reciprocal proportion. Dt 68–75 .. 177

Chapter 11. Duplicates and Outsiders. Dt 76–83 191

Chapter 12. Application of areas. II. Dt 84–85 207

Chapter 13. Intersecting Hyperbolas. H. G. Zeuthen and Dt 86 211

Chapter 14. Circles. Dt 87–94 225

Appendix A. Marinus' Commentary 241

Appendix B. Synopsis of Dt 39 and I.22 250

List of Definitions and Enunciations 253

Select Bibliography 267

Index 269

    Deductive Structure of Dt 1–9 55
    Deductive Structure of Dt 10–24 92
    Deductive Structure of Dt 25–38 114
    Deductive Structure of Dt 39–55 144
    Deductive Structure of Dt 56–67 175
    Deductive Structure of Dt 68–83 205
    Deductive Structure of Dt 84–94 238

Preface

With this book I present an interpretation of a work that kept me wondering for many years. Is it worthwhile to print another translation of Euclid's *Data* while there is a recent translation by George L. McDowell and Merle A. Sokolik (1993)? But for a few mistakes, which could be remedied by a single page of *errata*, their translation gives a fair impression of the *Data*. After reading it, however, one is left in the same bewilderment as that already expressed by the ancient commentators *Pappus* and *Marinus*: What is all this really about?

Therefore, what I defined as my main task was not to give another translation (which is here only for quick reference), but to disclose and explain what I think is the mathematical and philosophical content of the *Data*, – driven, not to say haunted, by two opposite judgements about Euclid's work:

"If no one had written Euclid's Data, I would never have missed it." (*Olaf Schmidt*, my teacher in Copenhagen, in a private conversation).

"The appropriate measure of the geometric researches conducted by Euclid and his contemporaries is to be sought not in the Elements, but in the Data .." (*Wilbur Richard Knorr* (1986, 102)).

Nor did I intend to present a new critical edition; after all, Heinrich Menge had a sound judgement, and little is gained by looking for variants in manuscripts outside his scope: there is, in fact, no doubt as to the content of the *Data* – the doubts are all about its interpretation. One might look for a different succession of theorems (which is bound to occur, concluding from the Arabic tradition), and some that Menge printed as alternative proofs might as well have been the 'original' ones. I will discuss that in my remarks to the actual propositions. But by and large I did not expect the benefit of a new recension to match the cost and time. Those who look for text variants will need Menge's edition, which is not so easy to come by in this century. I have used his apparatus constantly, but apart from minor deviations I follow his text.

The most recent commentary is by Clemens Thaer (Data 1962), a very brief and succinct interpretation in modern algebraic jargon, appended to his German translation. Needless to say (at least when you have read even part of *my* tale) I disagree with him at almost all points, so much so that one would think we were reading completely different texts. Thaer has some valuable references to the Arabic tradition.

My commentary is conceived as a continuous narrative, trying to give a coherent view of what was perhaps never really coherent. Being a so-called *internalist* I take the mathematics of the *Data* seriously as being worth analyzing, at the same time trying to be as much of a historian as the sources demand and my training allows me. I shall proceed as linearly as possible through the theorems of the *Data*; all along, I supply such lemmas as I find necessary for saving Euclid's reasonings or my own understanding. Some of the comments and explanations in what follows I have set forth earlier in my paper (Taisbak 1991). Since then I have paid more attention to the *Data* than to any other topic, and part of what I wrote then has been revised. And some (to me) difficult questions, which were left open, have to be answered, though a keen eye will still find loose ends.

Stylistic considerations and an overall sense of not being in the *Elements*, but wandering in a parallel world with *some* elements in the background raises the question whether, at any time of his career, Euclid the Στοιχειώτης, the writer of the *Elements*, was in fact the author of the treatise. And if he was not, who was? Perhaps Euclid of Alexandria did nothing to the work but collect and copy the theorems from older sources; that thesis may be the best explanation of the various layers in the *Data*. At least, no harm is done by thinking of Euclid as the editor.

I have tried to collect the propositions into groups with a common topic. Although this is not always possible, it is useful all the same: the grouping reveals repetitions and illogical separations, which suggest that the *Data* was not conceived as one coherent presentation of geometrical analysis, but rather is a compilation of theorems stated in the idiom of *Givens* – to which there is no natural ending. The author tried to conceal that fact by adding the formula ὅπερ ἔδει δεῖξαι, *quod erat demonstrandum*, at the end of theorem 94: surprisingly the only occurrence in the *Data* of that traditional full stop to propositions. So he, at least, thought that an end was reached.

Acknowledgements

After publishing my "Coloured Quadrangles" in 1982 I intended to do something similar to Euclid's *Data*, which had kept popping up on my desktop and in my mind for some ten years. Sandwiched between the incompatible attitudes of Wilbur Knorr and Olaf Schmidt I began to read the *Data* as a piece of serious mathematics without paying much attention to the secondary literature, trying to understand from scratch what is going on, though always with an eye in the Elements. In one of my rare lectures on Greek mathematics (I earned my living by teaching Latin grammar and prose composition) one bright student, Anders Hyldahl, made a commented translation into Danish - a useful work for both of us but still leaving us bewildered about the foundation and aim of the work.

I might have given up the project were it not for the invitation from Benno Artmann and Ian Mueller to participate in a small seminar in Oberwolfach in August 1990. That meeting and that group changed my intellectual life and made a homesick scholar, if not like, at least willing to bear travelling to meet them. Ian Mueller insisted that I published as much as I dared about the *Data* in the anthology *Peri ton Mathematon* in 1991, and Benno Artmann (in many ways my *alter ego*) kept me running by an intense correspondence that is rivalled only by that of Len Berggren. Both of them have keenly read and commented on my manuscript in its becoming, and Len undertook the burdensome task of reading carefully every single word and letter to save me from solecisms and logical blunders, and suggested many useful emendations and additions. My countryman and always loyal backer Jens Høyrup, with his usual nose for events, was also there in Oberwolfach, and then: that *enfant terrible* in our trade, Sabetai Unguru, whom until that day I had feared to encounter, but who at every possible occasion ever since has proved a faithful friend and encouraging colleague.

In short: participants in that group and many others to follow in the gatherings about ancient mathematics have formed in my mind the concept "Friends of the Profession" : not only sharing a common interest in deciphering the ideas of the ancients, but real friends in a present, not too promising world. One of them, Richard McKirahan, at the very last moment saved me as I was gaping helplessly at the commentary by Marinus Neapolitanus. I owe them my thanks for patiently waiting for the final outcome of my various attempts to initiate them into my interpretations of the *Data*. *Hypomeinantes* till the very end (S. Matthew 10.22)!

Lastly, I thank Providence for delaying my book by uprooting our household, *viris canibusque*, from "Villa Roberval" in the nothern precincts of Copenhagen (where, indeed, my wife and I had spent three eventful decades before the cataclysm of unbearable neighbours) and moving us far into the country of South Seeland. There we have found at the farm "Højgaard" the ideal surroundings for an active retirement, reminiscent of, although more pleasant and comfortable than, the smallholding of my childhood.

October 2002
Marinus III Hauniensis

Notation and Abbreviations

Def. 1	*Data*'s definition 1.
Dt 1–94	*Data*'s theorems are numbered Dt 1 – Dt 94.[1]
Θϵ 1–94	Numbers of theorems (Θεωρήματα) in the Greek text.
Dt 8*	My supplement to Dt 8 (if such a one occurs).
V.16	Proposition 16 in Book V of the *Elements*.
V.Def. 5	Definition 5 in Book V of the *Elements*.
C.N. 1	Common notion (κοινὴ ἔννοια) 1 in Book I of the *Elements*.
Axiom 0*	Axiom 0*, my supplement.
X => Y	If X then Y
∴	(And) therefore

RATIOS

A:B	The ratio of A to B.
A:B :: C:D	Translates "A is to B as C to D", meaning that the ratio A:B is the same as C:D or (in other words) A, B, C, and D are proportional (cf. *Elements* V.def.5).
A δ>: B	Denotes the relation defined in Def. 11, "A is by a given (magnitude) greater than to be in (a given) ratio to B" (cf. p. 57 ff.)

FIGURES

∠A	The angle with vertex A.
(AB,BC) or (ABC) or (AC)	the *parallelogram* contained by the lines AB and BC, in the Greek text often also denoted ABΓ or AΓ. The latter is the 'diagonal notation', very frequent in Greek geometrical texts, denoting squares, rectangles and parallelograms according to context. My parentheses are there to show that a *figure* and no line segment is meant.
⊏⊐AB,BC or ⊏⊐ABC or ⊏⊐(AC)	the rectangle contained by the lines AB and BC, in the Greek text often also denoted ABΓ or AΓ.
□AB	the square on the line AB.
□(AC)	the square ABCD, 'diagonal notation'.
△ABC	the triangle ABC

---

[1] The notation Dt allows me to forget that *Data* is plural in Latin, but singular when used about Euclid's book. If you must read the numbers aloud, *datum* is acceptable.

NOTATION AND ABBREVIATIONS

GIVENS
gvn.AB   the magnitude AB is given.[2]
gvn.A:B   the ratio of A to B is given.

While a ratio may simply be given, a geometric object (point, line, angle, figure) is given *in some respect*. That limitation is an important feature of the *Data*, with these predicates recurring over and over again, so that a shorthand notation will be useful. Thus, if a thing T is a geometric object, one of the following statements may hold:

pos.T   T is given in position (θέσει)
mag.T   T is given in magnitude (μεγέθει)
frm.T   T is given in form (εἴδει)

Transcription of Greek Letters in Diagrams

Α = A	Β = B	Γ = C	Δ = D	Ε = E	Ζ = Z	Η = H	Θ = Q
Ι = I	Κ = K	Λ = L	Μ = M	Ν = N	Ξ = X	Ο = O	Π = P
Ρ = R	Σ = S	Τ = T	Υ = U	Φ = F	Χ = G	Ψ = Y	Ω = W

In the Greek text the order of letters often seems to be quite arbitrary, the same line segment being called AB and BA indiscriminately in the same context. In my remarks and comments I try to be strictly alphabetical and consistent, to make things easier for a modern reader. Reviel Netz (1999, 85, result 2) thinks that the lettering follows the order in which the objects appear in the course of the proof.

The enunciations of the 94 propositions in the *Data* are listed on p. 253 ff. for a quick reference.

---

[2] This stenographical notation is chosen as the shortest possible. It should not convey any ideas of mathematical logic.

Introduction

Under the Name of *Givens*

Euclid's *Data* opens with the passive perfect participle δεδομένα (neuter plural), which means 'given'; its Latin form *data* remained the title in modern times; and even though I prefer to think of Euclid's work as 'Givens' (in view of the vulgarization of the word *data* in the era of computers) I might create too much confusion by changing its name. So *Data* it is.

In the *Data* Euclid proves deductively that if some items are given, some other items are also given, *into the bargain* so to speak. Many years ago, when first reading Ptolemy's *Almagest*, I wondered at the very strange use of the predicate 'is given', which is applied not only to the input of a problem, but also to the output (Ptolemy-Heiberg I:38):

> [The chords] AB and AC are both given in magnitude, measured in those units of which the diameter has 120; and let BC be joined. I say that BC is also given.

This assertion is translated by Manitius (Ptolemy-Manitius, 29, 14) into

> ... dass auch .. BC sich bestimmen lässt.

In the same chapter (I:42), Ptolemy says about the impossibility of trisecting an angle

> if a chord is given ..., the one that subtends a third of the same arc is in no way given by geometrical methods, διὰ τῶν γραμμῶν.

Again Manitius dodges and writes

> kann .. die das Drittel desselben Bogens unterspannende Sehne durch geometrische Konstruktion auf keinerlei Weise gefunden werden.

Manitius (and with him Toomer 70 years later, 'cannot *be found* by geometrical methods' (Ptolemy-Toomer, 54)) was convinced that hardly any contemporary mathematician would use the predicate 'is given' of what has been found or proved. But the Greeks did so, and that is one unfamiliar fact which must be understood, if Euclid's *Data* is to be assessed properly.

When I started to translate the *Data*, I found it very longwinded that a certain phrase kept popping up time and again, several times in every proposition: *if this item is given, that item is* also *given*. I decided to cancel all those *also*s and restore them only where they were absolutely necessary. But then I discovered that I was leaving out an essential feature of the *Data*: the Givens hang together in chains, the purpose of any proposition being to produce more links to them.

Mathematical Lacunas in the Opening of the *Elements*
The first item that presents itself to a novice in Greek mathematics is The Given Straight Line Segment of proposition I.1 of the *Elements*. Not only a segment, but The *Given* Segment AB. The problem is to construct on that segment an equilateral triangle; the trick consists in producing a point C which will serve as apex of the triangle. On the way there, the given segment is reproduced (twice), circles being the reproducing engines.

Another instance of reproducing a line segment is proposition I.2, which solves the problem, 'at a given point to place a line segment equal to a given one'. Let A be the given point and BC the given line segment. Thus it is required to place at the point A a line segment equal to the given segment BC.

Now the segment BC cannot simply be moved to the point A, as movements of given segments seem not to be allowed, [3] nor are there any vehicles to perform the movement. And who can tell that it should not be used for some purpose where it is? So BC is given *in position* and must stay there; another segment has to be provided and put in position at the point A, as made plain by the verb θέσθαι, to place, cognate with θέσις, position. I shall not go through the solution of I.2 (the building of Euclid's Copy Machine), but focus on the first sentence in the construction:

> From the point A to the point B let the straight line AB be joined; and on it let the equilateral triangle DAB be constructed.

But what we learned in I.1 was to construct a triangle on a *given* segment; and AB is not given explicitly. The point A is given, and we infer that the point B is given, although nothing is said to that effect. Euclid considers the endpoints B and C to be given because BC is given, and the line segment AB to be given because its endpoints are given. If these are not mathematical lacunas, they are at least noisy tacit assumptions.

---

[3] *Elements* I.4 being an apparent exception, but see *remark* on Dt 25, p. 94.

Tacit Assumptions

Some of the difficulties met in the *Data* are due to what is *not* said in that booklet. Many of the tacit assumptions may be classified as 'what everyone would know or understand' (*e.g.* Axiom 0*, *Any point or magnitude may be (taken and) appointed given*), only a few are really fatal omissions. That is why, I think, most readers of the *Data* never bothered about its foundation – everything looks so natural and self-evident. But precisely that attitude is bound to run into problems when it comes to *proving* statements about Givens. Quite the opposite complication confronts us when the author plunges into proving the improvable, *i.e.* statements whose axiomatic character seems indisputable to me (*e.g.* Dt 25).

Together, the two kinds of blemishes – Saccheri would have styled them *naevi* – reveal a mathematical and philosophical mind which is not as thorough as we are wont to believe that Euclid's was. After all, Euclid invented the fifth postulate, didn't he? One is tempted to ascribe (at least the foundational theorems and definitions of) the *Data* to someone else – or to Euclid as a young man. But if young Euclid wrote the piece, why did he stop so abruptly after Dt 94? He cannot have dropped dead, young Euclid, can he? Either his papyrus roll or his sources ran out at that point; and perhaps it was rather a question of style whether a theorem should go into the *Data* or into one of a few other geometric treatises, *e.g.* the *Loci* just to name one.

The Tool Box

The tacit assumptions will be treated at their due place. First, however, I present one item which is no real omission, rather an unmentioned prerequisite of historical significance, I call it by the name invented by Ken Saito:

> *The Tool Box:* Any proposition in the Elements I–VI may be used in the *Data*.

That does not necessarily imply that the editor's elements were those that we know as Euclid's. Wisdom called *Elements* was circulated before Euclid's edition. I have some doubt as to the editor's knowledge of Book V in its Eudoxan form; the treatment of ratio points to another (and therefore probably earlier) theory (cf. remarks on Def. 1 and Dt 1). The Tool Box, however, suggests a question: does any of the propositions in the *Elements* presuppose any theorem of the *Data*? *Elements* I.2 seems to do, as I hinted above; although it is more plausible that what looks to us as a *lacuna* was one of the tacit assumptions in the *Elements* which triggered the study of Givens. Perhaps the question is empty, to believe Wilbur Knorr's description of the *Data* (to which I more or less subscribe): The *Data* is a complement to the *Elements*, recast in a form more serviceable for the analysis of problems. ... The subject matter overlaps that of the *Elements*. ... Indeed, only in rare instances does the *Data* present a result without a parallel in the *Elements*, (Knorr 1986, 109), *cf.* also (Heath 1921, 421-22). In which sense it is a complement, and whether its form is more serviceable, may be discussed. In my remarks on Dt 41 I consider the idea of 'parallel results'.

Chapter 1. Definitions

Definitions 1–4: The Foundation

ὅρος 1   Δεδομένα τῷ μεγέθει λέγεται χωρία τε καὶ γραμμαὶ καὶ γωνίαι, οἷς δυνάμεθα ἴσα πορίσασθαι.

ὅρος 2   Λόγος δεδόσθαι λέγεται, ᾧ δυνάμεθα τὸν αὐτὸν πορίσασθαι.

ὅρος 3   Εὐθύγραμμα σχήματα τῷ εἴδει δεδόσθαι λέγεται, ὧν αἵ τε γωνίαι δεδομέναι εἰσὶ κατὰ μίαν καὶ οἱ λόγοι τῶν πλευρῶν πρὸς ἀλλήλας δεδομένοι.

ὅρος 4   Τῇ θέσει δεδόσθαι λέγονται σημεῖά τε καὶ γραμμαὶ καὶ γωνίαι, ἃ τόν αὐτὸν ἀεὶ τόπον ἐπέχει.

Def. 1   *Given in magnitude is said of figures and lines and angles for which we can provide equals.*

Def. 2   *A ratio is said to be given for which we can provide the same.*

Def. 3   *Rectilineal figures are said to be given in form if each [4] angle is given and the ratios of the sides to one another are given.*

Def. 4   *Given in position is said of points and lines and angles which always hold the same place.*

Remarks: [5]
The first four definitions are the basis of everything else in the *Data*. [6] Of those, Def. 2 stands alone because a given ratio is simply given, whereas a given point is given *in position*, and the so-called *magnitudes* (lines, angles, figures) can be given both *in position* and *in magnitude*. Rectilineal figures may even be given *in form*. To under-

---

[4] Literally 'one by one'. – I have voted for 'given in *form*', following (Heath 1956, II:254), although 'given in *shape*' would also convey the meaning.

[5] These are *collective remarks on* Def. 1–4. *Special remarks on* Def. 1 follow on p. 26; *on* Def. 2, p. 31; *on* Def. 3 & 4, p. 33.

[6] Def. 1 is invoked in 23, Def. 2 in 50 of the 94 propositions.

stand this limitation or splitting, which pervades the *Data*, we may for a start notice that the *Data is about geometry only*, not numbers (despite the fact that given numbers are legion in Books VII–IX of the *Elements*), and about *plane* geometry at that: the *Data* has a close symbiosis with Books I–VI, and I remind of *The Tool Box*, the tacit assumption that *any proposition in the Elements I–VI may be used in the Data* (see p. 15).[7]

The very first definition of the *Data* moves geometry from the thinking-shop (φροντιστήριον[8]) of the *Elements* to the workshop of the active mathematician, that is: to the class-room. The use of 'we' in the definitions is alien to Euclid's style; in the *Elements* no person is involved in constructions or proofs in any way, nor does anyone show up in the propositions in the *Data*, except the *I* in the assertion 'I say that ...' (λέγω ὅτι). But exactly the 'we' is the clue to the stage setting: we are the audience in the class-room of the mathematician at work, following what goes on on the blackboard (which, for authenticity's sake, we may call by the Greek word *pinax* irrespective of material).

There is no doubt that 'given' means what it means (*pace* Marinus Neapolitanus, see appendix A). That an object is given to us means that it is, in some relevant sense and scope, put at our disposal. In mathematics, however, the term *given* is used in an idiomatic way about conditions at the outset of a problem. Often the Given is, at the same time, thrust upon you, not to get rid of, so that you must use that object and obey that relation and no others. If we let ourselves be guided by what we expect to be the mathematician's meaning of 'given', some word like 'known' comes to our mind: What is wrong, then, with the word 'known'? Nothing, of course (and the mathematicians of medieval Islam used it in their analyses consistently), but Euclid for some reason or other did not use it.[9] In my comments I sometimes use the verb 'know' (reluctantly), because I can think of no better word to describe the information concerned. – What, then, are these definitions defining, if not the predicate 'given', for which no

---

[7] The *Data* stops a long way short of spatial problems. But then, such problems were reduced, if possible, to plane ones in Greek geometry. Nevertheless, the selective character of the propositions that occur in the Data rather points to an exercise book, not a systematic treatise.

[8] Aristophanes, *Clouds* 94.

[9] Your guess is as good as mine. Perhaps because Plato(nism) had problematized 'knowledge'; but rather, I think, because vital (numerical) properties of the items are *not* known even if the item is constructed, *e.g.* length and area. Cf. note 12. – In (Berggren/Van Brummelen 2000, 25 ff.) there is a presentation and discussion of 'Competing definitions of "known" ' in Islamic mathematics.

definition is needed? Def. 1 and 2 seem to claim that some other items are *also* given (besides those given in the obvious sense).

The Plane

We are supposed to understand what it means that an object is positioned in the Plane, according to what I call *Mueller's Single Plane Law* (Mueller 1981, 208): [10]

> Normally, when plane geometry is developed as an independent subject, it is taken for granted that all objects considered lie in a single plane which never has to be mentioned.

To stress that fact I write Plane with a capital P. The scene where the play of the *Data* takes place, is an abstraction from physical boards (πίνακες) where geometry may be illustrated. How do we know that there are any objects there at all to be considered? Again I may refer to Ian Mueller, who on a couple of pages (14–16, 27–29) succinctly describes the difference between Hilbert's and Euclid's treatment of geometrical *existence*. Let me confine myself to summarize Mueller's conclusions, which seem to fit the *Data* perfectly:

None of the postulates in Book I of the *Elements* provides means for establishing the existence of points; Euclid feels free to invoke points as he needs them (Axiom 0*). The Plane is supposed to be full of points, and one is free to choose among them. The same holds to a certain extent for *lines* and *line segments*. A line segment may be taken at random, that is: its endpoints may be chosen, and thanks to postulate 1 a straight line segment will join them. But then the next step will often be to show *by construction* that certain other line segments are there, brought about by the constructive operations defined in postulates 1, 2, and 3 in *Elements* I. To quote Mueller (p. 15):

> It is possible to see in Euclid's first three postulates something like an identification of existence and constructibility (as Zeuthen would have it). However, it is difficult to construe this identification as a matter of *conscious choice* (my emphasis) in the absence of an available alternative to it.

Three questions and one

As far as I understand the *Data*, questions of objects' existence are neither asked nor answered. What kind of questions, then, can be asked? I suggest three: Where is the

---

[10] Cf. scholium no. 3 (Menge 261): *The theory of Givens is supposed to be about objects lying in one plane, as* (is the case) *also in the first six books of the Elements.*

object? How great is it? What does it look like? About ratio only one is asked: Which ratio is this?

Where?

About points, only the first question can be asked: To know a point is to know its position. Def. 4 tells us that if it stays still at one and the same place, it is said to be *given in position*. The one interesting property of a point is its coordinates – and I use this anachronism on purpose to prepare our minds for the difficulties that arise from speaking of (given) positions of points without a coordinate system. In fact, the coordinate system cheats us; see p. 83.

How Great?

To know a magnitude is to know *how great* it is, to know its size. One difference between magnitudes and points, between entities with and without size, is that magnitudes can be compared, points cannot;[11] points are all alike, so what counts is *where* they are. Magnitudes can have equals and unequals; to know a magnitude is to know and recognize its equals. To tell how great an object is, one must *either* have some sort of measure *or* another object of the same kind to compare it with. The *Data* chose the latter alternative in accordance with the *Elements*:[12] If an equal object can be *provided*, the first object is said to be *given in magnitude* (Def. 1). Understanding the *Data* begins with an acceptable interpretation of that phrase. Something similar holds for ratio: to know a ratio is to know and recognize the same ratio. If the same ratio can be *provided*, the ratio at hand is said to be *given* (Def. 2).

What is it like?

This is the third question we may ask, in the case of rectilineal figures. The easiest way of telling that is to point to a similar figure; if such a one can be provided, the former is said to be *given in form*. Now, according to the Elements VI.Def. 1, similarity involves same ratios for corresponding sides, as well as equal angles, so we may look for such conditions instead, as demanded by Def. 3.

---

[11] Points have no size: *Elements* I, def. 1, σημεῖόν ἐστιν, οὗ μέρος οὐθέν, a point is that which has no part. But they do have positions: according to Proclus (Friedlein 95, 21) the Pythagoreans defined a point to be 'a monad having position as well', μονὰς προσλαβοῦσα θέσιν.

[12] Measuring is banished from Euclidean geometry, perhaps because of the incommensurability problem. Lines and areas can be *compared* and found to be equal or unequal.

CHAPTER 1. DEFINITIONS

Six parallel statements.
These questions and their connection with the concept *given* can be illustrated by presenting six 'parallel' statements from the *Elements* and the *Data*.

The first pair is about the 'How great' question and Def. 1, which can be interpreted as follows: *To say that A is given* (or: is a given magnitude, or: is given in magnitude) *is to say that there is* (?) *a magnitude C such that A = C.* Thus we get to know the size of A by knowing C. The idea is clear in Dt 3, which is 'proved' by appeal to C.N. 2:

Dt 3. *If any number of given magnitudes be added together, the magnitude composed of them will also be given.*
Instantiation: If gvn.A and gvn.B then gvn.A+B.

C.N. 2. *If equals be added to equals, the wholes are equal.*
Instantiation: If A = C and B = D then A+B = C+D.

The 'proof' of Dt 3 (which we shall evaluate in the comments on that proposition) says that since gvn.A means A = C, and gvn.B means B = D, and since A+B = C+D (C.N.2), therefore we may conclude gvn.A+B, because there is a magnitude C+D equal to A+B. Some would say that Dt 3 is but another way of expressing C.N. 2; I will come back to that in a moment when explaining the question mark after 'is'.

The second pair is about ratio and Def. 2, which I interpret as follows: *To say that the ratio A:B is a given ratio is to say that there is* (?) *a ratio D:E such that A:B and D:E are the same ratio.* Thus we get to know the ratio A:B by knowing D:E. An example is Dt 8, proved by appeal to V.22.:

Dt 8. [Magnitudes] *which have a given ratio to the same* [magnitude], *will also have a given ratio to one another.*
Instantiation: If gvn.A:B and gvn.B:C then gvn.A:C. [13]

V.22. *If there be any number of magnitudes whatever, and others equal to them in multitude, which taken together two and two are in the same ratio, they will also be in the same ratio* ex aequali (δι' ἴσου).
Instantiation: If A:B :: D:E and B:C :: E:F then A:C :: D:F.

The proof of Dt 8 says that since gvn.A:B means that there is a ratio D:E such that

---

[13] The *Data* uses symmetry of given ratio freely and tacitly: If gvn.C:B then gvn.B:C.

A:B :: D:E, and gvn.B:C means that there is a ratio E:F such that B:C :: E:F, and since (by V.22) A:C :: D:F, therefore we may conclude gvn.A:C, because there is a ratio D:F that is the same as A:C. Some would say that Dt 8 is but another way of expressing V.22; I will come back to that.

The third pair is about similarity: *To say that the triangle ABC is given in form is to say that there is (?) a triangle DEF such that the two are equiangular.* Def. 3 does not say that, but ensures givenness in form by means of Def. 1 and 2; however, in propositions proving that a triangle is given in form a similar triangle is brought in by construction. Thus Dt 40 is proved by appeal to VI.6.

Dt 40. *If a triangle have one given angle, and the sides about the given angle have a given ratio to one another, the triangle is given in form.*
Instantiation: If gvn.∡A and gvn.AB:AC then frm.△ABC.

VI.6. *If two triangles have one angle equal to one angle and the sides about the equal angles proportional, the triangles will be equiangular and will have those angles equal which the corresponding sides subtend.*
Instantiation: If ∡A = ∡D and AB:AC :: DE:DF
                  then △ABC is equiangular with △DEF.

Let us face a pertinent question: Which status do they have, the *extras* that are *provided* ('brought in', 'acquired', expressed by the verb πορίσασθαι, to purchase by some unmentioned operation, or by πεποιήσθω, 'let it be made')? What does the 'is' mean which I have noted by question marks? The obvious answer was hard to come by, for my part, but believe it or not: They are *given*.

What we have been taught are some concepts about given magnitudes, ratios and forms, but the very concept of *given* remains undefined (and undefinable, if I am not mistaken). Nor has the important concept of *position* been mentioned, although that will manifest itself as perhaps the most important one in the *Data*. Scrutiny of the propositions shows at least three kinds of 'Givens' in the *Data* (of which only the third is mentioned in the definitions): A geometric object is given

1) if it is given (or pointed to) by the *Master*;[14] so that it is possible, by Euclidean means, to provide its equal, or the same ratio, or a similar figure.
2) if it is shown to be equal to, or the same as, or similar to a given;
3) if it is *provided* as equal to or similar to or having a given ratio to a given.

---

[14] Please allow this broad title of the Giver.

CHAPTER 1. DEFINITIONS

Here the connection with the *Elements* becomes evident: *a geometric magnitude is given if it can be constructed by Euclidean means as defined and displayed in the Elements*. Let us look again at the First Pair (above): In C.N. 2 we know but that the magnitudes are respectively equal, whereas in Dt 3 they share the property of givenness in one (or more) of the types above. So Dt 3 is not simply another way of expressing C.N. 2. Similar remarks hold for the other pairs. More will be said about that; for the moment, however, the connection with the *Elements* can be further illustrated by answering another question: *What are Equalities Good for?* In Euclid's *Elements* three kinds of likeness play dominant roles:

1) *Equality of magnitudes*, e.g. congruent line segments, rectilineal figures of equal content, equal angles (a problematic kind of magnitudes).
2) *Sameness of ratios*: Two ratios may be 'the same' (say 15:24 and 40:64). They are said to be *representatives* of the same ratio (5:8), and taken together as a whole the four of them are also called a *proportion*, ἀναλογία, or said to be 'proportional', ἀνάλογον.[15] (No harm is done by using numerical examples.)
3) *Similarity of figures*: Two triangles or polygons may look alike, be similar. Book VI of the *Elements* (particularly VI.4–7) treats such figures thoroughly. Similarity can be conceived as a combination of equality (of angles) and ratio (of sides).

Def. 1 of the *Data* can be understood *via* equality of magnitudes. An example: In *Elements* I.6 we learn that if two angles in a triangle be equal to one another, two sides will also be equal to one another. Def. 1 ensures that if such a triangle is at hand (for some reason or other) and one of those sides is 'given', the other side is 'thrown in' as an extra 'given'. I do not know why Euclid chose not to use the term *prosdedoménon*.

To a modern mind, conversant with the 'equality sign' =, this means that if one side of an equality is known, the other side is also known. That seems to me to be the *raison d'être* of proving equalities: to get one item by means of another which is equal to it but easier (?) to come by. *Buy one, get one free*. A magnitude M that is not im-

---

[15] Ratios are spoken of like natural numbers: they may be οἱ αὐτοί 'the same', but are never said to be ἴσοι 'equal'; equality of magnitudes and sameness of ratios are quite different relations, and the Greeks seem to have been aware of that. In Dt 1 and 2, however, Euclid considered 'the same' as *clones*, (cf. footnote 32) which is quite anomalous. Cf. (Vitrac 1990, 1:502 ff.) 'Sur l'égalité'.

I use the term *representatives* about the several manifestations of one and the same ratio. Each representative has two terms, the *antecedent* (15) and the *consequent* (24).

mediately given in the obvious way, is proved to be given if there is another known magnitude N such that M = N.

The magnitude N must be known to make the equality work. *How* it is known is of lesser interest; it can be accessible in one of a couple of ways – handed out by the Master, constructed (vaguely called 'provided'), or simply taken (κείσθω, 'let it have been laid') – sometimes with the extra qualification or limitation 'given in magnitude'.

Def. 2 can be understood *via* sameness of ratios. It ensures that if one representative of the same ratio is 'given' ('provided' in some way and therefore at hand), any other representative is also 'given'.

Def. 3 can be understood *via* similarity of figures. It ensures that if the shape of one of a couple of similar figures is 'given', the other figure is also 'given in form'. As hinted above, there is some discrepancy between what Def. 3 says and how it is actually used; an explanation for that will be given when we discuss Dt 39 and the following propositions.

Def. 4 is more difficult to fathom. Apparently, no equality is at stake here; rather it is a question of uniqueness, of fixedness, a pledge that once an item is put in the Plane,[16] it will stay where it is put. Def. 4 implies intricate questions about the whereabouts of geometric items: where are they when or if they are not in the Plane? And the Plane itself, what kind of thing is that? The proofs involving Def. 4 (Dt 25–29) are among the weakest in the *Data* while being most important. In Dt 25 we will dive into the problem and dig out, not equality, but *assignment*, – a feature familiar to computer programmers – which will put *givenness in position* at the same conceptual level as *givenness in magnitude*.

Latent Co-actors
Def. 1–4 seem to be speaking of single objects, but our questions reveal that this impression is only superficial. In order to work, each of the definitions needs a co-actor, another magnitude or ratio or figure. This is pronounced in Def. 1 and 2, but is also necessary in Def. 3 (as will appear in the proofs). Only Def. 4 seems to define singles (points, lines, and angles) without co-actors, at the same time introducing methods and concepts which are alien to the Euclidean behaviour in the *Elements* and call for an unfamiliar understanding of the Plane and geometric objects. I postpone the interpretation of Def. 4 till we need it in Dt 25.

To find out what is said and what is unsaid in the *Data*, the following exposition may be helpful. Let w, x, y denote magnitudes, and z a point, all of them *under investigation*. Their status is at the outset undetermined, 'status' being used loosely to denote a property to be defined. Let a, b, f denote magnitudes, and p a point.

---

[16] Read about the capital P on p. 19.

The statement 'x is given in magnitude' means that 'there is a *given* magnitude a, and x is equal to a'.

The statement 'w is given in form' means that 'there is a *given* rectilineal figure f, and w is similar to f'.

The statement 'z is given in position' means that 'there is a *given* point p in the plane, and z coincides with p'.

The statement '$x{:}y$ is a given ratio' means that 'there is a *given* ratio $a{:}b$, and $x{:}y$ is the same ratio'.

These statements support the impression that 'given' is a primitive needing no definition. What is intended are some definitions of items being 'given in certain ways'. But do the statements do anything but shift the focus from the x, y, w, z to the a, f, b, p? What is gained by such a duplication? This might start an endless loop if we expect Euclid to be defining what it means to be 'given'. But we are saved because he is doing nothing of the kind. He defines that some objects are *also* given (in the said respects), besides ... Besides what? Besides those that are already given by the Master or ... taken. Let us make an axiom of this tacit assumption, the Axiom of Appointment:

Axiom 0* *Any point or magnitude in the Plane may be (taken and) appointed given.*

That is: it may be represented by an instantiation in the Plane. In practice, only *points* and *lines* are taken; by propositions from the *Elements* they will then produce any of the magnitudes that appear in the *Data*, and consequently we may draw any rectilineal figure and call it 'given'. This tacit allowance is of the utmost importance in the *Data*, used time and again from Dt 5 onwards. Why, then, is it not there explicitly? Because (I think) it was as obvious and self-evident to Euclid as the Single Plane concept; *the play of Givens simply presupposed the right to have something given* – or taken, if givers are scarce. What is important are the consequences of taking or giving: that some items are given in the bargain by equality.

Let us consider the statements 'x is equal to a', 'w is similar to f', 'z coincides with p', '$x{:}y$ is the same as $a{:}b$'. In statements about Givens, as we read them in the *Data*, the co-actors, $a, f, p, a{:}b$, are absent, and the relations are expressed by the statements 'given in magnitude', 'given in form', 'given in position', or simply 'given'. The co-actors are there only potentially, waiting in the wing, as it were. That means that we need not bother about sizes or shapes or coordinates (*i.e.* distances and directions); the mere possibility of providing these will suffice, and we may order them whenever we want them, that is, we may call for The Helping Hand (see p. 28).

Most moderns, when we are told that a certain item is *given in magnitude*, would look for constants, numbers and measure: *what* is its magnitude, its size? The simple statement is far from satisfactory and may be felt as lacking information; this suppression of distracting details is, however, rather practical if the purpose is to collect a lot of tools for geometrical analysis in a short treatise, and that seems to be one purpose of the *Data*.

It may be this feature that has lead some historians (*e.g.* Clemens Thaer) to compare the *Data* with formulae: when using a formula we profit from (someone) having done the cumbersome work (for us) once and for all. When using a *datum* we profit from having proved its givenness once and for all.

Most modern commentators import the latent co-actors when interpreting the *Data*, as if they were necessary in order to see the statements as masked 'equalities'? [17] Obviously, putting something into the text which is not immediately there may spoil the play, if in no other way than by making it into another play. We have no basis for assuming that Euclid thought of the latent constants in his proofs, and such an assumption often leads to absurdly complicated expressions; I will present some of these in my comments. Nevertheless, even I have found it helpful to introduce a few latent co-actors in Dt 10–21 to elucidate my explanations of the metamorphoses in those theorems; to ensure that they are not abused, I put them in braces {..}.

Remarks on Def. 1, πορίσασθαι, *to provide*:

Def. 1 (*Given in magnitude is said of figures and lines and angles for which we can provide equals*) does not define 'given' (which is a well-known predicate from colloquial language), but introduces the term 'in magnitude' and tells us that a geometric object can and will be *given in magnitude* if a certain condition is fulfilled: if two geometric objects are equal,[18] and it is possible to provide, πορίσασθαι, one of them, the other object is given in magnitude. To see what is implied by Def. 1, let us treat the definition as a *protasis* and make the corresponding *ekthesis*:

---

[17] Clemens Thaer treats the *Data* in that way. Zeuthen's interpretation of Dt 86 is a glaring example. Their idea was to translate the *Data* into a formal language familiar to the modern reader, but in fact they created a new genre. At best one might admit that this new genre puts us in a better position to see what the *Data* is not, *viz*. algebra.

[18] Used of figures 'equal' always means no more than 'of equal content', 'having the same area'. It does not mean 'congruent'. Cf. (Heath, 1956, I:327) comment on *Elem*. I.35 'Equality in a new sense'. About 'area' (Hartshorne, 40 f.) is valuable reading.

CHAPTER 1. DEFINITIONS

Let L be a straight line [segment],[19] and let it be possible to provide a straight line M equal to L. I say that L is given in magnitude.

What about M? Obviously M must have some property besides being equal to L or else L would do. And absurdly (*e.g.*) the legs of any isosceles triangle would be given (because they are equal).

The interpretation must focus on the procedure *to provide*. What does πορίσασθαι mean? It is vital to the interpretation of the *Data*, but for a start only, since it does not occur after Dt 5; then it is replaced by other verbs meaning 'make', 'bring about' (*e.g.* πεποιήσθω in Dt 6). Euclid obviously uses the word as being self-explanatory.[20] It seems to be imported from everyday market prose meaning 'procure', 'furnish', 'get (for oneself)': any kind of purchase without specification of method and currency. We may safely translate it by '*get* (something) *represented* (in the Plane)'. It is meant to comprise several specific operations, *e.g.* συνίστασθαι, 'put together', (the normal word for constructing a triangle), ἀγαγεῖν, 'draw' (a straight line), γράφεσθαι, 'describe' (a circle), ποιεῖν, 'make' or 'produce' (a ratio).

Mathematicians Unemployed or Admitted Tools ?

The main content of the concept *provide* seems (to me) to be that *mathematicians should not bother about 'making' or 'constructing' or 'creating' any of the given items*. To see (and accept) that interpretation it is necessary to revise one's ideas about constructions and tools in Greek mathematics. If it were true that 'the admitted tools were ruler and compasses', we should expect the words κανών and διαβήτης to appear in the *Elements*. In a certain sense Euclid and his contemporaries obeyed this law – but the attitude which it is essential to grasp as the most important prerequisite to understanding the *Data* is the following:

When mathematicians are doing geometry, describing circles, constructing triangles, producing straight lines, they are not really *creating* these items, but only *drawing pictures* of them. Plato describes them so vividly in the *Republic* VII, 527a6–b6:

---

[19] Throughout this book the meaning of *straight line* is *line segment* unless it is expressly called *infinite*.

[20] That the meaning of the cognate substantive πόρισμα is also cryptic to us, though it seems to have been self-evident to the Ancients, does not diminish our bewilderment; see (Mueller 1981, 180). Alexander Jones in his edition of Pappus, Collection Book 7 (Pappus-Jones, 547 ff.), gives a thorough treatment of Euclid's lost work Πορίσματα, leaving no hope for an easy interpretation. Marinus Neapolitanus uses ἐν πόρῳ "within our competence" (?) about possible constructions, cf. note 171.

'... this science is in direct contradiction with the language employed by its adepts. ... Their language is most ludicrous, though they cannot help it, for they speak as if they were doing something and as if all their words were directed towards action. For all their talk is of squaring and applying and adding and the like, whereas in fact the real object of the entire study is pure knowledge ... the knowledge of that which always is, and not of a something which at some time comes into being and passes away.'

The real play is played in the Realm of Intelligence. When Euclid in Book I of the *Elements* wrote (as Postulate 1)

Let it be demanded that a straight line be drawn from any point to any point,

I will understand it as

Whenever there are two points, there is also one [and only one] *straight line joining them,*

and we are permitted to behave accordingly, that is to conceive a picture of that line. (To exhibit it in practice, we would of course need a ruler, but that would be to work in another world than the Realm of Intelligence.) When he wrote (as postulate 3)

and [let it be demanded] *that a circle can be described with any centre and opening,*

I will understand it as

Whenever there is a point and an interval, there is also a circle having the point for centre and the interval as opening [of the compass, that is radius],[21]

and we can also conceive that circle; but to draw it in the picture would require some sort of compass, and that also would be to leave the Realm of Intelligence.

The Helping Hand

It may be appropriate to introduce *The Helping Hand*, a well-known factotum in Greek geometry, who takes care that lines are drawn, points are taken, circles described, perpendiculars dropped, etc. The perfect imperative passive is its verbal mask: 'Let a

---

[21] διαστήματι, 'stride, standing apart'; cf (Fowler & Taisbak 1999).

circle have been described with centre A and radius AB'; 'let it lie given' κείσθω δεδομένον. No one who has done the *Elements* in Greek will have missed it; never is there any of the commands or exhortations so familiar from our own class-rooms: 'Draw the median from vertex A', or 'If we cut the circle by that secant', or 'Let us add those squares together'. Always *The Helping Hand* is there first to see that things are done, and to keep the operations free from contamination by our mortal fingers.

There is no magic involved, though; *the Helping Hand* can do only such work as is warranted by postulates or propositions. Thus it can let circles be described by postulate 3 of Book I; by proposition I.1 it can let equilateral triangles be designed on given line segments; angles can be bisected thanks to I.10, whereas you cannot expect it to trisect an angle for you (see my remarks on Dt 29). Its main effect and interest is to keep *you* out of the doings. Greek geometry is not about 'what you can do', but about 'what can be done'.[22]

By geometrical methods, διὰ τῶν γραμμῶν.
I suggest that we understand the verb πορίσασθαι as meaning 'provide', 'furnish in a non-specified way' *by such axioms and methods as are ensured in the Elements*. Ptolemy gave the clue (cf. p. 13): a third of the same arc is in no way given *by geometrical methods'*, διὰ τῶν γραμμῶν. And I suggest that we accept the 'we' in Def. 1 and 2 as geometers and their audiences who watch what The Helping Hand hands out. That Def. 1, as well as the others, is meant as a biconditional will appear unambiguously from its use in the proofs. Accordingly we may rewrite it as

A figure or a line or an angle is given in magnitude if and only if we can provide its equal.

To avoid involving *ourselves* in the definition, I suggest a further 'amendment':

A figure or a line or an angle is given in magnitude if and only if it is possible to provide its equal.

But when and how is it possible to provide an equal? From Dt 39 onwards Euclid time and again uses an unstated argument, which must be the real meaning of Def. 1:

Def. 1* *A figure or a line or an angle is given in magnitude if and only if it is equal to one that is given.*

---

[22] Of course, this is still so without being so clearly expressed.

Whenever Def. 1 is invoked in a proof, def. 1* is meant, and the axiom of appointment (or assignment), Axiom 0* will save us from running out of Givens: *Any point or magnitude may be (taken and) appointed given.*

Magnitude

Geometric objects (except points) can be provided as 'given in magnitude' (μεγέθει, dative) just because they are 'magnitudes', μεγέθη.[23] The most significant property of a magnitude (cf. Aristotle *Metaphysics* 1020a) is that it is continuous, συνεχές, *i.e.* infinitely divisible; numbers (being discrete) are not magnitudes, nor are points, although they are very important objects. Points are explicitly ruled out by definition 1 of *Elements* I, 'A point is that which has no part', *i.e.* is indivisible.

The magnitudes considered in the *Data* are rectilineal figures (*choria*[24]), line segments, and (perhaps) angles. Angles are considered to be magnitudes in so far as they can be compared and are infinitely halvable; they may be equal, or one may be less than the other, often because one is part of the other, as in the *Elements* I.7. Care must be taken when angles appear in proportions (cf. Dt 2). As in Book V of the *Elements*, if a magnitude is not specified (as *e.g.* a triangle), it is represented by a line segment [25] as if what holds of a line segment will also hold of any magnitude. Sums and differences are represented naively as line segments in the obvious way.

Starting Points

Summing up our remarks on Def. 1, we may say that a line or an angle is given in magnitude

1) if it is given or taken at the outset, and therefore is at hand in the Plane; such a magnitude can be copied by procedures from the *Elements*.
2) if its equal is at hand, that is if there is an instantiation of it in the Plane.[26]

---

[23] Euclid uses the terms 'A is given in magnitude' and 'A is a given magnitude' indiscriminately, even in the same context, as *e.g.* in Dt 2. Some (Alexander Jones & *alii*) maintain that there is a difference: an angle can be given in magnitude, but is not considered a magnitude. I am not so sure of that; read my *remarks* on Dt 2.

[24] χωρία, not including circles, which are treated separately in Deff 5–8.

[25] With named endpoints if it is to take part in addition or subtraction, otherwise only marked with one letter. In the latter case there is little need of any figure at all.

[26] A standard illustration could be radii in a circle; they are given (postulate 3) if and only if one of them is given.

Whenever comments refer to Def. 1, Def. 1* is meant, if necessary with Axiom 0*.
3) Dt 2 renders a third criterion: A magnitude is given if it has a given ratio to a given magnitude.

To get started we may take or put a point or a line segment and call it 'given'. This can bear comparison with the *assignment* procedure in progamming language, often denoted by the syntax ':='. All Givens but those which are taken are deduced by definitions or propositions. But then, what does it mean to 'appoint' an item 'given'? That question is closely connected with another one: Where are geometric items when they are not given? Within the concept described above they are latent in the Plane waiting for appointment. In my remarks on Dt 25 (p. 96) you will see them coming from somewhere else.

◆

Remarks on Def. 2:
In Def. 1 two statements about a straight line L were considered, 'L is given', 'An equal to L can be provided', and it was laid down that the two are equivalent. – Def. 2 (*A ratio is said to be given for which we can provide the same*) seems to be doing something analogous for a ratio R: 'R is given' is equivalent to 'The same ratio as R can be provided'. But for an important reason the analogy is incomplete, since L is a magnitude to be handled in the Plane, whereas R is something impalpable; according to *Elements* V.Def. 3, λόγος is a σχέσις, a *habitus* or circumstance bound up with two magnitudes ('terms') of the same kind, but Def. 2 mentions no magnitudes.

However, without dwelling too long upon the meaning of the word we may safely agree that ratio is never there without its terms; to handle a ratio is always to handle at least one of its representatives (as defined below and in footnote 15). Therefore some (I, for one) allow themselves to consider a ratio as an equivalence class [27] of couples of magnitudes. Let us consider line segments in the Plane, denoted A, B, C, etc., and define

> The ratio of A to B (to be witten A:B) is a family of couples C,D such that the ratio C:D is the same as A:B according to *Elements* V.Def. 5 (or any other valid definition of ratio known to the Greeks). Every couple in the family is a *representative* of the ratio A:B. That A:B and C:D belong to the same family is denoted by A:B :: C:D.

---

[27] Because of the law of transitiviy (V.11). Reflexivity and symmetry were taken for granted by the ancients. To avoid objections about anachronisms let us speak of 'families' of couples.

Thus a ratio is always *magnitudinified*, represented and exhibited by a couple of homogeneous magnitudes. Therefore one would expect the concept of 'given ratio' to emerge immediately from that of given magnitudes, leading to a definition like

Def. 2X   *A ratio is given if its terms are given magnitudes.*

Whenever two [28] comparable magnitudes appear, they 'have' a ratio, and if such a ratio can be said to be given, one would expect it to be so particularly when the magnitudes are given. Nevertheless, Euclid has 2X as a theorem, Dt 1.

In fact, Def. 2 says less than 2X, but more than the Greek text: The arguments in Dt 1 and 2 show that the definitions are meant as biconditionals; Def. 2 should be read

Def. 2Y   *A ratio is given if and only if it is possible to provide the same.*

Let the ratio be represented by A:B, A and B being homogeneous magnitudes, say line segments. We will elucidate Def. 2 by the following bipartition:

1) A:B is given if there are *given* homogeneous magnitudes C and D such that A:B :: C:D (according to some definition of same ratio, *e.g. Elements* V.5).
2) If A:B is given then there are *given* homogeneous magnitudes C and D such that A:B :: C:D.

Dt 1 uses 1), Dt 2 uses 2). The said propositions show that Def. 2 depends as heavily as Def. 1 on the operation to 'provide', that is to furnish in a non-specified way by axioms and methods ensured in the *Elements*. In Dt 5 and *passim*, Euclid demonstrates what it means to provide the same ratio, by explicitly providing the same ratio *in given terms*: both the antecedent and the consequent provided are magnitudes given according to Def. 1. So Def. 2 should be strengthened to

Def. 2Z   *A ratio is given if it is possible to provide the same with given terms.*

---

[28] The widespread tendency to 'translate' a ratio into a single letter (often a Greek one), probably identifying it with a real number, is misleading if the goal is to represent Greek ideas. – In treatments of the *Elements* there is some controversy about the status of ratios: is there a two-place or only a four-place relation? Cf. (Mueller 1981, 66 and 118). In the *Data* there is no doubt: a ratio A:B is an individual item, however impalpable. A and B are magnitudes, most often (and least problematically) understood to be line segments; one may think of them as *positive real numbers*, that is as *lengths* of line segments, while remembering that the Greek geometers could not think like that, for want of such numbers.

Def. 2 is often used in a stronger form, implied by Def. 2Z; whenever Def. 2 is referred to in comments, the following is meant, and Axiom 0* will save us from circularity:

Def. 2* *A ratio is given if it is the same as a given ratio.*

◆

Remarks on Def. 3
Def. 3 (*Rectilineal figures are said to be given in form if each angle is given and the ratios of the sides to one another are given*) is in no way difficult if Def. 1 and 2 are understood. Among the first four definitions it has the closest affinity to the *Elements*, cf. VI.def.1, *Similar rectilineal figures are such as have their angles severally equal and the sides about the equal angles proportional.* When first needed, however, in Dt 39 (which is somewhat weakly proved), it seems to be supplanted by two un-proved lemmas involving position:

Def. 3*A' *A triangle is given in form if its vertices are given in position.*

Def. 3*B' *A triangle is given in form if it is similar to one that is given in form.*

3*A' saves 3*B' from circularity by shifting the burden to Def. 4. We know what a figure looks like if we can point to a similar figure (p. 20). More will be said about that from Dt 39 onwards (p. 115).

◆

Remarks on Def. 4:
Def. 4 (*Given in position is said of points and lines and angles which always hold the same place*) is applied in Dt 25–30: it is quite enigmatic, probably empty: an object is where it is. The 'place-holding' idiom is not used in the proofs, but replaced by an argument involving some kind of *movement*, which cannot but raise difficult questions. Read about μεταπίπτειν in the remarks on Dt 25.

Definitions 5–8: Circles

ὅρος 5   Κύκλος τῷ μεγέθει δεδόσθαι λέγεται, οὗ δέδοται ἡ ἐκ τοῦ κέντρου τῷ μεγέθει.

ὅρος 6   Τῇ θέσει δὲ καὶ τῷ μεγέθει κύκλος δεδόσθαι λέγεται, οὗ δέδοται τὸ μὲν κέντρον τῇ θέσει, ἡ δὲ ἐκ τοῦ κέντρου τῷ μεγέθει.

ὅρος 7   Τμήματα κύκλων τῷ μεγέθει δεδόσθαι λέγεται, ἐν οἷς αἱ γωνίαι δεδομέναι εἰσὶ καὶ αἱ βάσεις τῶν τμημάτων τῷ μεγέθει.

ὅρος 8   Τῇ θέσει δὲ καὶ τῷ μεγέθει τμήματα δεδόσθαι λέγεται, ἐν οἷς αἵ τε γωνίαι δεδομέναι εἰσὶ τῷ μεγέθει καὶ αἱ βάσεις τῶν τμημάτων τῇ θέσει καὶ τῷ μεγέθει.

Def. 5   *A circle is said to be given in magnitude if its radius is given in magnitude.*

Def. 6   *And a circle is said to be given in position and in magnitude if its centre is given in position and its radius in magnitude.*

Def. 7   *Segments of circles are said to be given in magnitude if the angles in them and the bases of the segments are given in magnitude.*

Def. 8   *And segments of circles are said to be given in position and in magnitude if the angles in them are given in magnitude and the bases of the segments are given in position and in magnitude.*

Remarks: Def. 5–8 are about circles; not being χωρία (*choria* in the *Data* always means 'rectilineal figures') and therefore not covered by Def. 1 they must have their own definitions. They are used in Dt 87–94 only. The predicate *given in form* does not apply to circles, because they are all similar. The particle δὲ in the Greek text shows that 5 and 6 are thought of as being one definition; the same holds for 7 and 8. – Segments, τμήματα, are defined in *Elements* III, Def. 6, *A segment of a circle is the figure contained by a straight line* (the 'base') *and an arc of a circle*. In def. 7 and 8 Euclid considers circles to be areas (as in the *Elements*), whereas in def. 5 and 6 they may also be lines.

◆

Definitions 9–12: Magnitudes by a Given Greater

ὅρος 9   Μέγεθος μεγέθους δοθέντι μεῖζόν ἐστιν, ὅταν, ἀφαιρεθέντος τοῦ δοθέντος, τὸ λοιπὸν τῷ αὐτῷ ἴσον ᾖ.

ὅρος 10   Μέγεθος μεγέθους δοθέντι ἔλασσόν ἐστιν, ὅταν, προστεθέντος τοῦ δοθέντος, τὸ ὅλον τῷ αὐτῷ ἴσον ᾖ.

Def. 9   *A magnitude [M] is by a given [G] greater than a magnitude [N] if, when the given magnitude [G] be subtracted [from M], the remainder is equal to the same*[N].

Def. 10   *A magnitude [M] is by a given [G] less than a magnitude [N] if, when the given magnitude [G] be added [to M], the sum is equal to the same [N].*

CHAPTER 1. DEFINITIONS 35

Remarks: Def. 9 and 10 deal with given differences; only one proposition (but worth studying), Dt 12, applies Def. 9. I have supplied individual names M and N to make things understandable, though it is quite unGreek.

◆

ὅρος 11   Μέγεθος μεγέθους δοθέντι μεῖζόν ἐστιν ἢ ἐν λόγῳ, ὅταν, ἀφαιρεθέντος τοῦ δοθέντος, τὸ λοιπὸν πρὸς τὸ αὐτὸ λόγον ἔχῃ δεδομένον.

ὅρος 12   Μέγεθος μεγέθους δοθέντι ἔλασσόν ἐστιν ἢ ἐν λόγῳ, ὅταν, προστεθέντος τοῦ δοθέντος, τὸ ὅλον πρὸς τὸ αὐτὸ λόγον ἔχῃ δεδομένον.

Def. 11   *A magnitude is by a given greater than in ratio to a magnitude if, when the given magnitude be subtracted, the remainder has a given ratio to the same.*

Def. 12   *A magnitude is by a given less than in ratio to a magnitude if, when the given magnitude be added, the sum has a given ratio to the same.*

Remarks: Those two definitions will be treated in Chapter 3 with Dt 10–21, from p. 57.

◆

Definitions 13–15: Lines and Direction

ὅρος 13   Κατηγμένη ἐστιν ἡ ἀπὸ δεδομένου σημείου ἐπὶ θέσει εὐθεῖαν ἀγομένη εὐθεῖα ἐν δεδομένῃ γωνίᾳ.

ὅρος 14   Ἀνηγμένη ἐστιν ἡ ἀπὸ δεδομένου σημείου πρὸς θέσει εὐθείᾳ ἀγομένη εὐθεῖα ἐν δεδομένῃ γωνίᾳ.

ὅρος 15   Παρὰ θέσει ἐστὶν ἡ διὰ δεδομένου σημείου θέσει εὐθεία παράλληλος ἀγομένη.

Def. 13   *A straight line is dropped if it is drawn in a given angle from a given point to a straight line given in position.*

Def. 14   *A straight line is raised if it is drawn in a given angle from a given point on a straight line given in position.*

Def. 15   *A straight line is drawn parallel in position if it is drawn through a given point parallel to a straight line given in position.*

Remarks: Def. 13–15 are not used in the *Data*; a scholiast ascribed them to Apollonius, and the terms defined ('dropped', κατηγμένη, and 'raised', ἀνηγμένη) surely point in his direction. He uses παρά freely for παράλληλος ,– an abbreviation which is seen often in the *Data* as well as in the *Elements* (*e.g.* VI.2). That Apollonius was well-trained in using the *Data* is patent in the *Conica*, whenever he is to find or construct something (a good example is II.49, translated by McDowell/Sokolik p. 185–203). I do not believe that Apollonius wrote the *Data*: some parts of that booklet are far too primitive to suit him. But then, some parts are not.

◆

Chapter 2. Magnitudes and Ratio I
Dt 1–9

Dt 1–4. Four Axiomatic (?) Fundamentals.

Θε 1. Τῶν δεδομένων μεγεθῶν ὁ λόγος ὁ πρὸς ἄλληλα δέδοται.

ἔστω δεδομένα μεγέθη τὰ Α, Β· λέγω, ὅτι τοῦ Α πρὸς τὸ Β λόγος ἐστὶ δοθείς.
 ἐπεὶ γὰρ δέδοται τὸ Α, δυνατόν ἐστιν αὐτῷ ἴσον πορίσασθαι. πεπορίσθω καὶ ἔστω τὸ Γ. πάλιν, ἐπεὶ δεδομένον ἐστὶ τὸ Β, δυνατόν ἐστιν αὐτῷ ἴσον πορίσασθαι. πεπορίσθω καὶ ἔστω τὸ Δ. ἐπεὶ οὖν ἴσον ἐστὶ τὸ μὲν Α τῷ Γ, τὸ δὲ Β τῷ Δ, ἔστιν ἄρα ὡς τὸ Α πρὸς τὸ Γ, οὕτως τὸ Β πρὸς τὸ Δ· ἐναλλὰξ ὡς τὸ Α πρὸς τὸ Β, οὕτως τὸ Γ πρὸς τὸ Δ. τοῦ Α ἄρα πρὸς τὸ Β λόγος ἐστὶ δοθείς· ὁ αὐτὸς γὰρ αὐτῷ πεπόρισται ὁ τοῦ Γ πρὸς τὸ Δ.

Dt 1. *The ratio of given magnitudes to one another is given.*

Let A and B [29] be given magnitudes; I say that the ratio A:B is given.

For, since A is given, it is possible to provide a [magnitude] equal to it [Def. 1]. Let it have been provided, and let it be C. Again, since B is given, it is possible to provide a [magnitude] equal to it. Let it have been provided, and let it be D.

| A_____ | C_____ |
| B_____ | D_____ |

Then, since A = C and B = D, therefore A:C :: B:D. *Enallax* [30] A:B :: C:D [V.16]. Therefore the ratio A:B is given, for the ratio C:D has been provided the same as it [Def. 2]. ∎

---

[29] τὰ Α, Β. The definite article is normally left out in translations into English. One could spell it out: let the [magnitudes] A and B be given magnitudes.

[30] I borrow the term ἐναλλάξ instead of the traditional 'alternately'. See a survey of *metamorphoses* of proportions in the remarks to Dt 5.

Remarks: gvn.A & gvn.B => gvn.A:B.[31]

My first reaction to Dt 1 was 'How can such an obvious truth be proved?' and then 'Must it be proved? Is it not an axiom?' One would think that the ratio between given magnitudes must be given *par excellence*. I have voted for that point of view by Def. 2X (p. 32).

We interpreted Def. 2 as a biconditional saying that a necessary and sufficent condition for a ratio to be given is that (Def. 2*) *some representative is at hand in given terms*. But then Dt 1 is not a theorem, but simply another way of stating the definition: Let A and B be the given magnitudes; then, by Def. 1, they have equals C and D, which (by the stronger Def. 1*) must be *given* equals. Therefore A:B :: C:D (see comments below), and C:D is provided the same as A:B *in given terms* (so as to suit Def. 2*). But we had such a representative from the very outset, since A and B are given. Thus Dt 1 is not a theorem, but says the same as Def. 2*.

My claim is that Def. 1 and 2 must be sharpened as 1* and 2* to save
the concept 'given' – and they are supported by the doings in the Data.
Let us follow Euclid's 'proof' of Dt 1: After providing a ratio C:D which is not only another representative of the ratio A:B, but has its terms respectively equal [32] to the given magnitudes A and B, Euclid concludes immediately that A:C :: B:D, and commentators refer to *Elements* V.Def. 5, rightly in so far as the conclusion is in accordance with that definition; but the argument would be known under any theory of *logos*.

The next step uses the *enallax*-property (V.16) to get A:B :: C:D. *This is much too heavy ammunition* for proving facts that hardly need a proof; after all A:B :: C:D can be inferred along the lines of V.7 [33] by immediate recourse to any definition of same ratio; and as the *enallax*-theorem is rather difficult to prove (and probably must use arguments as V.7), this proof seems a bit supererogatory. Perhaps, however, it is unfair to adopt such minimalistic points of view if *The Tool Box* (p. 15) is open.

◆

---

[31] Read *If A is given and B is given then the ratio A:B is given*. This notation is meant to be stenographic, not to be understood as symbolic logic. In Dt 1–4 given magnitudes are denoted by capitals A, B, C, etc. From Dt 5 onward I denote undetermined magnitudes by late letters x, y, z, while given ones are early ones a, b, c, often g. In the spirit of the *Data* you had better see them as line segments, cf. note 28.

[32] We may call it a *clone* if A and B are line segments; areas can be equal without being 'clones', *i.e.* congruent; line segments cannot.

[33] Applied twice it says: If A = C then A:B :: C:B; if B = D then C:B :: C:D. Transitivity (V.11) ensures A:B :: C:D.

CHAPTER 2. MAGNITUDES & RATIO I. (1–9)

Θϵ 2. Ἐὰν δεδομένον μέγεθος πρὸς ἄλλο τι μέγεθος λόγον ἔχῃ δεδομένον, δέδοται κἀκεῖνο τῷ μεγέθει.

δεδομένον γὰρ μέγεθος τὸ Α πρὸς ἄλλο τι μέγεθος τὸ Β λόγον ἐχέτω δεδομένον· λέγω, ὅτι δέδοται καὶ τὸ Β τῷ μεγέθει.
 ἐπεὶ γὰρ δέδοται τὸ Α, δυνατόν ἐστιν αὐτῷ ἴσον πορίσασθαι. πεπορίσθω καὶ ἔστω τὸ Γ. καὶ ἐπεὶ δέδοται ὁ τοῦ Α πρὸς τὸ Β λόγος· {οὕτως γὰρ ὑπόκειται·} δυνατόν ἐστιν αὐτῷ τὸν αὐτὸν πορίσασθαι. πεπορίσθω καὶ ἔστω ὁ τοῦ Γ πρὸς τὸ Δ λόγος. καὶ ἐπεί ἐστιν ὡς τὸ Α πρὸς τὸ Β, οὕτως τὸ Γ πρὸς τὸ Δ, ἐναλλὰξ ἄρα ἐστὶν ὡς τὸ Α πρὸς τὸ Γ, οὕτως τὸ Β πρὸς τὸ Δ. ἴσον δὲ τὸ Α τῷ Γ· ἴσον ἄρα καὶ τὸ Β τῷ Δ· δέδοται ἄρα τὸ Β μέγεθος· ἴσον γὰρ αὐτῷ πεπόρισται τὸ Δ.

Dt 2. *If a given magnitude have a given ratio to some other magnitude, the other is also given in magnitude.* [34]

For, let the given magnitude A have a given ratio to some other magnitude B; I say that B is given in magnitude.
 For, since A is given, it is possible to provide a [magnitude] equal to it [Def. 1]. Let it have been provided, and let it be C. And since the ratio A:B is given – for that has been premised –[35] it is possible to provide the same as it [Def. 2]. Let it have been provided, and let it be the ratio C:D.

A _____   C _____
B _____   D _____

And since A:B :: C:D, *enallax* A:C :: B:D [V.16]; and A = C; therefore B = D [see *remarks*]. Therefore the magnitude B is given, for the [magnitude] D has been provided equal to it [Def. 1]. ∎

Remarks: gvn.A:B & gvn.A => gvn.B (partial converse of Dt 1).
Suppose that the given magnitude is some line segment A. Since A is given, another line segment C equal to A can be 'provided' and therefore is given (Def. 1).[36] Is A given *in position*, then? Ever since Pappus it is customary to think of propositions

---

[34] μέγεθος, magnitude, is used, even (as here) in the same context, both in a concrete sense 'geometric magnitude' and (in the dative only) meaning 'with respect to magnitude', *i.e.* size, as opposed to 'in form' and 'in position'.

[35] This argument is missing in ms a, may be a gloss. It is not in his normal style.

[36] *Elements* I.1 shows how to provide a line equal to a given.

1–24 as treating magnitudes without position; but how could the geometer speak of τὸ A ('the magnitude A') without pointing to a line fixed in the Plane? Nor do I see any way of copying A (to provide C) if A is not in position. However, since *the actual position is not important*, and any position will do, A is not in a *given* position.

By hypothesis A has a given ratio to some line segment B; therefore a fourth proportional D can be provided such that A:B :: C:D. But the length of B is undetermined at the outset; this raises a pertinent question: can the ratio A:B be given without its actual terms being given? It certainly can if only one representative, one couple of terms is given. That is to say, since the ratio A:B is given, there are given magnitudes, say line segments P and Q in the given ratio, and A:B :: P:Q.

Not until now can D be found to make C:D the given ratio, *i.e.* C:D :: P:Q. P and Q are the first '*latent co-actors*' we meet in the *Data* (read about them p. 24). They must be there, since their job, to represent the given ratio, cannot be done by C and D. Are P and Q given in position, then? Again, since it does not matter where in the Plane they be put, we say no, although they must stay still in order to be copied. They do have positions, albeit they are unseen phantoms. I mention these puzzles in order to bring out the weakness of the concept 'given in position', which (to my view) is the weakest part of the *Data* while being of utmost importance.

The *assumption of the fourth proportional* is a heavy one and will not hold for all kinds of magnitudes, not even for all kinds of two-dimensional magnitudes, say angles (if we choose to treat angles as magnitudes [37]). If the magnitudes are line segments, however, no harm is done, as is proved in VI.12. But let me quote (Mueller 1981, 139):

> It seems clear that no Greek ever questioned the 'assumption of the existence of a fourth proportional', perhaps because the use was not noticed, but more probably because the existence of such a proportional to three given geometrical objects was considered obvious on the basis of intuitive ideas about continuity.

How far are we? We have P:Q :: C:D, and therefore (by some transitivity proposition, *e.g.* V.11) A:B :: C:D. The last step, which conludes that B = D since A = C, is taken, not by V.14, but by means of the *enallax* theorem (V.16), which (as in Dt 1) seems to be too strong a tool for establishing what will follow immediately from any definition of same ratio.

---

[37] Glen van Brummelen suggested in private correspondence how to use Dt 2 to 'prove' that it is possible to trisect any angle. We shall kill that idea in the remarks on Dt 29.

The discussion between Jean-Louis Gardies and Ken Saito [38] about *Elements* V.14 is instructive when we try to evaluate Dt 1 and 2. Let us recapitulate V.14, which is proved by strong means, V.8. [39] We quote the equality-part only, skipping what is proved about inequalities:

V.14 If A:B :: C:D and A = C then B = D.

For convenience, Ken Saito uses the name V.14a for the following proposition, which is stated nowhere, but is used implicitly in many proofs in the *Elements*; again we need only the equality part:

V.14a If A:B :: C:D and A = B then C = D.

The role of V.14 in the *Elements* V is restricted to proving the *enallax* property (V.16) and V.18;[40] in many appropriate cases (as Gardies points out) Euclid does not use V.14, but makes a *détour* applying 14a together with the *enallax* proposition.

Ken Saito proposes to understand this *détour* as belonging in an earlier stratum of the *Elements*, which did not include V.14 (because it did not include V.8, I suggest) and (consequently) had to prove the *enallax* theorem on another basis than Eudoxus' definition V.Def. 5. The proceedings in Dt 1 and 2 to my view corroborate Ken Saito's interpretation: in Dt 1 Euclid uses what amounts to the partial converse of V.14a,

V.14a part. conv. If A = B and C = D then A:B :: C:D

followed by the *enallax* proposition. In Dt 2 he invokes the *enallax* property, then applies V.14a. I conclude, with Ken Saito, that he had neither V.Def. 5 nor V.14 at his disposal.

The peculiarities in Dt 2 can be listed as follows:
1) To represent the given ratio A:B while B is undetermined the unmentioned (latent) magnitudes (line segments) P and Q must be introduced and put in position (though in no explicitly given position) in order to furnish (VI.12) the ratio C:D.
2) Transitivity of same ratio is assumed tacitly, but hardly noticed because P and Q are latent.
3) The *enallax* property is involved supererogatorily.
4) The fourth proportional is assumed, harmlessly as long as line segments are concerned.

◆

---

[38] *Revue d'Histoire des Sciences*, 1991, XLIV/3-4, 457-467; 1994, XLVII/2, 273-284.

[39] V.8, If A > B then A:C > B:C and C:B > C:A. It uses V.def.4 (and therefore in a sense 'Archimedes' axiom'); cf. (Mueller 1981, 139-144), and (Vitrac 1990, 85 ff.)

[40] V.18, If A:B :: C:D then (A+B):B :: (C+D):D.

Θε 3. Ἐὰν δεδομένα μεγέθη ὁποσαοῦν συντεθῇ, καὶ τὸ ἐξ αὐτῶν συγκείμενον δεδομένον ἔσται.

συγκείσθω γὰρ ὁποσαοῦν δεδομένα μεγέθη τὰ AB, ΒΓ· λέγω, ὅτι καὶ τὸ ἐκ τῶν AB, ΒΓ συγκείμενον τὸ AΓ δεδομένον ἐστίν.
ἐπεὶ γὰρ δέδοται τὸ AB, δυνατόν ἐστιν αὐτῷ ἴσον πορίσασθαι. πεπορίσθω καὶ ἔστω τὸ ΔE. πάλιν, ἐπεὶ δέδοται τὸ ΒΓ, δυνατόν ἐστιν αὐτῷ ἴσον πορίσασθαι. πεπορίσθω καὶ ἔστω τὸ EZ. ἐπεὶ οὖν ἴσον ἐστὶ τὸ μὲν AB τῷ ΔE, τὸ δὲ ΒΓ τῷ EZ, ὅλον ἄρα τὸ AΓ ὅλῳ τῷ ΔZ ἐστιν ἴσον· δέδοται ἄρα τὸ AΓ : ἴσον γὰρ αὐτῷ πεπόρισται τὸ ΔZ.

Dt 3. *If any number of given magnitudes be added together, the magnitude composed of them will also be given.*

For, let any number of given magnitudes AB and BC have been added together; I say that the [magnitude] AC composed of AB and BC is also given.

A   B   C D   E   Z

For, since AB is given, it is possible to provide a [magnitude] equal to it [Def. 1]. Let it have been provided, and let it be DE. Again, since BC is given, it is possible to provide a [magnitude] equal to it. Let it have been provided, and let it be EZ.

Then, since AB = DE and BC = EZ, the whole AC = the whole DZ [C.N. 2]. Therefore AC is given; for the [magnitude] DZ equal to it has been provided [Def. 1].
■

Remarks: gvn.P & gvn.Q => gvn.P+Q.

Dt 3 seems to be a *petitio principii*: gvn.P means that P is equal to a given R (which can be provided, that is: instantiated in the Plane); and gvn.Q means that Q is equal to a given S (which can be provided, that is: instantiated in the Plane); according to Common Notion 2, P+Q = R+S. Is R+S given? At least, it is (with some copying, perhaps) instantiated in the Plane as a continuous magnitude. But to conclude that because R and S are both given, their sum R+S is also given, is to conclude what we are supposed to prove. We would happily accept Dt 3 as an axiom.[41]

As in the *Elements*, 'any number' of objects means 'two' if two is enough to ensure

---

[41] Len Berggren objects: 'But Euclid does not say it is given, only that it is *provided*. It is you who seem to want to conflate these two notions (for the best of reasons, I admit, but still it is you and not Euclid).' But let me remind of my remarks on Dt 1, that Def. 1 and 2 must be sharpened as 1* and 2* to save the concept 'given'.

universality.[42] The magnitudes involved in this proposition (in fact, in Dt 1 – Dt 23) are meant to be general but (as in *Elements* V) illustrated by line segments; their 'sum' AC is represented in the normal Euclidean way by a concatenation of the two segments in one straight line,[43] and their difference in the next proposition are also represented in the naive way. This limitation can have no serious consequences for the theory as long as the magnitudes involved satisfy Common notion 2 and 3,[44] which say nothing about the actual making of wholes and differences.

Euclid probably reflected no longer on this matter than on the existence of the fourth proportional: a magnitude is a continuous and limited object; a line segment is a continuous and limited object; why should we hesitate to represent the one by the other, as long as we do not use any properties of line segments which are not shared by magnitudes in general? More remarkable is, in fact, the use of line segments to illustrate *numbers* in the arithmetical books of the *Elements*.

◆

Θε 4. ’Εὰν ἀπὸ δεδομένου μεγέθους δεδομένον μέγεθος ἀφαιρεθῇ, τὸ λοιπὸν δεδομένον ἔσται.

ἀπὸ γὰρ δεδομένου μεγέθους τοῦ ΑΒ δεδομένον μέγεθος ἀφῃρήσθω τὸ ΑΓ· λέγω, ὅτι τὸ λοιπὸν τὸ ΓΒ δεδομένον ἐστίν.
ἐπεὶ γὰρ δέδοται τὸ ΑΒ, δυνατόν ἐστιν αὐτῷ ἴσον πορίσασθαι. πεπορίσθω καὶ ἔστω τὸ ΔΖ. πάλιν, ἐπεὶ δέδοται τὸ ΑΓ, δυνατόν ἐστιν αὐτῷ ἴσον πορίσασθαι. πεπορίσθω καὶ ἔστω τὸ ΔΕ. ἐπεὶ οὖν ἴσον ἐστὶ τὸ μὲν ΑΒ τῷ ΛΖ, τὸ δὲ ΑΓ τῷ ΔΕ, λοιπὸν ἄρα τὸ ΒΓ λοιπῷ τῷ ΕΖ ἐστιν ἴσον· δέδοται ἄρα τὸ ΒΓ· ἴσον γὰρ αὐτῷ πεπόρισται τὸ ΕΖ.

Dt 4. *If a given magnitude be subtracted from a given magnitude, the remainder will be given.*

For, let the given magnitude AC have been subtracted from the given magnitude AB. I say that the remainder CB is given.

---

[42] B. Vitrac comments on this Euclidean *non*-induction (Vitrac 1990) I, 278, found in I.45 and *passim*.

[43] A 'sum' (συγκείμενον) of two segments may be defined as a line segment cut into two parts, either of which is congruent with either of the original segments.

[44] *Common notion* 2: if equals be added to equals, the sums are equal. *Common notion* 3: if equals be subtracted from equals, the remainders are equal.

A   C   B D   E   Z
─────────── ──── ── ───────── ────

For, since AB is given, it is possible to provide a [magnitude] equal to it [Def. 1]. Let it have been provided, and let it be DZ. Again, since AC is given, it is possible to provide a [magnitude] equal to it Let it have been provided, and let it be DE.

Then, since AB = DZ and AC = DE, the remainder BC is equal to the remainder EZ [C.N. 3]. Therefore BC is given; for EZ has been provided equal to it [Def. 1]. ∎

Remarks: gvn.P & gvn.Q => gvn.P–Q.
Another stenograph, gvn.P+Q & gvn.P => gvn.Q, would show that Dt 4 is a partial converse to Dt 3; as the ancients' rhetorical mathematics had no 'signs' for plus and minus, the symmetry of such statements tends to disappear. In our symbols it stands out clearly. Dt 4 suffers from the same logical defect as Dt 3, *petitio principii*, assuming that 'provided' means 'given' (Def. 1*).

In modern jargon the two would be: *Givenness in magnitude is preserved under addition and subtraction.* They are to the *Data* what Common Notions 2 and 3 are to the *Elements*, simply substituting 'given' for 'equal'. They are simpler statements than Dt 1 and 2, as they involve no ratios, so systematically Dt 3 and 4 could have been the first propositions of the *Data*. Dt 1 is invoked in 14, Dt 2 in 31, Dt 3 in 23, and Dt 4 in 29 out of the 94 propositions.

◆

Dt 5–9. Metamorphoses of Ratios

Theorems 5–9 can be viewed as constituting a series with the common hypothesis that a ratio of homogeneous magnitudes P:Q is given; whenever it matters, P is supposed to be greater than Q,[45] which condition is secured by making Q a part of P (in the sense of *Elements* Book I, C.N. 8).

The proofs involve some of the *metamorphoses* of ratios, defined in *Elements* V. Def. 12–16 and (partly) proved in V.16–18: how to get new proportions from given ones by interchanging or adding or subtracting the terms; I refer you to B. Vitrac's comments, particularly on V.Def. 12 (Vitrac 1990, II:50 f.) In the *Data* they are signalled by the Greek word (in the dative) for the operation or actor, which is often rendered by its Latin counterpart. I will stick to the Greek, so you may look to the following survey for a translation. We met the *enallax* metamorphosis in Dt 1 and 2.

---

[45] In the Greek theory of *logos*, the antecedent is often greater than the consequent; a slight evidence that *logoi* were not considered to be fractions. I follow Benno Artmann, who considers this as evidence for an earlier anthyphairesis definition.

From		P:Q :: R:S	
we get			
enallax	alternando	P:R :: Q:S	V.16
anapalin	invertendo	Q:P :: S:R	evident from V.Def. 5
synthenti	componendo	(P+Q):Q :: (R+S):S [46]	V.18

The following apply only if $P > Q$, and therefore (V.14) $R > S$:

dielonti	separando	(P–Q):Q :: (R–S):S	V.17
anastrepsanti,	convertendo	P:(P–Q) :: R:(R–S)	

From	P:Q :: R:S		
and	Q:U :: S:V	we get by 'skipping equal distances'	
di' isou, ex aequali	P:U :: R:V [47]		V.22

Addition and subtraction is associated with ratios by V.12, 19, and 24.

If P:Q :: R:S then (P+R):(Q+S) :: P:Q	V.12
If P:Q :: R:S then (P–R):(Q–S) :: P:Q	V.19
If P:Q :: R:S and U:Q :: V:S then (P+U):Q :: (R+V):S [48]	V.24

Dt 5 to 9 [49] prove the following statements about metamorphoses of a given ratio:

Dt 5	If gvn.P:Q then gvn.P:(P–Q)	*anastrepsanti*
Dt 6	If gvn.P:Q then gvn.(P+Q):P and gvn.(P+Q):Q	(*anapalin &*) *synthenti*
Dt 7	If gvn.P:Q and gvn.P+Q then gvn.P and gvn.Q	
Dt 8	If gvn.P:Q and gvn.R:Q then gvn.P:R	*di'isou* via *symmetry, compounding*
Dt 9	If gvn.P:Q:R:... and gvn.P:S, gvn.Q:T, gvn.R:U, ... then gvn.S:T:U: ...	
		(Givenness of ratio is *epidemic*)
Dt 22	If gvn.P:R and gvn.Q:R then gvn.(P+Q):R	
Dt 23	If gvn.(P+Q):(R+S) and gvn.P:R and gvn.Q:S then gvn.P:Q:R:S	

◆

---

[46] If transformations of *logoi* were algebra, we would be adding 1 to both sides; but it is quite unlikely that a Greek mathematician would associate this operation with the unit, μονάς. Clemens Thaer nevertheless does so in Dt 6 and elsewhere.

[47] V.22 proves an important property of compound ratios: Compounding is indifferent to the choice of representatives. No such truth is explicitly stated in the *Elements*, and Euclid did not exploit the idea of compounding so far as he might have; see (Saito1986).

[48] V.24 can be interpreted as having something to do with adding fractions that have a common denominator. But we should be wary of any interpretation that makes fractions out of *logoi*.

[49] Dt 22 and 23 should be included in this group; we will treat them in due course.

46 EUCLID'S *DATA*

Θε 5. ’Εὰν μέγεθος πρὸς ἑαυτοῦ τι μέρος λόγον ἔχῇ δεδομένον, καὶ πρὸς τὸ λοιπὸν λόγον ἕξει δεδομένον.

μέγεθος γὰρ τὸ ΑΒ πρὸς ἑαυτοῦ τι μέρος τὸ ΑΓ λόγον ἐχέτω δεδομένον· λέγω, ὅτι καὶ πρὸς τὸ λοιπὸν τὸ ΒΓ λόγον ἔχει δεδομένον.
κείσθω γὰρ δεδομένον μέγεθος τὸ ΔΖ. καὶ ἐπεὶ λόγος ἐστὶ δοθεὶς ὁ τοῦ ΒΑ πρὸς τὸ ΑΓ, ὁ αὐτὸς αὐτῷ πεπορίσθω ὁ τοῦ ΖΔ πρὸς ΔΕ. λόγος ἄρα ἐστὶν ὁ τοῦ ΖΔ πρὸς ΔΕ δοθείς. δοθὲν δὲ τὸ ΖΔ. δοθὲν ἄρα καὶ τὸ ΔΕ· καὶ λοιπὸν ἄρα τὸ ΕΖ δοθέν ἐστιν. {ἔστι δὲ καὶ τὸ ΔΖ δοθέν·} λόγος ἄρα τοῦ ΔΖ πρὸς τὸ ΖΕ δοθείς. καὶ ἐπεί ἐστιν ὡς τὸ ΔΖ πρὸς ΔΕ, οὕτως καὶ τὸ ΑΒ πρὸς ΑΓ, ἀναστρέψαντι ἄρα ἐστὶν ὡς τὸ ΔΖ πρὸς τὸ ΖΕ, οὕτως τὸ ΑΒ πρὸς τὸ ΒΓ. {λόγος δὲ τοῦ ΔΖ πρὸς ΖΕ δοθείς, ὡς δέδεικται}· λόγος ἄρα καὶ τοῦ ΑΒ πρὸς τὸ ΒΓ δοθείς.

Dt 5. *If a magnitude have a given ratio to some part of itself, it will also have a given ratio to the remainder.*

Let a magnitude AB have a given ratio to some part of itself AC. I say that it has also a given ratio to the remainder BC.

For, let a given magnitude DZ have been set out [50] [Axiom 0*].

A_____C_____B D____E____Z

And since the ratio BA:AC is given, let ZD:DE have been provided the same as it [Def. 2]. Then the ratio ZD:DE is given. And ZD is given; therefore DE is given [Dt 2]; therefore the remainder EZ is given [Dt 4]; {and DZ is given;} therefore the ratio DZ:ZE is given [Dt 1].

And since DZ:DE :: AB:AC, *anastrepsanti*[51] [DZ:(DZ–DE) :: AB:(AB–AC), that is] DZ:ZE :: AB:BC.

{And the ratio DZ:ZE is given, as has been proved.} Therefore the ratio AB:BC is also given [Def. 2*]. ∎

Remarks: gvn.x:y => gvn.x:$(x-y)$.
   Concordance: x = AB, y = AC, $x-y$ = CB, g = DZ, h = DE, $g-h$ = EZ.

---

[50] Literally 'let it have been laid given', κείσθω δεδομένον.

[51] V.19 corollary. The dative ἀναστρέψαντι means literally 'for the person who turns [the ratio] upside down'. English translations of Greek mathematics mostly use the Latin form *convertendo*, 'by turning around'. See the survey of these *metamorphoses* on the previous pages.

Dt 5 is the first proposition of some substance, and we may hope to improve our understanding of the *Data* by studying it carefully. Euclid begins his proof in a shocking way with the imperative κείσθω δεδομένον μέγεθος, 'let a given magnitude have been laid'. He is in need of a given magnitude, so he helps himself (or rather is helped by *The Helping Hand*, cf. p. 28) to a given line segment DZ, meant to represent a one-, two-, or three-dimensional magnitude. We infer that *if a given magnitude is needed, it can be conjured up with no ado*. The imperative *keistho* (or *ekkeistho*) means 'let it have been positioned', from τίθημι, cognate verb to *thesis*, position. We would be perfectly happy with the following license, *Any magnitude may be* (taken and) *appointed given*, that is: represented by an instantiation in the Plane. Probably Euclid considered Axiom 0* as a matter of course, (as did the editor of the Book I of the *Elements*, cf. above, p. 19), since 'taking' and 'appointing' are fundamental, undefined operations in the game of Givens.

The argument in Dt 5 supplies some extra Givens in order to be able to use *Elements* Book V. A given magnitude DZ (g) is *demanded* and split in the same ratio as AB (*e.g.* by VI.12);[52] we summarize the proof in our notation:

Take [Axiom *0]	gvn.g	1
Make [VI.12][53]	g:h :: x:y	2
[Def. 2*]	gvn.g:h	3
[Dt 2, Dt 4]	gvn.h, gvn.g–h	4
[Dt 1]	gvn.g:(g–h)	5
[2]	g:h :: x:y	6
[V.19 corr]	g:(g–h) :: x:(x–y)	7 *anastrepsanti*
[Def. 2*]	gvn.x:(x–y)	8

Euclid expands the given ratio x:y into a proportion (2) by 'providing' g:h as a representative of the ratio x:y (by means of *latent coactors*, see next paragraph). However, g:h is not only the same ratio, but the same ratio *in given terms*: since g is taken for given, h will also be given according to Def. 2. (That accounts for my supplement to Def. 2, 2*, 'A ratio is said to be given for which we can provide the same *in given magnitudes.*') The *anastrepsanti* metamorphosis (see survey on p. 44) leads from (2) to (7), thus providing a ratio g:(g–h) which is the same as x:(x–y). Since g:(g–h) is in given terms, the latter is also entitled to be called given.

---

[52] When the text demands a (third) given magnitude to be there and a fourth to be provided in a given ratio, I refer to VI.12 as the constructing engine. Of course that will hold for line segments only, but it can be made to work for rectangles too (VI.1), and thus for any rectilineal figure.

[53] VI.12 will serve for line segments and areas. Cf. Mueller, quoted on p. 40.

But how can he provide g:h in the same ratio as x:y if x and y are not given in magnitude? While the magnitudes DZ, DE, EZ are given (that is, either taken or proved by Dt 2 and 4), the status of AB, AC, and BC is undetermined. Only the ratio is given, not the magnitudes constituting it; they may be any sizes in the given ratio. How is that possible? Can a ratio be given without its terms being given? In order to find a fourth proportional (VI.12) there must be three *given* line segments. *Either* AB and AC are meant to be given in magnitude, *or* (more likely) a representative of the ratio AB:AC, say a ratio of given line segments U:V, must be lurking in the wing, materialized in line segments but not mentioned. The *Data* certainly employs more craft than is listed: *The Latent Co-actors.*

Dt 5 surprises by treating the *anastrepsanti* case, which is a complex conversion. In the *Elements* it is not proved, but merely mentioned as a porism to V.19; in terms of Book V it certainly takes some roundabout reasoning to prove it (see Vitrac 1990, II:113 f.). In the *Data* one would expect a simpler proposition to be the first, say the *anapalin* case: if gvn.P:Q then gvn.Q:P, but that one is used without proof as a matter of course. Also the *dielonti* case is used without proof (*e.g.* in Dt 10): If gvn.P:Q then gvn.(P–Q):Q.

However, from a certain point of view Dt 5 *is* simpler: The *anastrepsanti* conversion is among the first to present itself if these metamorphoses are applied to chords and harmonies. If (in figure 5.1) a string of length 4 is divided in the ratio 4:3 (to sound the concord of the fourth), it is at the same point divided in the ratio 4:1 (the double octave); thus the latter ratio is the first one turned 'upside down', an ἀναστροφή.

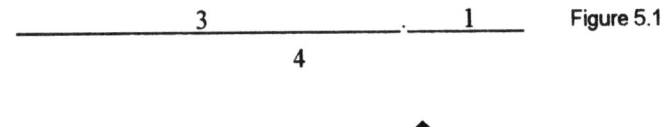

Figure 5.1

◆

Θε 6. ' Ἐὰν δύο μεγέθη συντεθῇ πρὸς ἄλληλα λόγον ἔχοντα δεδομένον, καὶ τὸ ὅλον πρὸς ἑκάτερον αὐτῶν λόγον ἕξει δεδομένον.

συγκείσθω γὰρ δύο μεγέθη τὰ ΑΓ, ΓΒ, πρὸς ἄλληλα λόγον ἔχοντα δεδομένον· λέγω, ὅτι καὶ ὅλον τὸ ΑΒ πρὸς ἑκάτερον τῶν ΑΓ, ΓΒ λόγον ἔχει δεδομένον.
 ἐκκείσθω γὰρ δεδομένον μέγεθος τὸ ΔΕ. καὶ ἐπεὶ λόγος ἐστὶ τοῦ ΑΓ πρὸς ΓΒ δοθείς, ὁ αὐτὸς αὐτῷ πεποιήσθω ὁ τοῦ ΔΕ πρὸς ΕΖ. ὁ ἄρα τοῦ ΔΕ πρὸς ΕΖ λόγος ἐστὶ δοθείς· δοθὲν δὲ τὸ ΔΕ· δοθὲν ἄρα καὶ τὸ ΕΖ· καὶ ὅλον ἄρα τὸ ΔΖ δοθέν ἐστιν. ἔστι δὲ ἑκάτερον τῶν ΔΕ, ΕΖ δοθέν· λόγος ἄρα τοῦ ΔΖ πρὸς ἑκάτερον τῶν ΔΕ, ΕΖ δοθείς. καὶ ἐπεί ἐστιν ὡς τὸ ΑΓ πρὸς ΓΒ, οὕτως τὸ ΔΕ πρὸς ΕΖ, συνθέντι ὡς τὸ ΑΒ πρὸς τὸ ΒΓ, οὕτως τὸ ΔΖ πρὸς ΖΕ· καὶ ἀναστρέψαντι ὡς τὸ ΑΒ πρὸς τὸ ΑΓ, οὕτως τὸ ΔΖ πρὸς ΔΕ. καὶ ἐπεὶ ὡς τὸ ΔΖ πρὸς ἑκάτερον τῶν ΔΕ, ΕΖ, οὕτως τὸ ΑΒ πρὸς ἑκάτερον τῶν ΑΓ, ΓΒ, λόγος ἄρα καὶ τοῦ ΑΒ πρὸς ἑκάτερον τῶν ΑΓ, ΓΒ δοθείς.

Dt 6. *If two magnitudes having a given ratio to one another be added together, the whole will also have a given ratio to each of them.*

For, let the two magnitudes AC, CB, having a given ratio to one another, have been added together; I say that the whole AB also has a given ratio to each of AC, CB.

For, let a given magnitude DE have been set out [Axiom 0*].

A   C   B D E Z

And since the ratio AC:CB is given, let DE:EZ have been made the same;[54] then the ratio DE:EZ is given [Def. 2*]; and DE is given; therefore EZ is also given [Dt 2], and so the whole DZ is given [Dt 3].

Each of DE, EZ is given; therefore [Dt 1] the ratio of DZ to each of DE, EZ is given. And since AC:CB :: DE:EZ, *synthenti* [(AC+CB):BC :: (DE+EZ):ZE, that is] AB:BC :: DZ:ZE [55] [V.18], and *anastrepsanti* [AB:(AB–BC) :: DZ:(DZ–EZ), that is] AB:AC :: DZ:DE [V.19].

And since, as DZ is to each of DE, EZ, so is AB to each of AC, CB, therefore the ratio of AB to each of AC, CB is given [Def. 2*]. ∎

Remarks: gvn.x:y => gvn.$(x+y)$:x & gvn.$(x+y)$:y.
    Concordance: x = AC, y = CB, $x+y$ = AB, g = DE, h = EZ, $g+h$ = DZ.

Dt 6 uses the *synthenti* and *anastrepsanti* metamorphoses, as defined above. After the *synthenti* metamorphosis he could have used his result from Dt 5, but in the context it was perhaps easier to repeat the *anastrepsanti* argument. Here is the proof of Dt 6 in our notation:

        gvn.x:y
Take [Axiom 0*] gvn.g
Make [VI.12, Def. 2*] g:h :: x:y, whence gvn.g:h

[54] From now on Euclid uses πεποιήσθω 'let have been made' (which is the best idiom I can think of, even if bad English), instead of πεπορίσθω; astonishingly neither term is seen in the *Elements* - as if the author of the *Data* were trained in another terminology than Euclid. Or did he simply follow his sources slavishly? In Dt 5 the Vatican ms (Menge's apparatus. p. 10, li. 11) has a nine letters' vaccuum beside πεπορίσθω – would that be πεποιήσθω wiped out?

[55] The dative συνθέντι means literally 'for the person who adds together'. The Latin counterpart is *componendo* (see the survey of metamorphoses).

[Dt 2, Dt 3]	gvn.h and gvn.$g+h$
synthenti	$g:h :: x:y \Rightarrow (g+h):h :: (x+y):y$
[Dt 1, Def. 2*]	gvn.$(x+y):y$
anastrepsanti	$(g+h):g :: (x+y):x$
[Dt 1, Def. 2*]	gvn.$(x+y):x$

As in Dt 5 we may ask for the status of AC (x) and CB (y). Their ratio is given, but they are not given in magnitude, for if they were, the proposition would follow immediately from Dt 3 and 1. To make up for that, two other magnitudes (straight lines) in the given ratio are imported, one of them (DE = g) taken, the other (EZ = h) provided by VI.12 to make the given ratio. Again, how is that ratio given? In order to provide it, a representative must be latent, say U:V :: AC:CB. The magnitude AB and its parts remain *not given in magnitude* throughout the proposition.

◆

Θε 7. Ἐὰν δεδομένον μέγεθος εἰς δεδομένον λόγον διαιρεθῇ, ἑκάτερον τῶν τμημάτων δεδομένον ἐστίν.

δεδομένον γὰρ μέγεθος τὸ ΑΒ εἰς δεδομένον λόγον διῃρήσθω τὸν τοῦ ΑΓ πρὸς ΓΒ· λέγω, ὅτι ἑκάτερον τῶν ΑΓ, ΓΒ δοθέν ἐστιν.

ἐπεὶ γὰρ λόγος ἐστὶ τοῦ ΑΓ πρὸς ΓΒ δοθείς, λόγος ἄρα καὶ τοῦ ΑΒ πρὸς ἑκάτερον τῶν ΑΓ, ΓΒ δοθείς. δοθὲν δὲ τὸ ΑΒ· δοθὲν ἄρα καὶ ἑκάτερον τῶν ΑΓ, ΓΒ.

Dt 7. *If a given magnitude be divided in a given ratio, each of the parts is given.*

For, let the given magnitude AB have been divided in the given ratio AC:CB; I say that each of AC, CB is given.

A———————C————B

For, since the ratio AC:CB is given, the ratio of AB to each of AC, CB is given [Dt 6]. And AB is given; therefore each of AC, CB is given [Dt 2]. ∎

Remarks: gvn.$x+y$ & gvn.$x:y$ $\Rightarrow$ gvn.x & gvn.y.
    Concordance x = AC, y = CB, $x+y$ = AB.

It is easy to see that if, in Dt 6, furthermore $x+y$ is given then x and y are given by Dt 2. In our notation:

[Dt 6] gvn.(x+y):x AND (x+y):y
[Dt 2] gvn.x AND gvn.y

Dt 7 is the first proposition in the *Data* which does not refer to axioms or propositions from the *Elements*, but stands as a sort of corollary (to Dt 6). Such *internal* propositions are not frequent. [56] Furthermore, they rarely need to bring in *latent co-actors*. How little information is contained in Dt 7 can be seen if we introduce the latent co-actors and make algebra of it (*a* and *b* being Givens):

$$x+y = a \text{ AND } x/y = b \implies x = ab/(b+1) \text{ AND } y = a/(b+1).$$

None of these steps or values are found in the text, *pace* Clemens Thaer; we are told that x and y are *given* and nothing more. [57] – Dt 7 is used only once (in Dt 58), but Menge seems not to have noticed it, since the relevant parts are halves, and the half of a Given must be given by Def. 1.

After Dt 6 a proof of the *dielonti* case (gvn.P:Q => gvn.(P–Q):Q) was to be expected, but that one is proved nowhere in the *Data*; it seems to be taken for granted – or happened not to be used.

Θε 8. Τὰ πρὸς τὸ αὐτὸ λόγον ἔχοντα δεδομένον, καὶ πρὸς ἄλληλα λόγον ἕξει δεδομένον.

ἐχέτω γὰρ ἑκάτερον τῶν Α, Γ πρὸς τὸ Β λόγον δεδομένον· λέγω, ὅτι καὶ τὸ Α πρὸς τὸ Γ λόγον ἕξει δεδομένον.
ἔστω γὰρ δεδομένον μέγεθος τὸ Δ. καὶ ἐπεὶ λόγος ἐστὶ τοῦ Α πρὸς τὸ Β δοθείς, ὁ αὐτὸς αὐτῷ πεποιήσθω ὁ τοῦ Δ πρὸς τὸ Ε. δοθὲν δὲ τὸ Δ· δοθὲν ἄρα καὶ τὸ Ε. πάλιν, ἐπεὶ λόγος ἐστὶ τοῦ Β πρὸς τὸ Γ δοθείς, ὁ αὐτὸς αὐτῷ πεποιήσθω ὁ τοῦ Ε πρὸς τὸ Ζ. δοθὲν δὲ τὸ Ε· δοθὲν ἄρα καὶ τὸ Ζ. ἔστι δὲ καὶ τὸ Δ δοθέν· λόγος ἄρα τοῦ Δ πρὸς τὸ Ζ ἐστι δοθείς. καὶ ἐπεί ἐστιν ὡς μὲν τὸ Α πρὸς τὸ Β, οὕτως τὸ Δ πρὸς τὸ Ε, ὡς δὲ τὸ Β πρὸς τὸ Γ, οὕτως τὸ Ε πρὸς τὸ Ζ, διίσου ἄρα ἐστὶν ὡς τὸ Α πρὸς τὸ Γ, οὕτως τὸ Δ πρὸς τὸ Ζ. λόγος δὲ τοῦ Δ πρὸς τὸ Ζ δοθείς· λόγος ἄρα καὶ τοῦ Α πρὸς τὸ Γ δοθείς.

---

[56] 13 out of 94, *viz.* Dt 7, 9. 10, 17, 22, 25–27, 29, 47, 49, 53, 63.

[57] (Thaer 1962, 66, §7). I do not understand his comment:˙die Annahme, dass die Teile vorweggegeben sind, [ist] hier sinnlos' (the assumption that the parts are given in advance makes no sense here). But Euclid does not suppose the parts to be given in advance. Thaer could be thinking of *latent co-actors*, but his text is too brief to verify that. – In all the propositions 7–24 he reads a lot of extras into the text, makes products out of rectangles or ratio times line segment, handles all kind of fractions and ratios of ratios, etc.

Dt 8. [Magnitudes] [58] *which have a given ratio to the same, will also have a given ratio to one another.*

For, let each of A, C have a given ratio to B; I say that A will also have a given ratio to C.

```
A_____    D_____
B_____    E_____
C_____        Z_____
```

Let D [59] be a given magnitude [Axiom 0*]. And since the ratio A:B is given, let D:E have been made the same as it [Def. 2]. And D is given; therefore E is given [Dt 2].

Again, since the ratio B:C is given, let E:Z have been made the same. And E is given; therefore Z is given. And D is given; therefore the ratio D:Z is given [Dt 1].

And since A:B :: D:E and B:C :: E:Z, therefore *di' isou* [60] A:C :: D:Z [V.22]. And the ratio D:Z is given; therefore the ratio A:C is also given [Def. 2*]. ∎

Remarks: gvn.A:B & gvn.C:B => gvn.A:C. [61]

This important theorem proves that *Givenness is contagious*, and joins a series of *transitivity* statements in the *Elements*:

- *C.N.1* Magnitudes which are equal to the same are equal.
- *I.30* Straight lines parallel to the same straight line are also parallel to one another.
- *V.11* Ratios which are the same as the same ratio are also the same as one another.
- *X.12* Magnitudes commensurable with the same magnitude are commensurable with one another also.

All of them take symmetry of the relevant relationsship for granted.

---

[58] Like Common notion 1, Dt 8 does not express the word 'magnitude', but uses the plural of the article, τὰ.

[59] The Greek says τὸ Δ, 'let the magnitude D be a given one'.

[60] Latin *ex aequali*. See remarks on Dt 5, *metamorphoses* of ratio.

[61] I do not render A, B, and C as x, y, and z here since the letters denote magnitudes and not endpints of line segments (as in Dt 5–7).

The effective lemma in Dt 8 is V.22, known as the *ex aequali* or *di' isou* relation. To make that theorem work he must import three given magnitudes in the given ratio, D taken, E and Z provided.[62] Since A:C is the 'compound ratio'[63] of A:B and B:C, we may formulate Dt 8 briefly:

Dt 8* *Given ratios compound into a given ratio.*

I will refer to Dt 8 by 'transitivity' or 'compounding'. Here it is in my notation:

take [Axiom 0*]	gvn.D
find E [VI.12]	D:E :: A:B
[Def. 2*, Dt 2]	gvn.D:E and gvn.E
find Z [VI.12]	E:Z :: B:C
[Def. 2*, Dt 2]	gvn.E:Z and gvn.Z
[Dt 1]	gvn.D:Z
Since	A:B :: D:E and B:C :: E:Z
di'isou [V.22]	A:C :: D:Z
[Def. 2*]	gvn.A:C.

◆

Θε 9. Ἐὰν δύο ἢ πλείονα μεγέθη πρὸς ἄλληλα λόγον ἔχῃ δεδομένον, ἔχῃ δὲ τὰ αὐτὰ μεγέθη πρὸς ἄλλα τινὰ μεγέθη λόγους δεδομένους, [εἰ καὶ] μὴ τοὺς αὐτούς, κἀκεῖνα τὰ μεγέθη πρὸς ἄλληλα λόγους ἕξει δεδομένους.

δύο γὰρ ἢ πλείονα μεγέθη τὰ Α, Β, Γ πρὸς ἄλληλα λόγον ἐχέτω δεδομένον, ἐχέτω δὲ τὰ αὐτὰ μεγέθη τὰ Α, Β, Γ πρὸς ἄλλα τινὰ μεγέθη τὰ Δ, Ε, Ζ λόγους δεδομένους, μὴ τοὺς αὐτοὺς δέ· λέγω, ὅτι καὶ τὰ Δ, Ε, Ζ, μεγέθη πρὸς ἄλληλα λόγον ἕξει δεδομένον.

ἐπεὶ γὰρ λόγος ἐστὶ τοῦ Α πρὸς τὸ Β δοθείς, τοῦ δὲ Α πρὸς τὸ Δ λόγος ἐστὶ δοθείς, καὶ τοῦ Δ ἄρα πρὸς τὸ Β λόγος ἐστὶ δοθείς. ἀλλὰ τοῦ Β πρὸς τὸ Ε λόγος ἐστὶ δοθείς· καὶ τοῦ Δ ἄρα πρὸς τὸ Ε λόγος ἐστὶ δοθείς. πάλιν, ἐπεὶ λόγος ἐστὶ τοῦ Β πρὸς τὸ Γ δοθείς, τοῦ δὲ Β πρὸς τὸ Ε λόγος ἐστὶ δοθείς, καὶ τοῦ Ε ἄρα πρὸς τὸ Γ λόγος ἐστὶ δοθείς. τοῦ δὲ Γ πρὸς τὸ Ζ λόγος ἐστὶ δοθείς· καὶ τοῦ Ε ἄρα πρὸς τὸ Ζ λόγος ἐστὶ δοθείς· τὰ Δ, Ε, Ζ ἄρα πρὸς ἄλληλα λόγον ἔχει δεδομένον.

---

[62] Provided by VI.12 if they are line segments or rectilinear figures. The *Data* does not care about methods: if a magnitude is given, it can be provided.

[63] A ratio a:c is said to be compounded from the ratios k:m and p:q, if there is a magnitude b such that k:m :: a:b and p:q :: b:c. Read more about compounding ratios in (Mueller 1981) passim (see his index p. 377), and in (Saito 1986, 25–59).

Dt 9. *If two or more magnitudes have a given ratio to one another, and if the same magnitudes have given, but not the same, ratios to some other magnitudes, then those [other] magnitudes will have given ratios to one another.*

For, let two or more magnitudes A, B, C have to one another a [64] given ratio, and let the same magnitudes A, B, C have to some other magnitudes D, E, Z given ratios, but not the same. I say that the magnitudes D, E, Z will have a given ratio to one another.

```
A_____    D_____
B_____          E_____
C_____                Z_____
```

For, since the ratio A:B is given, and the ratio A:D is given, therefore the ratio D:B is given [Dt 8]. But the ratio B:E is given; therefore the ratio D:E is given.

Again, since the ratio B:C is given, and the ratio B:E is given, therefore the ratio E:C is given. But the ratio C:Z is given; therefore the ratio E:Z is given; therefore D, E, Z have a given ratio to one another. ∎

Remarks: After Dt 8 it is meaningful to speak of three magnitudes having a given ratio, in symbols: gvn.A:B:C. With the transitivity of Dt 9 givenness of ratio is expanded to any number of magnitudes, gvn.A:B:C:...:N. When I tried to invent a short term for it and realized that Dt 8 proves a sort of contagion, I could think of nothing better than *Givenness of ratio is epidemic*. The ratios must be 'not the same' if the proposition is to have any non-trivial content; for if gvn.A:B:C and A:B:C::D:E:Z, then of course gvn.D:E:Z. – I have seen no instance of Dt 9 being used, let alone wanted. The same holds of Dt 7 (though possibly used in Dt 58), whereas Dt 5, 6, and particularly 8 are used frequently.

◆

---

[64] The text has λόγον δεδομένον in the singular, probably to suit the case of two magnitudes; if there be three or more magnitudes, the ratios may (and most often will) be different, so that the plural would be the proper form.

Deductive Structure of Dt 1–9

		Dt 1	2	3	4	5	6	7	8	9	use
Data											
Ax.	0*	.	.	.	.	+	+	.	+	.	
Def.	1	+	+	+	+	.	.	.	.	.	
	2	+	+	.	.	+	+	.	+	.	
Dt	1	.	.	.	.	+	+	.	+	.	
	2	.	.	.	.	+	+	+	+	.	
	3	.	.	.	.	.	+	.	.	.	
	4	.	.	.	.	+	.	.	.	.	
	6	.	.	.	.	.	.	+	.	.	
	8	.	.	.	.	.	.	.	.	+	
Elem											
C.N.	2	.	.	+	.	.	.	.	.	.	
C.N.	3	.	.	.	+	.	.	.	.	.	
V.def.	5	?	.	.	.	.	.	.	.	.	
	14	.	+	.	.	.	.	.	.	.	
	16	+	+	.	.	.	.	.	.	.	
	18	.	.	.	.	.	+	.	.	.	
	19	.	.	.	.	+	+	.	.	.	
	22	.	.	.	.	.	.	.	+	.	
		Dt 1	2	3	4	5	6	7	8	9	

The diagram shows that (*e.g.*) Dt 5 uses Axiom 0*, Def 2, Dt 1, 2, and 4, and also proposition V.19 from the *Elements*. Dt 5 itself is not used within this group, but turns up in Dt 10, 11, 23, and never again. Dt 7 is but a corollary to 6, Dt 9 to 8. The group involves some important theorems from Book V of the *Elements*.

Chapter 3. By a Given Greater than in Ratio
Dt 10–21

Definition 11

I have given the wording of Definition 11 a slight turn to make the idea stand out more clearly. Perhaps I had better not, since the expression in Greek is more syntax than meaning, as I shall explain below. Literally it says, quite meaninglessly:

> A magnitude [M] is by a given greater than a magnitude [N] than in ratio if, when the given magnitude be subtracted, the remainder has a given ratio to the same [N],

the first 'than' being expressed as a genitive of comparison, the second by the normal particle ἤ.

In the theory presented in Dt 10–21, the doings reveal that the phrase should be understood as *M is by the given G greater than the magnitude L which has to N a given ratio*. (So M = L+G, and L:N is a given ratio). By an abbreviating shift of syntax (which points to some sort of idiolect) this was standardized into the rather confusing formula *A magnitude (M) is by a given greater than a magnitude (N) than in ratio* – confusing because M is not greater than N but greater than the magnitude which has to N the given ratio. The apparent nonsense does not affect the correct use of the relation in the actual propositions.

While investigating this peculiar relation we shall need V.12 and 19:

V.12   P:Q :: R:S => (P+R):(Q+S) :: P:Q
V.19   P:Q :: R:S => (P–R):(Q–S) :: P:Q (if P > R and so Q > S),

that is, when magnitudes are added to (or subtracted from) the antecedent and consequent of some ratio then the new ratio is the same as the first one *if* the magnitudes added (or subtracted) are in that ratio.[65]

One may ask if the condition is also a necessary one, and Dt 14 and 15 can be read as showing that the operation leads to the same ratio *only if* the magnitudes are in the given ratio.

According to our analysis of Def. 2, two magnitudes x and y have a given ratio if and only if there are two given magnitudes a and b such that $x{:}y :: a{:}b$. Dt 14 considers

---

[65] Geometrically, these truths are eqivalent with VI.24, *The parallelograms around the diameter of any parallelogram are similar to the whole and to one another*, cf. p. 156.

a given ratio $x{:}y$, adds given magnitudes c and d to antecedent and consequent respectively, and examines the ratio produced $(x+c){:}(y+d)$. From V.12 follows that if the ratio $c{:}d$ is the same as $x{:}y$ then $(x+c){:}(y+d)$ is also the same as $x{:}y$. But what happens if $c{:}d$ is not the same ratio as $x{:}y$?

Suppose that $c{:}d > x{:}y$ (as defined in V.Def. 7 and elaborated in V.8 and 10). Assuming continuity and a fourth proportional there will be a magnitude g such that if it be subtracted from c, $(c-g){:}d$ is the same as $x{:}y$. Since d is given and the ratio is given, $c-g$ is also given (Dt 2), and so g (Dt 3). We now face a situation which can be aptly described as *c is by the given magnitude g too great to be in the given ratio to d.* Def. 11 says as much.

Whether this relation has any mathematical relevance remains to be seen, but it has a close affinity to the relation *x has a given ratio to y.* Graphically (and anachronistically) the affinity can be described as follows: while the proportion $x{:}y :: a{:}b$ can be represented by a straight line from (but not including) the origin (0,0) with the slope b/a, the ratio $(x-g){:}y :: a{:}b$ can be represented by a straight line from (but not including) the point $(g,0)$ with the same slope b/a. A comparison between the indeterminate equations
$$bx - ay = 0$$
and
$$bx - ay = bg$$
also illustrates the situation, although I do not see why one should investigate these equations particularly.[66]

As I mentioned, this relation is used in Dt 14 and 15 to prove a necessary condition in connexion with V.12 and 19. It has properties in common with the concept *given ratio*, e.g. (some of) the metamorphoses (see p. 44), as proved in Dt 10 and 11. From the wording of it, it is obviously an order relation, and Dt 13 and 16–19 show that. Ordered ratios, as defined in V.def.7, are absent from the arithmetical books of the *Elements* (VII–IX) and might therefore be considered as one of Eudoxus' new approaches.[67] But even if this criteria of greater ratio were new, the concept must undoubtedly be older; and one way of handling two magnitudes having a greater ratio than two others could be that defined in Def. 11, with the special feature that the 'ratio-difference' is a *given* magnitude.

I shall refer to the relation as R'11 with reference to Def. 11, and use the stenograph $\qquad$ A $\delta{>}{:}$ B $\qquad$ for the statement

A is by a given [magnitude] *greater than in* [a given] *ratio to B.*

---

[66] Algebraists (like Clemens Thaer) will like the formula $x = \frac{a}{b} y + g$, with a, b, and g given.

[67] Def. 5 and 7 of Book V appear to have been conceived at the same time, Def. 7 being the logical negative of Def. 5.

CHAPTER 3. BY A GIVEN GREATER THAN IN RATIO. (10–21)

hoping to make the structure stand out more clearly; it means that there is a given magnitude C such that (A–C):B is a *given* ratio. In the *Data*, apart from theorems 10, 11, 13–21, it is used only in theorem 86. Outside the *Data* I know few occurrences of the relation, the generalized Apollonius' circle being one of them.[68] But even there it is but a smart way to designate a complex relation, without any real mathematical significance. – In Def. 12 Euclid defines the negative counterpart to R'11, which is not used in the *Data*, and it is in fact superfluous, since it can easily be proved that if A δ>: B then B δ<: A, read *B is by a given less than A than in ratio*.

Synopsis of Dt 10–21

Dt 10, 11, 13–21 (12 is an intruder) deal with various properties of R'11, some of them analogues to the metamorphoses of ratios (cf. p. 44), so I will use the names 'antecedent' and 'consequent' for the terms involved. A synopsis will illustrate what it is all about, and at the same time intimate that some of the statements are mathematically trivial; in the running notes to each theorem I will try to explain the procedures. In the Greek text everything is illustrated geometrically, and some of it may be more easily understood from the diagrams (simple straight lines) than from my shorthand. If you *must* think of the magnitudes as real numbers, remember that they are positive. The symbol ~ (in Dt 12) denotes the positive difference between two magnitudes irrespective of which is greater. If the minus symbol – is used, the subtrahend is supposed to be the lesser magnitude.

Dt 10.1 The *synthenti* (*componendo*) metamorphosis (adding antecedent and consequent while keeping the consequent):
x δ>: y => $x+y$ δ>: y.

Dt 10.2 The *dielonti* (*subtrahendo*) metamorphosis (the converse of 10.1; it works only if x is greater than the given magnitude; if not, something else is proved):
$x+y$ δ>: y => x δ>: y OR (gvn.$x+z$ AND gvn.$y:z$)

Dt 11 A biconditional proving some important metamorphoses:
x δ>: y <=> x δ>: $x+y$.

---

[68] See my paper *Discovering Apollonius' Circle* in Delphi, August 1996, printed in the proceedings from that seminar (Third International Conference on Ancient Mathematics, Delphi, Greece July 30 - August 3, 1996, Ed. by Tasoula Berggren, Simon Fraser University.) 'With two given points in a plane and a given ratio of unequal lines, it is possible to describe in the plane a circle such that straight lines from the given points, inflected on the circumference of the circle, have (to one another) the given ratio.'

Dt 12 has nothing to do with R'11 (not even serving as a lemma for the following):
gvn.$x+y$ AND gvn.$y+z$ => $x = z$ OR gvn.$x \sim z$

Dt 13 Transitivity of a mixed relation:
gvn.x:y AND y δ>: z => x δ>: z.
(compare $x = y$ AND $y > z$ => $x > z$).

Dt 14 Addition of given magnitudes to antecedent and consequent:
gvn.x:y AND gvn.c AND gvn.d =>
gvn.$(x+c)$:$(y+d)$ OR $x+c$ δ>: $y+d$ [OR $y+d$ δ>: $x+c$]. [69]
(compare $x = y$ => $x+c = y+d$ OR $x+c > y+d$ OR $x+c < y+d$).

Dt 15 Subtraction of given magnitudes from antecedent and consequent:
gvn.x:y AND gvn.c AND gvn.d =>
gvn.$(x-c)$:$(y-d)$ OR $x-c$ δ>: $y-d$ [OR $y-d$ δ>: $x-c$].
(compare $x = y$ => $x-c = y-d$ OR $x-c > y-d$ OR $x-c < y-d$).

Dt 16 A mixture of propositions 14 and 15:
gvn.x:y AND gvn.c AND gvn.d => $x+c$ δ>: $y-d$.
(compare $x = y$ => $x+c > y-d$).

Dt 17 x δ>: y AND z δ>: y => gvn.x:z OR x δ>: z [OR z δ>: x].
(compare $x > y$ AND $z > y$ => $x = z$ OR $x > z$ OR $x < z$).

Dt 18 x δ>: y AND x δ>: z => gvn.y:z OR y δ>: z [OR z δ>: y].
(compare $x > y$ AND $x > z$ => $y = z$ OR $y > z$ OR $y < z$).

Dt 19 Transitivity: x δ>: y AND y δ>: z => x δ>: z.
(compare $x > y$ AND $y > z$ => $x > z$).

Dt 20 Analogue of Dt 15:
gvn.a AND gvn.b AND gvn.x:y =>
gvn.$(a-x)$:$(b-y)$ OR $a-x$ δ>: $b-y$ [OR $b-y$ δ>: $a-x$].

Dt 21 Superfluous analogue of Dt 14:
gvn.a AND gvn.b AND gvn.x:y =>
gvn.$(a+x)$:$(b+y)$ OR $a+x$ δ>: $b+y$ [OR $b+y$ δ>: $a+x$].

◆

---

[69] Read 'one of them is by a given magnitude greater than in ratio to the other'.

CHAPTER 3. BY A GIVEN GREATER THAN IN RATIO. (10–21)

Θε 10. ' Εὰν μέγεθος μεγέθους δοθέντι μεῖζον ᾖ ἢ ἐν λόγῳ, καὶ τὸ συναμφότερον τοῦ αὐτοῦ δοθέντι μεῖζον ἔσται ἢ ἐν λόγῳ· καὶ ἐὰν τὸ συναμφότερον τοῦ αὐτοῦ δοθέντι μεῖζον ᾖ ἢ ἐν λόγῳ, καὶ τὸ λοιπὸν τοῦ αὐτοῦ ἤτοι δοθέντι μεῖζόν ἐστιν ἢ ἐν λόγῳ, ἢ τὸ λοιπὸν μετὰ τοῦ ἑξῆς, πρὸς ὃ τὸ ἕτερον λόγον ἔχει δεδομένον, δοθέν ἐστιν.

μέγεθος γὰρ τὸ ΑΒ μεγέθους τοῦ ΒΓ δοθέντι μεῖζον ἔστω ἢ ἐν λόγῳ· λέγω, ὅτι καὶ τὸ συναμφότερον τὸ ΑΓ τοῦ αὐτοῦ τοῦ ΓΒ δοθέντι μεῖζόν ἐστιν ἢ ἐν λόγῳ.

ἐπεὶ γὰρ τὸ ΑΒ τοῦ ΒΓ δοθέντι μεῖζόν ἐστιν ἢ ἐν λόγῳ, ἀφῃρήσθω τὸ δοθὲν μέγεθος τὸ ΑΔ· λοιποῦ ἄρα τοῦ ΔΒ πρὸς τὸ ΒΓ λόγος ἐστὶ δοθείς· καὶ συνθέντι τοῦ ΔΓ πρὸς τὸ ΒΓ λόγος ἐστὶ δοθείς. καί ἐστι δοθὲν τὸ ΑΔ· τὸ ΓΑ ἄρα τοῦ ΓΒ δοθέντι μεῖζόν ἐστιν ἢ ἐν λόγῳ.

πάλιν δὴ τὸ ΑΓ τοῦ ΓΒ δοθέντι μεῖζον ἔστω ἢ ἐν λόγῳ· λέγω, ὅτι τὸ λοιπὸν τὸ ΑΒ τοῦ αὐτοῦ τοῦ ΒΓ ἤτοι δοθέντι μεῖζον ἔσται ἢ ἐν λόγῳ, ἢ τὸ ΑΒ μετὰ τοῦ ἑξῆς, πρὸς ὃ τὸ ΒΓ λόγον ἔχει δοθέντα, δοθέν ἐστιν.

ἐπεὶ γὰρ τὸ ΑΓ τοῦ ΓΒ δοθέντι μεῖζόν ἐστιν ἢ ἐν λόγῳ, ἀφῃρήσθω τὸ δοθὲν μέγεθος. τὸ δὴ δοθὲν ἤτοι ἔλασσόν ἐστι τοῦ ΑΒ ἢ μεῖζον. ἔστω πρότερον ἔλασσον, καὶ ἔστω τὸ ΑΔ· λοιποῦ ἄρα τοῦ ΔΓ πρὸς ΓΒ λόγος ἐστὶ δοθείς· διελόντι ἄρα τοῦ ΔΒ πρὸς ΒΓ λόγος ἐστὶ δοθείς. καί ἐστι δοθὲν τὸ ΑΔ· τὸ ΑΒ ἄρα τοῦ ΒΓ δοθέντι μεῖζόν ἐστιν ἢ ἐν λόγῳ.

ἀλλὰ δὴ τὸ δοθὲν μεῖζον ἔστω τοῦ ΑΒ, καὶ κείσθω αὐτῷ ἴσον τὸ ΑΕ· λόγος ἄρα λοιποῦ τοῦ ΕΓ πρὸς τὸ ΓΒ ἐστι δοθείς· ὥστε καὶ ἀνάπαλιν τοῦ ΒΓ πρὸς τὸ ΕΓ λόγος ἐστὶ δοθείς· καὶ ἀναστρέψαντι ὁ τοῦ ΒΓ πρὸς ΒΕ λόγος ἐστὶ δοθείς. καί ἐστι τὸ ΕΒ μετὰ τοῦ ΒΑ δοθέν· ὅλον γὰρ τὸ ΑΕ δοθέν ἐστιν· τὸ ΒΑ ἄρα μετὰ τοῦ ἑξῆς, πρὸς ὃ τὸ ΒΓ λόγον ἔχει δοθέντα, δοθέν ἐστιν.

Dt 10. *If a magnitude* [M] *be by a given* [magnitude] *greater than in* [a given] *ratio to a magnitude* [N],[70] *the sum* [M+N] *will be by a given* [magnitude] *greater than in* [a given] *ratio to the same* [N]. – *And if the sum* [M+N] *be by a given* [magnitude] *greater than in* [a given] *ratio to the same* [N], *the remainder* [M] *is either by a given* [magnitude] *greater than in* [a given] *ratio to the same* [N], *or the remainder together with the adjacent to which the other one* [N] *has a given ratio, is given.*

For, let the magnitude AB be by a given [magnitude] greater than in [a given] ratio to the magnitude BC; I say that the sum AC is by a given greater than in ratio to CB.

A   D B   C

---

[70] It is quite un-Greek to supply names of variables in the *protasis*, but I do so nevertheless to make things understandable at once. I have given my translation a slight interpretative turn.

[1] For, since AB is by a given greater than in ratio to the magnitude BC, let AD [equal to] the given magnitude have been subtracted; then the ratio of the remainder DB to BC is given [Def. 11]; and *synthenti* the ratio [(DB+BC):BC, *i.e.*] DC:BC is given [Dt 6]. And AD is given; therefore CA is by a given greater than in ratio to CB.

[2] Again, let AC be by a given greater than in ratio to CB; I say that the remainder AB either will be by a given greater than in ratio to BC, or AB plus the *adjacent*, to which BC has a given ratio, is given.

For, since AC is by a given greater than in ratio to CB, let the given magnitude have been subtracted. The given magnitude is either less than AB or greater.

[2.a] Let it first be less, and let it be [equal to] AD; then the ratio of the remainder DC to CB is given [Def. 11]. *Dielonti* the ratio [(DC−CB):CB *i.e.*] DB:CB is given; and AD is given; therefore AB is by a given greater than in ratio to CB.

<u>A   B   E   C</u>

[2.b] But then, let the given magnitude be greater than AB, and let AE be set out equal to it; then the ratio of the remainder EC to CB is given [Def. 11]; so that *anapalin* the ratio BC:EC is given; and *anastrepsanti* the ratio [BC:(BC−EC) *i.e.*] BC:BE is given [Dt 5]; and EB plus BA is given; for the whole AE is given; therefore BA together with the *adjacent* [BE], to which BC has a given ratio, is given. ∎

Remarks: Dt 10, first part, proves the *synthenti* metamorphosis for R'11. The second part is split in two, proving the *dielonti* metamorphosis, if it works; if it does not, something less comes out.

> Note that the ratios below in { } are expressed only – and sufficiently – by the idiom *is given*, and have no counterparts in the Greek text. It remains an unanswered question whether Euclid thought of those *latent co-actors* (see p. 24) or not. Anyway, they *must* be there if the ratios are to be reproduced. Many distracting and irrelevant constants and operations are suppressed in the Greek idiom. – In the following I summarize the proofs, renaming the magnitudes to make a (to me) more readable notation. First I present my own style, followed by a transcription from the Greek text, where the alphabetic order is somewhat surprising.

CHAPTER 3. BY A GIVEN GREATER THAN IN RATIO. (10–21)

10.1) $\quad x\ \delta{>}: y \Rightarrow x{+}y\ \delta{>}: y$
Concordance: $\quad AB = x, BC = y, AC = x{+}y, AD = g, DB = x{-}g$.
The first part is very straightforward, using nothing but Def. 11 and the *synthenti* metamorphosis (Dt 6).

Since $x\ \delta{>}: y$		AB $\delta{>}$: BC,
there is [Def. 11] a given g		gvn.AD
so that gvn.$(x{-}g){:}y$	{:: $a{:}b$ }	gvn.DB:BC.
[Dt 6] gvn.$(y{+}x{-}g){:}y$	{:: $(a{+}b){:}b$}	gvn.DC:BC. (*synthenti*)
[Def. 11] $x{+}y\ \delta{>}: y$		CA $\delta{>}$: CB.

Dt 6, the *synthenti* theorem (gvn.P:Q => gvn.(P+Q):Q) is the effective ingredient in the proof. As I said above, the magnitudes and ratio in braces {a, b, and $a{:}b$} do not occur in the text, but are latent in the statement *is a given ratio*. Beware: we are not told the value of the final given ratio {$(a{+}b){:}b$}, only that it is given.. The *Data* speaks about properties, not quantities; an interpretation by algebraic notation brings forth how much information is *not* there, and that is about the best that can be said for such translations – although in this and the following propositions it may help a modern reader to follow the arguments.

10.2) $\quad x{+}y\ \delta{>}: y \Rightarrow x\ \delta{>}: y$
The second part is the converse of the first part, starting with AC and BC as initial magnitudes, AC still called τὸ συναμφότερον, the sum, and BC τὸ λοιπόν, the remainder; but now Euclid may run into difficulties from lack of negative magnitudes, and so he has to split the proof into two cases:

10.2.a) In the first case, the given AD (g) is less than AB (x), and the proof uses the same figure as the first part, applying Def. 11 and the *dielonti* metamorphosis (gvn.P:Q => gvn.(P–Q):Q), which is not proved in the *Data*. [71]

Since $x{+}y\ \delta{>}: y$		AC $\delta{>}$: CB,
there is [Def. 11] a given g		gvn.AD
so that gvn.$(x{+}y{-}g){:}y$	{:: $a{:}b$}	gvn.DC:CB.
Suppose that $g < x$		AD < AB.
Dielonti gvn.$(x{-}g){:}y$	{:: $(a{-}b){:}b$}.	gvn.DB:BC.
[Def. 11] $x\ \delta{>}: y$		AB $\delta{>}$: BC.

---

[71] Menge (and after him McDowell/Sokolik) wrongly refer to Dt 5, which proves the *anastrepsanti* operation (gvn.P:Q leads to gvn.P:(P–Q)).

10.2.b) Concordance: AB = x, BC = y, AC = $x+y$, AE = h, BE = $h-x$.

The second case occurs if the given, now named AE (h), is greater than AB (x), so that $x+y-h < y$ and the *dielonti* operation (to subtract the consequent from the antecedent) does not work; Euclid proves that the ratio $(h-x):y$ is given, (now Dt 5 is at work), and that $x+(h-x)$ is given, – the latter being as obvious to Euclid in the diagram as to us by our notation, since it is equal to the given line AE (h).

Since $x+y$ δ>: y		AC δ>: BC,	
there is [Def. 11] a given h		gvn.AE	
such that gvn.$(x+y-h):y$	{:: a:b}	gvn.EC:CB.	
Suppose that $h > x$		AE > AB.	
Anapalin gvn.$y:(x+y-h)$	{:: b:a}	gvn.BC:EC	
[Dt 5] gvn.$y:(h-x)$	{:: b:($b-a$)}	gvn.BC:BE	
Let $h-x = z$ be named 'The adjacent'		BE is 'The neighbour'	
Since $z+x = h$, $z+x$ is given		EB+BA = gvn.AE	

After applying Dt 5, *anastrophe*, the proof gives up the idiom and considers the given AE as a composed magnitude which is given (in fact, it is *the* given magnitude h from the *protasis*), namely AB (x) *plus* an extra BE ($h-x$). The Greek text calls it τὸ ἑξῆς, which means 'the adjacent one', 'the neighbour' (*viz.* of $x =$ AB); a curious term at first sight, [72] but BE manifestly is a neighbour when a drawing is made. Figures were integral parts of the arguments. – The crux is, of course, that AB cannot be by a *negative* magnitude greater than BC than in ratio; we might restate the two cases of the second part of the proposition as follows:

> **Dt 10*** If a sum (AC) is greater than a part (BC) than in ratio, then the remainder (AB) will either be by a given greater than the part (BC) than in ratio, or the remainder (AB) will be less than a given by a magnitude which has to the part (BC) a given ratio.

◆

Θε 11. Ἐὰν μέγεθος μεγέθους δοθέντι μεῖζον ᾖ ἢ ἐν λόγῳ, τὸ αὐτὸ καὶ συναμφοτέρου δοθέντι μεῖζον ἔσται ἢ ἐν λόγῳ, καὶ ἐὰν τὸ αὐτὸ συναμφοτέρου δοθέντι μεῖζον ᾖ ἢ ἐν λόγῳ, τὸ αὐτὸ καὶ τοῦ λοιποῦ δοθέντι μεῖζον ἔσται ἢ ἐν λόγῳ.

μέγεθος γὰρ τὸ ΑΒ τοῦ ΒΓ δοθέντι μεῖζον ἔστω ἢ ἐν λόγῳ· λέγω, ὅτι καὶ τοῦ ΑΓ δοθέντι μεῖζόν ἐστιν ἢ ἐν λόγῳ.

---

[72] And McDowell/Sokolik gave up translating it and wrote 'that'.

ἐπεὶ γὰρ τὸ ΑΒ τοῦ ΒΓ δοθέντι μεῖζόν ἐστιν ἢ ἐν λόγῳ, ἀφῃρήσθω τὸ δοθὲν μέγεθος τὸ ΑΔ· λοιποῦ ἄρα τοῦ ΔΒ πρὸς τὸ ΒΓ λόγος ἐστὶ δοθείς. ἀνάπαλιν καὶ συνθέντι λόγος ἐστὶ τοῦ ΓΔ πρὸς τὸ ΔΒ δοθείς· ὁ αὐτὸς αὐτῷ γεγονέτω ὁ τοῦ ΑΔ πρὸς τὸ ΔΕ· λόγος ἄρα καὶ τοῦ ΑΔ πρὸς τὸ ΔΕ δοθείς· δοθὲν δὲ τὸ ΑΔ· δοθὲν ἄρα καὶ τὸ ΔΕ· ὥστε καὶ λοιπὸν τὸ ΕΑ δοθέν ἐστιν. ἔστι δὲ καὶ ὅλου τοῦ ΑΓ πρὸς ὅλον τὸ ΕΒ λόγος δοθείς· ὥστε καὶ τοῦ ΕΒ πρὸς ΑΓ λόγος ἐστὶ οθείς. καί ἐστι δοθὲν τὸ ΑΕ· τὸ ΒΑ ἄρα τοῦ ΑΓ δοθέντι μεῖζόν ἐστιν ἢ ἐν λόγῳ.

ἀλλὰ δὴ τὸ ΒΑ συναμφοτέρου τοῦ ΑΓ δοθέντι μεῖζον ἔστω ἢ ἐν λόγῳ· λέγω, ὅτι τὸ αὐτὸ τὸ ΑΒ καὶ τοῦ λοιποῦ τοῦ ΒΓ δοθέντι μεῖζον ἔσται ἢ ἐν λόγῳ.

ἐπεὶ γὰρ τὸ ΑΒ τοῦ ΑΓ δοθέντι μεῖζόν ἐστιν ἢ ἐν λόγῳ, ἀφῃρήσθω τὸ δοθὲν μέγεθος τὸ ΑΕ· λοιποῦ ἄρα τοῦ ΕΒ πρὸς τὸ ΑΓ λόγος ἐστὶ δοθείς· ὥστε καὶ τοῦ ΑΓ πρὸς τὸ ΕΒ λόγος ἐστὶ δοθείς· ὁ αὐτὸς αὐτῷ γεγονέτω ὁ τοῦ ΑΔ πρὸς ΕΔ· καὶ τοῦ ΔΑ ἄρα πρὸς ΕΔ λόγος ἐστὶ δοθείς· καὶ ἀναστρέψαντι τοῦ ΔΑ πρὸς ΑΕ λόγος δοθείς· καὶ ἀνάπαλιν τοῦ ΕΑ πρὸς τὸ ΑΔ λόγος ἐστὶ δοθείς. καὶ δοθὲν τὸ ΑΕ· δοθὲν ἄρα καὶ ὅλον τὸ ΑΔ. καὶ ἐπεὶ ὅλου τοῦ ΑΓ πρὸς ὅλον τὸ ΕΒ λόγος ἐστὶ δοθείς, ὦν τοῦ ΑΔ πρὸς τὸ ΔΕ λόγος ἐστὶ δοθείς, ἔσται καὶ λοιποῦ τοῦ ΓΔ πρὸς λοιπὸν τὸ ΔΒ λόγος δοθείς· καὶ διελόντι τοῦ ΓΒ πρὸς τὸ ΔΒ λόγος ἐστὶ δοθείς· ὥστε καὶ τοῦ ΔΒ πρὸς τὸ ΒΓ λόγος ἐστὶ δοθείς. καί ἐστι δοθὲν τὸ ΔΑ· τὸ ΑΒ ἄρα τοῦ ΒΓ δοθέντι μεῖζόν ἐστιν ἢ ἐν λόγῳ.

Dt 11. *If a magnitude be by a given greater than in ratio to a magnitude, the same will be by a given greater than in ratio to the sum; and if the same be by a given greater than in ratio to the sum, the same will be by a given greater than in ratio to the remainder.*

[1] Let the magnitude AB be by a given [magnitude] greater than in [a given] ratio to the magnitude BC; I say that it is by a given greater than in ratio to AC.

A   E   D   B   C

For, since AB is by a given greater than in ratio to the magnitude BC, let the given magnitude AD have been subtracted; then the ratio of the remainder DB to BC is given [Def. 11]; *anapalin* and *synthenti* the ratio [(CB+BD):DB, *i.e.*] CD:DB is given [Dt 6].

Let AD:DE have become [73] the same ratio [as CD:DB]. Then the ratio AD:DE is given [Def. 2*]; and AD is given; therefore DE is given [Dt 2], so that the remainder EA is given [Dt 4].

And the ratio of the sum [AD+DC] AC to the sum [ED+DB] EB is given, so that the ratio EB:AC is given [V.12; Def. 2*]. And AE is given; therefore BA is by a given greater than in ratio to the magnitude AC [Def. 11].

---

[73] γεγονέτω; whether it is different from πεποιήσθω I am not sure - but as a means of distinguishing different layers in the *Data* those forms may be significant.

66   EUCLID'S *DATA*

[2] But then, let BA be by a given greater than in ratio to the sum AC; I say that the same AB is by a given greater than in ratio to the remainder BC.

For, since AB is by a given greater than in ratio to AC, let the given magnitude AE have been subtracted; then the ratio of the remainder EB to AC is given [Def. 11]; so that the ratio AC:EB is given.

Let AD:ED have become the same; then the ratio DA:ED is given; and *anastrepsanti* the ratio DA:AE is given [Dt 5]; and *anapalin* the ratio EA:AD is given; and AE is given; therefore the whole AD is given [Dt 2].

And since the ratio of the whole AC to the whole EB is given, whereof the ratio AD:DE is given, the ratio of the remainder DC to the remainder DB will be given [V.19, Def. 2]; and *dielonti* the ratio [(DC–DB):DB, that is] CB:DB is given; so that the ratio DB:BC is given; and DA is given; therefore AB is by a given greater than in ratio to the remainder BC.   ∎

Remarks: Dt 11 is in many ways the most instructive of the series 10–21. Again it will be helpful to remember V.12 and V.19 (see p. 57). They play important roles in the following propositions, some of which can be interpreted as supplements to them. Dt 11 proves an equivalence, x δ>: y <=> x δ>: $x+y$, the first part of which is straightforward.

Concordance: AB = x, BC = y, AD = g, AE = h, ED = $g-h = w$, BD = $x-g$, BE = $x-h$.

11.1 x δ>: y => x δ>: $x+y$

Heuristics:

x δ>: y means that there is a given g such that $(x-g):y$ is a given ratio {$a:b$}.

x δ>: $x+y$ means that there is a given h such that $(x-h):(x+y)$ is a given ratio {$c:d$}.

So the problem is to find a given h from the given g, and to find a (unseen) given ratio $c:d$ from the (unseen) given ratio $a:b$, by means of appropriate metamorphoses and the elementary propositions. To get $x+y$ as the new consequent he may invert the ratio (*anapalin*, $y:(x-g)$) and use the *synthenti* operation by adding $(x-g)$ to y, then turn it round again and try to get rid of the g afterwards. – To a Greek mathematician skilled in handling proportions this was perhaps easy, but I am sure that any of my readers will have to think twice and try trice before succeeding. Here is what Euclid did:

(Again we admit the hidden co-actors {$a:b$} in brackets; to a modern reader they will explain – more easily than long notes – what is going on. But never forget that they are not in the Greek text.)

CHAPTER 3. BY A GIVEN GREATER THAN IN RATIO. (10–21)

Since $x \ \delta{>}: y$		AB $\delta{>}$: BC	1
there is [Def. 11] a given g		gvn. AD	2
so that gvn.$(x{-}g){:}y$	{:: $a{:}b$}	gvn.DB:BC.	3
Anapalin &			
synthenti [Dt 6] gvn.$(y{+}x{-}g){:}(x{-}g)$	{:: $(b{+}a){:}a$}	gvn.CD:DB	4
Get w (fourth proportional)		DE	5
so that $g{:}w :: (y{+}x{-}g){:}(x{-}g)$		AD:DE :: CD:DB.	6
[Def. 2*, Dt 2] gvn.$g{:}w$, gvn.w		gvn. AD:DE, gvn.DE	7
[Dt 4, naming] gvn.$g{-}w$ = gvn.h		gvn.EA	8
[[5, V.12] $(x{+}y){:}(x{-}h) :: g{:}w$		AC:EB :: AD:DE]	9
[Def. 2*] gvn.$(x{+}y){:}(x{-}h)$	{:: $(b{+}a){:}a$}	gvn.AC:EB	10
Anapalin gvn.$(x{-}h){:}(x{+}y)$	{:: $a{:}(b{+}a)$}	gvn.EB:AC	11
Since gvn.h, this means [Def. 11] $x \ \delta{>}: x{+}y$		BA $\delta{>}$: AC.	12

Explanations: Lines 1–3 use Def. 11 to expand the hypothesis. Line 4 inverts the ratio (*anapalin*) and applies Dt 6 (*synthenti*, adding consequent to antecedent). To get rid of the subtrahend g in the antecedent we get $g{:}w$ (*i.e.* find the fourth proportional w) in the given ratio just found; whence w is given and less than g; then V.12 is applied implicitly to yield $(x{+}y){:}(x{-}g{+}w)$, *i.e.* $(x{+}y){:}(x{-}h)$; finally the ratio is inverted again, and Def. 11 ensures the assertion.

It pays to compare with Clemens Thaer's result to have a glimpse of a different world: 'If $x = \kappa y + c$, $z = y + x$, then $x = z\kappa/(\kappa{+}1) + c/(\kappa{+}1)$'. – κ denotes the given ratio, my $a{:}b$.

11.2 $x \ \delta{>}: x{+}y \Rightarrow x \ \delta{>}: y$

The converse of 11.1 can be proved by reversing the steps. However, from line 6 Euclid seems to loose his grip.

Since $x \ \delta{>}: x{+}y$		BA $\delta{>}$: AC	1
there is [Def. 11] a given h		gvn.AE	2
so that gvn.$(x{-}h){:}(x{+}y)$	{:: $a{:}b$}	gvn.EB:AC	3
Anapalin gvn.$(x{+}y){:}(x{-}h)$	{:: $b{:}a$}	gvn.AC:EB	4
[get w and $g = h{+}w$		DE and AD]	5
so that $g{:}w :: (x{+}y){:}(x{-}h)$	{:: $b{:}a$}	AD:ED :: AC:EB	6
[Def. 2*] gvn.$g{:}w$	{:: $b{:}a$}	gvn.DA:ED	7

[The ratio is given; but are AD and DE given? Euclid proves this, by way of the *anastrepsanti* metamorphosis in Dt 5, so he is not loosing any grip. Algebraically he is finding g and w from the 'equations' $g{-}w = h$ (a given magnitude) and $g{:}w :: a{:}b$ (a given ratio). How DE (and therefore AD) can be constructed from the given AE is illustrated on figure 11.1, with AH = AC and HK = BE.]

[Dt 5] gvn.g:(g–w) i.e. gvn.g:h   {∷ b:(b–a)}   gvn.DA:AE.   8
Anapalin gvn.h:g   {∷ (b–a):b}   gvn.EA:AD   9
[invert to be able to use Dt 2 literally !]
[Dt 2] gvn.g   gvn.AD   10
[[line 6, V.19] (x+y–g):(x–g) ∷ g:w   CD:DB ∷ AD:DE
because x–h–w = x–(h+w) = x–g   ]   11
[Def. 2*] gvn.(x+y–g):(x–g)   {∷ b:a}   gvn.DC:DB   12
Dielonti gvn.y:(x–g)   {∷ (b–a):a}   gvn.CB:DB   13
Anapalin gvn(x–g):y   {∷ a:(b–a)}   gvn.DB:BC   14
[Def. 11] x δ>: y   AB δ>: BC.   15

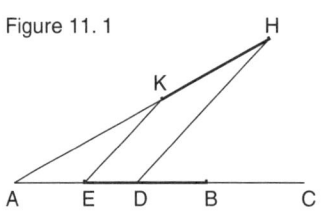

Figure 11. 1

Clemens Thaer (Thaer 1962, 66), suggests that 11.2, which is 'mathematically valuable', is a later addition, because it is not in the Arabic translation, and because of an omission (in step 11) which is not really one: a Greek geometer would see immediately that V.19 is applied. The conspicuous 'mathematical value' may be the construction in lines 5–6 (Figure 11.1).

◆

Θε 12. ' Ἐὰν ᾖ τρία μεγέθη καὶ τὸ μὲν πρῶτον μετὰ τοῦ δευτέρου ᾖ δοθέν, ᾖ δὲ καὶ τὸ δεύτερον μετὰ τοῦ τρίτου δοθέν, τὸ πρῶτον τῷ τρίτῳ ἤτοι ἴσον ἐστίν, ἢ τὸ ἕτερον τοῦ ἑτέρου δοθέντι μεῖζόν ἐστιν.

ἔστω τρία μεγέθη τὰ ΑΒ, ΒΓ, ΓΔ, καὶ τὸ μὲν ΑΒ μετὰ τοῦ ΒΓ δοθὲν ἔστω τὸ ΑΓ, τὸ δὲ ΒΓ μετὰ τοῦ ΓΔ δοθὲν ἔστω τὸ ΒΔ· λέγω, ὅτι τὸ ΑΒ τῷ ΓΔ ἤτοι ἴσον ἐστίν, ἢ τὸ ἕτερον τοῦ ἑτέρου δοθέντι μεῖζόν ἐστιν.
 ἐπεὶ γὰρ δοθέν ἐστιν ἑκάτερον τῶν ΑΓ, ΒΔ, τὰ δὴ δοθέντα ἤτοι ἴσα ἐστὶν ἢ ἄνισα. ἔστω πρότερον ἴσα· ἴσον ἄρα ἐστὶ τὸ ΑΓ τῷ ΒΔ. κοινὸν ἀφῃρήσθω τὸ ΒΓ· λοιπὸν ἄρα τὸ ΑΒ λοιπῷ τῷ ΓΔ ἴσον ἐστίν.
 μὴ ἔστω δὴ ἴσα, ἀλλ' ἔστω μεῖζον τὸ ΑΓ τοῦ ΒΔ, καὶ κείσθω τῷ ΒΔ ἴσον τὸ ΓΕ· δοθὲν δὲ τὸ ΒΔ· δοθὲν ἄρα καὶ τὸ ΓΕ. ἔστι δὲ καὶ ὅλον τὸ ΑΓ δοθέν· καὶ λοιπὸν τὸ ΑΕ δοθέν ἐστιν. καὶ ἐπεὶ ἴσον ἐστὶ τὸ ΕΓ τῷ ΒΔ, κοινὸν ἀφῃρήσθω τὸ ΒΓ· λοιπὸν ἄρα τὸ ΒΕ λοιπῷ τῷ ΓΔ ἴσον ἐστίν. καί ἐστι δοθὲν τὸ ΑΕ· τὸ ΑΒ ἄρα τοῦ ΓΔ δοθέντι μεῖζόν ἐστιν.

Dt 12. *If there be three magnitudes, and the first plus the second be given in magnitude, and the second plus the third be given in magnitude, the first is either equal to the third, or the one is greater than the other by a given magnitude.*

CHAPTER 3. BY A GIVEN GREATER THAN IN RATIO. (10–21)

Let AB, BC, CD be three magnitudes, and let AB plus BC be the given magnitude AC, and let BC plus CD be the given magnitude BD; I say that AB is either equal to CD, or the one is greater than the other by a given magnitude.

For, since each of AC, BD is given, the given magnitudes are either equal or unequal.

```
A        B      C        D
```

[1] First, let them be equal; then AC is equal to BD. Let the common magnitude BC have been subtracted; then the remainder AB is equal to the remainder CD [C.N. 3].

```
A      E     B      C       D
```

[2] Now let them be *not equal*, but let AC be greater than BD, and let CE have been set out equal to BD; and BD is given; therefore CE is given [Def. 1]. And the whole AC is given; [therefore] the remainder AE is given [Dt 4].

And since EC is equal to BD, let the common magnitude BC have been subtracted; then the remainder BE is equal to the remainder CD. And AE is given; therefore AB is greater than CD by a given magnitude [Def. 9]. ∎

Remarks: gvn.P+Q – gvn.Q+R => P = R or gvn.P–R.

Dt 12 has nothing to do with the theory of R'11; it probably crept in because of its verbal likeness to Dt 13 and 17..19, *If there be three magnitudes ...* The Arabic translation puts it before Dt 10. – The statement is an extension of Dt 4, *If a given magnitude be subtracted from a given magnitude, the remainder will be given.*

Dt 4 ensures that the difference (P+Q)–(Q+R) is given; so the problem is reduced to showing that the said difference is P–R. If P+Q = Q+R, the difference P–R is zero; to cope with that situation the Greeks had to divide the proposition in two, since there is no zero-magnitude.

Geometrically, in order to determine the difference between the two magnitudes, say line segments, (a magnitude equal to) the lesser must be put as part of the greater. That is why he must involve EB to see the difference AE. Algebraically, Dt 12 is quite elementary. We should not, however, (on the basis of symbols developed afterwards) underrate the usefulness of the theorem. Students are not always happy when asked to 'cancel a minus parenthesis' even as simple as $x+y-(y+z)$.

The text could be presented as a conditional string:

Hypotheses	gvn. AC = AB+BC	else	if AC > BD
	gvn. BD = BC+CD.		then put CE = BD
Assertion	AB = CD or		whence gvn.CE;
	gvn. AB–CD.		and AE = AC–CE,
Proof			whence gvn. AE;
if	AC = BD		and CE–BC = BD–BC,
then	AC–BC = BD–BC,		that is BE = CD,
	∴ AB = CD;		∴ AB–CD = AB–BE = AE.

Thus AB is greater than CD by the given magnitude AE.

Θε 13. 'Εὰν ᾖ τρία μεγέθη, καὶ τὸ μὲν πρῶτον πρὸς τὸ δεύτερον λόγον ἔχῃ δεδομένον, τὸ δὲ δεύτερον τοῦ τρίτου δοθέντι μεῖζον ᾖ ἢ ἐν λόγῳ, καὶ τὸ πρῶτον τοῦ τρίτου δοθέντι μεῖζον ἔσται ἢ ἐν λόγῳ.

ἔστω τρία μεγέθη τὰ ΑΒ, ΓΔ, Ε, καὶ τὸ μὲν ΑΒ πρὸς τὸ ΓΔ λόγον ἐχέτω δεδομένον, τὸ δὲ ΓΔ τοῦ Ε δοθέντι μεῖζον ἔστω ἢ ἐν λόγῳ· λέγω, ὅτι καὶ τὸ ΑΒ τοῦ Ε δοθέντι μεῖζόν ἐστιν ἢ ἐν λόγῳ.

ἐπεὶ γὰρ τὸ ΓΔ τοῦ Ε δοθέντι μεῖζόν ἐστιν ἢ ἐν λόγῳ, ἀφῃρήσθω τὸ δοθὲν μέγεθος τὸ ΓΖ· λοιποῦ ἄρα τοῦ ΔΖ πρὸς τὸ Ε λόγος ἐστὶ δοθείς. καὶ ἐπεὶ λόγος ἐστὶ δοθεὶς τοῦ ΑΒ πρὸς τὸ ΓΔ, ὁ αὐτὸς αὐτῷ γεγονέτω ὁ τοῦ ΑΗ πρὸς τὸ ΓΖ· λόγος ἄρα καὶ τοῦ ΑΗ πρὸς τὸ ΓΖ δοθείς. δοθὲν δὲ τὸ ΓΖ· δοθὲν ἄρα καὶ τὸ ΑΗ· καὶ λοιποῦ τοῦ ΗΒ πρὸς λοιπὸν τὸ ΔΖ λόγος ἐστὶ δοθείς. τοῦ δὲ ΔΖ πρὸς τὸ Ε λόγος ἐστὶ δοθείς· καὶ τοῦ ΗΒ ἄρα πρὸς τὸ Ε λόγος ἐστὶ δοθείς. καί ἐστι δοθὲν τὸ ΑΗ· τὸ ΑΒ ἄρα τοῦ Ε δοθέντι μεῖζόν ἐστιν ἢ ἐν λόγῳ.

Dt 13. *If there be three magnitudes, and the first have to the second a given ratio, while the second be greater than in ratio to the third by a given* [magnitude], *the first will also be greater than in ratio to the third by a given* [magnitude].

Let AB, CD, E be three magnitudes, and let AB have to CD a given ratio, and let CD be greater than in ratio to E by a given magnitude; I say that AB is also greater than in ratio to E by a given magnitude.

```
A        H                                    B
C   Z                           D
            E
```

For, since CD is greater than in ratio to E by a given magnitude, let the given magnitude CZ have been subtracted; then the ratio of the remainder DZ to E is given.

CHAPTER 3. BY A GIVEN GREATER THAN IN RATIO. (10-21)

And since the ratio AB:CD is given, let the same ratio have become AH:CZ; then the ratio AH:CZ is given [Def. 2*]. And CZ is given; therefore AH is also given [Dt 2]; and the ratio of the remainders HB:DZ is given [V.19].

And the ratio DZ:E is given; therefore the ratio HB:E is given [Dt 8]. And AH is given; therefore AB is by a given greater than in ratio to E [Def. 11]. ∎

Remarks: gvn.x:y & y δ>: z => x δ>: z.

Dt 13 proves a mixed 'transitivity'. The problem can be analyzed as follows: The second premiss y δ>: z means (Def. 11) that there is a given magnitude g such that the ratio (y–g):z is given (say a:b, which is not seen). Suppose that the conclusion is true, x δ>: z; then there is a given magnitude h such that the ratio (x–h):z is also given; by Dt 8 follows that the ratio (x–h):(y–g) must be a given ratio. If this were the same as x:y, the one given ratio mentioned, the ratio h:g will also be that one (because of V.19) and therefore given. Thus an analysis suggests that we try to get h:g the same as the given ratio x:y.

Since y δ>: z		CD δ>: E	
there is [Def. 11] a given g		gvn.CZ	
so that gvn.(y–g):z	{:: b:c}	gvn.DZ:E	
Get h such that h:g :: x:y		AH:CZ :: AB:CD	
[Def. 2*] gvn.h:g	{:: a:b}	gvn.AH:CZ	
[Dt 2] gvn.h		gvn.AH	
[[V.19] (x–h):(y–g) :: x:y		HB:DZ :: AB:CD]	
[Def. 2*] gvn.(x–h):(y–g)	{:: a:b}	gvn.HB:DZ	
[Dt 8, transitivity] gvn.(x–h):z	{:: a:c}	HB:E	
which means [Def. 11] x δ>: z.		AB δ>: E.	

Dt 13 should be compared with Dt 19, the transitivity property for the R'11 relation; *mutatis mutandis* the proofs run parallel.

◆

Θε 14. Ἐὰν δύο μεγέθη πρὸς ἄλληλα λόγον ἔχῃ δεδομένον, καὶ προστεθῇ ἑκατέρῳ αὐτῶν δεδομένον μέγεθος, τὰ ὅλα πρὸς ἄλληλα ἤτοι λόγον ἕξει δεδομένον, ἢ τὸ ἕτερον τοῦ ἑτέρου δοθέντι μεῖζόν ἐστιν ἢ ἐν λόγῳ.

δύο γὰρ μεγέθη τὰ ΑΒ, ΓΔ πρὸς ἄλληλα λόγον ἐχέτω δεδομένον, καὶ προσκείσθω ἑκατέρῳ αὐτῶν δεδομένον μέγεθος, τό τε ΑΕ καὶ τὸ ΓΖ· λέγω, ὅτι τὰ ὅλα τὰ ΕΒ, ΖΔ πρὸς ἄλληλα ἤτοι λόγον ἔχει δεδομένον, ἢ τὸ ἕτερον τοῦ ἑτέρου δοθέντι μεῖζόν ἐστιν ἢ ἐν λόγῳ.

ἐπεὶ γὰρ δοθέν ἐστιν ἑκάτερον τῶν ΕΑ, ΖΓ, λόγος ἄρα τοῦ ΕΑ πρὸς τὸ ΖΓ δοθείς. καὶ εἰ μὲν ὁ αὐτὸς τῷ τοῦ ΑΒ πρὸς ΓΔ, ἔσται καὶ ὅλου τοῦ ΕΒ πρὸς ὅλον τὸ ΖΔ λόγος δοθείς.

μὴ ἔστω δὴ ὁ αὐτὸς καὶ πεποιήσθω ὡς τὸ ΑΒ πρὸς ΓΔ, οὕτως τὸ ΗΑ πρὸς ΓΖ· λόγος ἄρα καὶ τοῦ ΗΑ πρὸς τὸ ΖΓ δοθείς. δοθὲν δὲ τὸ ΖΓ· δοθὲν ἄρα καὶ τὸ ΗΑ. ἔστι δὲ καὶ τὸ ΕΑ δοθέν· καὶ λοιπὸν ἄρα τὸ ΕΗ δοθέν ἐστιν. καὶ ἐπεὶ ὡς τὸ ΑΒ πρὸς τὸ ΓΔ, οὕτως τὸ ΗΑ πρὸς τὸ ΖΓ, λόγος ἄρα καὶ τοῦ ΗΒ πρὸς ΖΔ δοθείς. καί ἐστι δοθὲν τὸ ΕΗ· τὸ ΕΒ ἄρα τοῦ ΖΔ δοθέντι μεῖζόν ἐστι ἢ ἐν λόγῳ.

Dt 14. *If two magnitudes have a given ratio to one another, and a given magnitude be added to each of them, the wholes will either have a given ratio to one another, or one of them is by a given magnitude greater than in ratio to the other.*

For, let the two magnitudes AB, CD have a given ratio to one another, and let a given magnitude be added to each of them, namely AE and CZ. I say that the wholes EB, ZD either have a given ratio to one another, or one of them is greater than in ratio to the other by a given magnitude.

```
B          A                    H           E
D              C                            Z
```

For, since both EA and ZC are given, the ratio EA:ZC is given [Dt 1].

[1] And if it is the same as AB:CD, the ratio of the whole EB to the whole ZD will be given [V.12; Def. 2*].

[2] But then let it not be the same, and let have been made HA:CZ :: AB:CD; then the ratio HA:ZC is given. And ZC is given; therefore HA is given [Dt 2]. EA is given; therefore the remainder EH is given [Dt 4].

And since AB:CD :: HA:ZC, the ratio HB:ZD is given [V.12; Def. 2*]. And EH is given; therefore EB is by a given greater than in ratio to ZD [Def. 11]. ∎

Remarks: gvn.x:y AND gvn.c AND gvn.d =>
    gvn.$(x+c)$:$(y+d)$ OR $x+c$ δ>: $y+d$ [OR $y+d$ δ>: $x+c$].

Elements V.12 and 19 (cf. p. 57) prove that addition (or subtraction) of given magnitudes to (or from) the antecedent and consequent of a given ratio leads to the same ratio if the magnitudes added are in the given ratio. Dt 14 and 15 establish a necessary condition, that the operation leads to the same ratio *only if* the magnitudes are in the given ratio. As such the propositions are of some importance, although they seem to 'prove' a trivial trichotomy.

CHAPTER 3. BY A GIVEN GREATER THAN IN RATIO. (10–21)

Their immediate purpose is to serve as lemmas for Dt 17 and 18, which also seem to 'prove' a trivial trichotomy of doubtful use. The trichotomies are not spelled out in the proofs, but are hidden in the verbal phrasing as in Dt 14, 'they will either have a given ratio to one another or one of them is greater than in ratio to the other ...' – without any discussion of which of them will be the greater. Dt 14 is (superfluously) repeated in Dt 21 and Dt 15 in Dt 20, the latter being justified by the lack of negative magnitudes.

Since gvn.c, gvn.d.		gvn.AE, gvn.CZ	
∴ [Dt 1]	gvn.c:d	gvn.EA:ZC	
If c:d :: x:y		If EA:ZC :: AB:CD	
then [V.12]	$(x+c)$:$(y+d)$ :: c:d	then EB:ZD :: AE:CZ.	
[Def. 2*]	gvn.$(x+c)$:$(y+d)$	gvn.EB:ZD	
else get e so that e:d :: x:y		else get AH so that HA:CZ :: AB:CD	
[Def. 2*]	gvn.e:d	gvn.HA:ZC	
[Dt 2]	gvn.e	gvn.HA	
[tacit assumption $c>e$]		[H between A and E]	
[Dt 4, name h]	gvn.$c-e$ (= h)	gvn.EH	
[6, V.12	$(x+e)$:$(y+d)$ :: e:d	BH:DZ]	
Def. 2*	gvn.$(x+e)$:$(y+d)$		
that is	gvn.$(x+c-h)$:$(y+d)$	gvn.HB:ZD	
[Def. 11]	$x+c$ δ>: $y+d$	EB δ>: ZD.	

What happens if c is less than e so that H falls beyond E? Euclid does not treat that case,[74] but would probably make AE:CH :: AB:CD with H between C and Z, and follow the same track, ending with DZ δ>: EB, in our notation $y+d$ δ>: $x+c$. It is still true that *one is by a given magnitude greater than the other than in ratio*.

Θε 15. Ἐὰν δύο μεγέθη πρὸς ἄλληλα λόγον ἔχῃ δεδομένον καὶ ἀφαιρεθῇ ἀπὸ ἑκατέρου αὐτῶν δεδομένον μέγεθος, τὰ λοιπὰ πρὸς ἄλληλα ἤτοι λόγον ἕξει δεδομένον, ἢ τὸ ἕτερον τοῦ ἑτέρου δοθέντι μεῖζόν ἐστιν ἢ ἐν λόγῳ.

δύο γὰρ μεγέθη τὰ ΑΒ, ΓΔ πρὸς ἄλληλα λόγον ἐχέτω δεδομένον, καὶ ἀφῃρήσθω ἀφ' ἑκατέρου αὐτῶν δεδομένον μέγεθος, ἀπὸ μὲν τοῦ ΑΒ τὸ ΕΑ, ἀπὸ δὲ τοῦ ΓΔ τὸ ΓΖ· λέγω, ὅτι τὰ λοιπὰ τὰ ΕΒ, ΖΔ πρὸς ἄλληλα ἤτοι λόγον ἕξει δεδομένον, ἢ τὸ ἕτερον τοῦ ἑτέρου δοθέντι μεῖζόν ἐστιν ἢ ἐν λόγῳ.

---

[74] Nor does he treat all cases of theorems in the *Elements*.

ἐπεὶ γὰρ ἑκάτερον τῶν ΑΕ, ΓΖ δοθέν ἐστι, λόγος ἄρα τοῦ ΑΕ πρὸς ΓΖ δοθείς. καὶ εἰ μὲν ὁ αὐτός ἐστι τῷ τοῦ ΑΒ πρὸς ΓΔ, ἔσται καὶ λοιποῦ τοῦ ΕΒ πρὸς λοιπὸν τὸ ΖΔ λόγος δοθείς.

μὴ ἔστω δὴ ὁ αὐτός, καὶ πεποιήσθω ὡς τὸ ΑΒ πρὸς ΓΔ, οὕτως τὸ ΑΗ πρὸς τὸ ΓΖ. λόγος δὲ τοῦ ΑΒ πρὸς τὸ ΓΔ δοθείς· λόγος ἄρα καὶ τοῦ ΑΗ πρὸς τὸ ΓΖ δοθείς· δοθὲν δὲ τὸ ΓΖ· δοθὲν ἄρα καὶ τὸ ΑΗ. ἔστι δὲ καὶ τὸ ΑΕ δοθέν· καὶ λοιπὸν ἄρα τὸ ΕΗ δοθέν ἐστιν. καὶ ἐπεὶ ὡς τὸ ΑΒ πρὸς τὸ ΓΔ, οὕτως τὸ ΑΗ πρὸς τὸ ΓΖ, λοιποῦ ἄρα τοῦ ΗΒ πρὸς λοιπὸν τὸ ΖΔ λόγος ἐστὶ δοθείς. καί ἐστι δοθὲν τὸ ΕΗ· τὸ ΕΒ ἄρα τοῦ ΖΔ δοθέντι μεῖζόν ἐστιν ἢ ἐν λόγῳ.

Dt 15. *If two magnitudes have a given ratio to one another, and a given magnitude be subtracted from each of them, the remainders either will have a given ratio to one another, or one of them is by a given greater than in ratio to the other.*

Let the two magnitudes AB, CD have a given ratio to one another, and let a given magnitude have been subtracted from each of them, AE from AB and CZ from CD. I say that the remainders EB, ZD either will have a given ratio to one another, or one of them is by a given greater than the other than in ratio.

A_____E_____H_____B

C_____Z_____D

For, since both AE and CZ are given, the ratio AE:CZ is given [Dt 1].
[1] And if it is the same as AB:CD, the ratio of the remainder EB to the remainder ZD will also be given [V.19, Def. 2*].

[2] But then let it not be the same, and let have been made AH:CZ :: AB:CD. And the ratio AB:CD is given; therefore the ratio AH:CZ is also given. And CZ is given; therefore AH is given [Dt 2]. AE is given; therefore the remainder EH is given [Dt 4].

And since AB:CD :: AH:CZ, the ratio of the remainder HB to the remainder ZD is also given [V.19, Def. 2*]. And EH is given; therefore EB is by a given greater than ZD than in ratio [Def. 11]. ∎

Remarks: gvn.*x*:*y* AND gvn.*c* AND gvn.*d* =>
gvn.(*x–c*):(*y–d*) OR *x–c* δ>: *y–d* [OR *y–d* δ>: *x–c*]

Dt 15 is *mutatis mutandis* the subtractive counterpart of Dt 14 and as such deserves few remarks. If, however, the magnitudes to be subtracted from the antecedent and consequent are greater than those, Euclid must formulate the problem otherwise, and does so in Dt 20, since negative magnitudes are not defined. Apart from that, Dt 20

CHAPTER 3. BY A GIVEN GREATER THAN IN RATIO. (10–21)

runs exactly as Dt 15. But then he seems to have felt a need to duplicate Dt 14 also, although the negative problem does not affect that one; so Dt 21 repeats Dt 14, again with almost the same wording and syntax. What do we make of this redundancy? It points rather to an unfinished draft than to a thoroughly elaborated edition, unless it is the work of a later editor. Did Euclid understand the kind of relation he was treating? Is Dt 15 a trivial statement? Compare the trichotomy

$$x = y \Rightarrow x{-}c = y{-}d \quad \text{OR} \quad x{-}c > y{-}d \quad \text{OR} \quad x{-}c < y{-}d.$$

◆

Θε 16. Ἐὰν δύο μεγέθη πρὸς ἄλληλα λόγον ἔχῃ δεδομένον, καὶ ἀπὸ μὲν τοῦ ἑνὸς αὐτῶν δεδομένον μέγεθος ἀφαιρεθῇ, τῷ δὲ ἑτέρῳ αὐτῶν δεδομένον μέγεθος προστεθῇ, τὸ ὅλον τοῦ λοιποῦ δοθέντι μεῖζον ἔσται ἢ ἐν λόγῳ.

δύο γὰρ μεγέθη τὰ ΑΒ, ΓΔ λόγον ἐχέτω δεδομένον, καὶ ἀπὸ μὲν τοῦ ΓΔ δεδομένον μέγεθος ἀφῃρήσθω τὸ ΓΕ, τῷ δὲ ΑΒ δεδομένον μέγεθος προσκείσθω τὸ ΖΑ. λέγω, ὅτι ὅλον τὸ ΖΒ τοῦ λοιποῦ τοῦ ΕΔ δοθέντι μεῖζόν ἐστιν ἢ ἐν λόγῳ.

ἐπεὶ γὰρ λόγος ἐστὶ τοῦ ΑΒ πρὸς ΓΔ δοθείς, ὁ αὐτὸς αὐτῷ γεγονέτω τοῦ ΑΗ πρὸς τὸ ΓΕ· λόγος ἄρα καὶ τοῦ ΑΗ πρὸς τὸ ΓΕ δοθείς· δοθὲν δὲ τὸ ΓΕ· δοθὲν ἄρα καὶ τὸ ΑΗ. ἔστι δὲ καὶ τὸ ΑΖ δοθέν· ὅλον ἄρα τὸ ΖΗ δοθέν ἐστιν. καὶ ἐπεὶ ὡς τὸ ΑΒ πρὸς τὸ ΓΔ, οὕτως τὸ ΑΗ πρὸς ΓΕ, καὶ λοιποῦ τοῦ ΗΒ πρὸς λοιπὸν τὸ ΕΔ λόγος ἐστὶ δοθείς. καί ἐστι δοθὲν τὸ ΗΖ· τὸ ΖΒ ἄρα τοῦ ΕΔ δοθέντι μεῖζόν ἐστιν ἢ ἐν λόγῳ.

Dt 16. *If two magnitudes have a given ratio to one another, and from the one a given magnitude be subtracted, while to the other a given magnitude be added, the whole will be greater than in ratio to the remainder by a given magnitude.*

For, let two magnitudes AB, CD have a given ratio to one another, and from CD let the given magnitude CE have been subtracted, while to AB the given magnitude ZA be added. I say that the whole ZB is greater than in ratio to the remainder ED by a given magnitude.

```
Z_____A_____H_____B
C_____E____D
```

For, since the ratio AB:CD is given, let the same ratio as that have become AH:CE. Then the ratio AH:CE is given [Def. 2]; and CE is given. Therefore AH is also given [Dt 2]. But AZ is given; therefore the whole ZH is given [Dt 3].

And since AH:CE :: AB:CD, the ratio of the remainder HB to the remainder ED is also given [V.19; Def. 2]. And HZ is given; therefore ZB is by a given greater than in ratio to ED [Def. 11]. ∎

Remarks: gvn.*x*:*y* AND gvn.*c* AND gvn.*d* => *x*+*c* δ>: *y*–*d*.

A mixture of Dt 14 and 15; adding a magnitude to the antecedent and subtracting another from the consequent of a given ratio inevitably leads to the R'11 relation. Trivial? Compare $x = y \Rightarrow x+c > y-d$.

Since gvn.*x*:*y*   {:: *a*:*b*}   gvn.AB:CD
get *e* (4'th prop) *e*:*d* :: *x*:*y*   get AH, AH:CE :: AB:CD
[Def. 2*] gvn.*e*:*d*   gvn.AH:CE
[Dt 2]   gvn.*e*   gvn.AH
[Dt 3]   gvn.*c*+*e*   gvn.ZH
[line 2, V.19 (*x*–*e*):(*y*–*d*) :: *x*:*y*   HB:ED :: AB:CD   ]
[Def. 2*] gvn.(*x*–*e*):(*y*–*d*)   {:: *a*:*b*}   gvn.HB:DE
[*x*–*e* = *x*+*c*–(*c*+*e*)]
[Def. 11] (*x*+*c*) δ>: (*y*–*d*).   ZB δ>: ED

Putting $a:b = \kappa$ (cf note 28) Clemens Thaer renders this result as 'if $x = \kappa y$ then $(x+c) = \kappa(y-d) + (c+\kappa d)$'. Rightly, of course (since $e = \kappa d$), but does it represent the text of the *Data*? Similar remarks can be made about all his algebraizations. – About the following four theorems (Dt 17–20) he maintains that they prove that all linear transformations, $\{ax + b \mid a,b \in \mathbb{R}\}$ form a group ('dass die ganzen linearen Substitutionen einer Veränderlichen eine Gruppe bilden'). I am not sure that I understand what he means to say, and the *Data* certainly does not help me, – so probably Euclid would not understand either.

◆

Θε 17. Ἐὰν ᾖ τρία μεγέθη, καὶ τὸ πρῶτον τοῦ δευτέρου δοθέντι μεῖζον ᾖ ἢ ἐν λόγῳ, ᾖ δὲ καὶ τὸ τρίτον τοῦ αὐτοῦ δοθέντι μεῖζον ἢ ἐν λόγῳ, τὸ πρῶτον πρὸς τὸ τρίτον ἤτοι λόγον ἕξει δεδομένον, ἢ τὸ ἕτερον τοῦ ἑτέρου δοθέντι μεῖζον ἔσται ἢ ἐν λόγῳ.

ἔστω τρία μεγέθη τὰ ΑΒ, Γ, ΔΕ, καὶ ἑκάτερον τῶν ΑΒ, ΔΕ τοῦ Γ δοθέντι μεῖζον ἔστω ἢ ἐν λόγῳ· λέγω, ὅτι τὰ ΑΒ, ΔΕ ἤτοι πρὸς ἄλληλα λόγον ἔχει δεδομένον ἢ τὸ ἕτερον τοῦ ἑτέρου δοθέντι μεῖζόν ἐστιν ἢ ἐν λόγῳ.

ἐπεὶ γὰρ τὸ ΔΕ τοῦ Γ δοθέντι μεῖζόν ἐστιν ἢ ἐν λόγῳ, ἀφῃρήσθω τὸ δοθὲν μέγεθος τὸ ΔΗ· λοιποῦ ἄρα τοῦ ΗΕ πρὸς τὸ Γ λόγος ἐστὶ δοθείς. διὰ τὰ αὐτὰ δὴ καὶ τοῦ ΖΒ πρὸς τὸ Γ λόγος ἐστὶ δοθείς· καὶ τοῦ ΖΒ ἄρα πρὸς τὸ ΗΕ λόγος ἐστὶ δοθείς. καὶ πρόσκειται αὐτοῖς δεδομένα μεγέθη τὰ ΑΖ, ΔΗ· τὰ ὅλα ἄρα τὰ ΑΒ, ΔΕ πρὸς ἄλληλα ἤτοι λόγον ἔχει δεδομένον, ἢ τὸ ἕτερον τοῦ ἑτέρου δοθέντι μεῖζόν ἐστιν ἢ ἐν λόγῳ.

CHAPTER 3. BY A GIVEN GREATER THAN IN RATIO. (10–21) 77

Dt 17. *If there be three magnitudes, and the first be greater than in ratio to the second by a given magnitude, while the third be greater than in ratio to the same by a given magnitude, then the first will either have a given ratio to the third, or one of them will be greater than in ratio to the other by a given magnitude.*

Let there be three magnitudes AB, C, DE,[75] and let each of AB and DE be greater than in ratio to C by a given magnitude. I say that AB and CD either have a given ratio to one another, or one of them is greater than in ratio to the other by a given magnitude.

```
A        Z                              B
C_____
D           H                    E
```

For, since DE is by a given greater than in ratio to C, let the given magnitude DH have been subtracted. Then the ratio of the remainder HE to C is given [Def. 11]. For the same reason the ratio of ZB to C is given. Therefore ZB:HE is a given ratio [Dt 8]. And to them are added given magnitudes AZ and DH. Therefore the wholes AB and DE either have to one another a given ratio, or one of them is greater than in ratio to the other by a given magnitude [Dt 14]. ■

Remarks: x δ>: y AND z δ>: y => gvn.x:z OR x δ>: z [OR z δ>: x].

Dt 19 proves (in modern terms) that R'11 is an order relation, and Dt 17 proves that no conclusion can be drawn about two magnitudes having this relation to one and the same magnitude. The proof uses Dt 14, with the same tacit assumption.
Compare $x > y$ AND $z > y => x = z$ OR $x > z$ OR $z > x$.

Θε 18. Ἐὰν ᾖ τρία μεγέθη, ἓν δὲ αὐτῶν ἑκατέρου τῶν λοιπῶν δοθέντι μεῖζον ᾖ ἢ ἐν λόγῳ, τὰ λοιπὰ δύο πρὸς ἄλληλα ἤτοι λόγον ἕξει δεδομένον, ἢ τὸ ἕτερον τοῦ ἑτέρου δοθέντι μεῖζόν ἐστιν ἢ ἐν λόγῳ.

ἔστω τρία μεγέθη τὰ ΑΒ, ΓΔ, ΕΖ, ἓν δὲ αὐτῶν τὸ ΓΔ ἑκατέρου τῶν λοιπῶν τῶν ΑΒ, ΕΖ δοθέντι μεῖζον ἔστω ἢ ἐν λόγῳ. λέγω, ὅτι τὸ ΑΒ πρὸς τὸ ΕΖ ἤτοι λόγον ἔχει δεδομένον, ἢ τὸ ἕτερον τοῦ ἑτέρου δοθέντι μεῖζόν ἐστιν ἢ ἐν λόγῳ.

---

[75] Economy of names: if a magnitude must be split in the proof, it must have names to its extremities; if not, a single letter will do. Cp. Dt 13 and 19.

ἐπεὶ γὰρ τὸ ΓΔ τοῦ ΑΒ δοθέντι μεῖζόν ἐστιν ἢ ἐν λόγῳ, ἀφῃρήσθω τὸ δοθὲν μέγεθος τὸ ΓΗ. λοιποῦ ἄρα τοῦ ΗΔ πρὸς τὸ ΑΒ λόγος ἐστὶ δοθείς. ὁ αὐτὸς αὐτῷ γεγονέτω ὁ τοῦ ΓΗ πρὸς τὸ ΑΘ. λόγος ἄρα καὶ τοῦ ΓΗ πρὸς τὸ ΑΘ δοθείς. δοθὲν δὲ τὸ ΓΗ. δοθὲν ἄρα καὶ τὸ ΑΘ. καὶ ὅλου τοῦ ΓΔ πρὸς ὅλον τὸ ΘΒ λόγος ἐστὶ δοθείς.

πάλιν, ἐπεὶ τὸ ΓΔ τοῦ ΕΖ δοθέντι μεῖζόν ἐστιν ἢ ἐν λόγῳ, ἀφῃρήσθω τὸ δοθὲν μέγεθος τὸ ΓΚ. λοιποῦ τοῦ ΚΔ πρὸς ΕΖ λόγος ἐστὶ δοθείς. ὁ αὐτὸς αὐτῷ γεγονέτω ὁ τοῦ ΓΚ πρὸς ΛΕ. λόγος ἄρα καὶ τοῦ ΓΚ πρὸς ΛΕ δοθείς. δοθὲν δὲ τὸ ΓΚ. δοθὲν ἄρα καὶ τὸ ΛΕ. καὶ ὅλου τοῦ ΓΔ πρὸς ὅλον τὸ ΛΖ λόγος ἐστὶ δοθείς.

τοῦ δὲ ΓΔ πρὸς ΘΒ λόγος ἐστὶ δοθείς. καὶ τοῦ ΘΒ ἄρα πρὸς ΛΖ λόγος ἐστὶ δοθείς. καὶ ἀφῄρηται ἀπ' αὐτῶν δεδομένα μεγέθη τὰ ΘΑ, ΛΕ. τὰ ΑΒ, ΕΖ ἄρα ἤτοι πρὸς ἄλληλα λόγον ἕξει δεδομένον, ἢ τὸ ἕτερον τοῦ ἑτέρου δοθέντι μεῖζόν ἐστιν ἢ ἐν λόγῳ.

Dt 18. *If there be three magnitudes, and one of them be greater than in ratio to each of the remaining ones by a given magnitude, the remaining ones either will have a given ratio to one another, or one of them is greater than in ratio to the other by a given magnitude.*

For, let there be three magnitudes AB, CD, EZ, and let one of them CD be greater than in ratio to each of the remaining ones, AB, EZ, by a given magnitude. I say that either AB has a given ratio to EZ, or one of them is greater than in ratio to the other by a given magnitude.

```
Q         A                              B
C    H    K               D
L         E                    Z
```

For, since CD is greater than in ratio to AB by a given magnitude, let the given magnitude CH have been subtracted. Then the ratio of the remainder HD to AB is given [Def. 11].

Let the same ratio have become CH:AQ. Then the ratio CH:AQ is given [Def. 2]. And CH is given; therefore AQ is given [Dt 2], and the ratio of the whole of CD to the whole of QB is given [V.12, Def. 2].

Again, since CD is greater than in ratio to EZ by a given magnitude, let the given magnitude CK have been subtracted. [Then] the ratio of the remainder KD to EZ is given [Def. 11].

Let the same ratio have become CK:LE. Then the ratio CK:LE is given [Def. 2]. And CK is given; therefore LE is given, [Dt 2], and the ratio of the whole of CD to the whole of LZ is given [V.12, Def. 2].

And the ratio CD:QB is given; therefore the ratio QB:LZ is given [Dt 8]. And the given magnitudes QA, LE have been subtracted from them. Therefore AB, EZ either will have a given ratio to one another, or one of them is greater than in ratio to the other by a given magnitude [Dt 15]. ∎

CHAPTER 3. BY A GIVEN GREATER THAN IN RATIO. (10–21)   79

Remarks: x δ>: y AND x δ>: z => gvn.y:z OR y δ>: z [OR z δ>: y].
Another trivial statement, elaborately proved.
Compare x > y AND x > z => y = z OR y > z OR y < z.

◆

Θε 19. Ἐὰν ᾖ τρία μεγέθη, καὶ τὸ μὲν πρῶτον τοῦ δευτέρου δοθέντι μεῖζον ᾖ ἢ ἐν λόγῳ, ᾖ δὲ καὶ τὸ δεύτερον τοῦ τρίτου δοθέντι μεῖζον ἢ ἐν λόγῳ, καὶ τὸ πρῶτον τοῦ τρίτου δοθέντι μεῖζον ἔσται ἢ ἐν λόγῳ.

ἔστω τρία μεγέθη τὰ ΑΒ, ΓΔ, Ε, καὶ τὸ μὲν ΑΒ τοῦ ΓΔ δοθέντι μεῖζον ἔστω ἢ ἐν λόγῳ, τὸ δὲ ΓΔ τοῦ Ε δοθέντι μεῖζον ἔστω ἢ ἐν λόγῳ. λέγω, ὅτι καὶ τὸ ΑΒ τοῦ Ε δοθέντι μεῖζόν ἐστιν ἢ ἐν λόγῳ.

ἐπεὶ γὰρ τὸ ΓΔ τοῦ Ε δοθέντι μεῖζόν ἐστιν ἢ ἐν λόγῳ, ἀφῃρήσθω τὸ δοθὲν μέγεθος τὸ ΓΖ· λοιποῦ ἄρα τοῦ ΖΔ πρὸς τὸ Ε λόγος ἐστὶ δοθείς. πάλιν, ἐπεὶ τὸ ΑΒ τοῦ ΓΔ δοθέντι μεῖζόν ἐστιν ἢ ἐν λόγῳ, ἀφῃρήσθω τὸ δοθὲν μέγεθος τὸ ΑΗ· λοιποῦ ἄρα τοῦ ΗΒ πρὸς τὸ ΓΔ λόγος ἐστὶ δοθείς. ὁ αὐτὸς αὑτῷ γεγονέτω τοῦ ΗΘ πρὸς τὸ ΓΖ· λόγος ἄρα καὶ τοῦ ΗΘ πρὸς τὸ ΓΖ δοθείς. δοθὲν δὲ τὸ ΓΖ· δοθὲν ἄρα καὶ τὸ ΗΘ. ἔστι δὲ καὶ τὸ ΗΑ δοθέν· καὶ ὅλον ἄρα τὸ ΘΑ δοθέν ἐστιν. καὶ ἐπεὶ ὡς τὸ ΗΒ πρὸς τὸ ΓΔ, οὕτως τὸ ΗΘ πρὸς τὸ ΓΖ, καὶ λοιποῦ τοῦ ΘΒ πρὸς λοιπὸν τὸ ΖΔ λόγος ἐστὶ δοθείς. τοῦ δὲ ΖΔ πρὸς τὸ Ε λόγος ἐστὶ δοθείς· καὶ τοῦ ΘΒ ἄρα πρὸς τὸ Ε λόγος ἐστὶ δοθείς. καὶ δοθὲν τὸ ΘΑ· τὸ ΒΑ ἄρα τοῦ Ε δοθέντι μεῖζόν ἐστιν ἢ ἐν λόγῳ.

Dt 19.   *If there be three magnitudes, and the first be greater than in ratio to the second by a given magnitude, and the second be greater than in ratio to the third by a given magnitude, the first will be by a given greater than in ratio to the third by a given magnitude.*

Let there be three magnitudes AB, CD, E, and let AB be by a given greater than in ratio to CD, and let CD be by a given greater than in ratio to E. I say that AB is by a given greater than in ratio to E.

For, since CD is by a given greater than in ratio to E, let the given magnitude CZ be subtracted. Then the ratio of the remainder ZD to E is given [Def. 11].

```
A        H        Q                    B
C    Z                    D
E
```

Again, since AB is by a given greater than in ratio to CD, let the given magnitude AH have been subtracted. Then the ratio of the remainder HB to CD is given [Def. 11].

Let the same ratio have become HQ:CZ. Then the ratio HQ:CZ is given. And CZ is given; therefore HQ is given [Dt 2]. And HA is given; therefore the whole QA is given [Dt 3].

And since HB:CD :: HQ:CZ, the ratio of the remainder QB to the remainder ZD is given [V.19, Def. 2]. And the ratio ZD:E is given. Therefore the ratio of QB to E is given [Dt 8]. And QA is given; therefore BA is by a given greater than in ratio to E [Def. 11]. ∎

Remarks: x δ>: y AND y δ>: z => x δ>: z..

Since y δ>: z	CD δ>: E
there is [Def. 11] a given g	gvn.CZ
such that gvn.$(y-g)$:z {:: b:c}	gvn.DZ:E
Since x δ>: y	AB δ>: CD
there is [Def. 11] a given h	gvn.AH
such that gvn.$(x-h)$:y {:: a:b}	gvn.BH:CD
get k such that k:g :: $(x-h)$:y	get HQ:CZ :: BH:CD
[Def. 2*] gvn.k:g {:: a:b}	gvn.HQ:CZ
[Dt 2] gvn.k	gvn.HQ
[Dt 3] gvn.$h+k$	gvn.AQ
[V.19 $(x-(h+k))$:$(y-g)$ :: k:g]	
gvn.$(x-(h+k))$:$(y-g)$ {:: a:b}	gvn.BQ:DZ
[Dt 8] gvn.$(x-(h+k))$:z {:: a:c}	gvn.BQ:E
[Def. 11] x δ>: z.	AB δ>: E.

Compare $x > y$ AND $y > z$ => $x > z$. And remember the proof of 'mixed' transitivity Dt 13.

At the beginning of his appendix (p. 190) Menge prints an alternative proof which (according to Thaer) is also found in the Arabic tradition. It is a bit shorter and may be called smarter and more systematic since it uses Dt 13, the other transitive statement:

```
A           E      Z                      B
C
D
```

Dt 19 alternative

Since x δ>: y	AB δ>: C
there is [Def. 11] a given.g	gvn.AE
such that gvn.$(x-g)$:y	gvn.EB:C.

CHAPTER 3. BY A GIVEN GREATER THAN IN RATIO. (10-21)   81

Since gvn($x-g$):y & y δ>: z gvn.EB:C & C δ>: D
[Dt 13]   ($x-g$) δ>: z EB δ>: D
therefore there is [Def. 11] a given h gvn.EZ
such that gvn.(($x-g$)–h):z gvn.ZB:D
[Dt 3]   gvn.$g+h$ gvn.AZ
[Def. 11] x δ>: z AB δ>: D.

◆

As I mentioned in the remarks on Dt 15, the next propositions 20 and 21 are repetitions of Dt 14 and 15, even down to words and syntax:

Θε 20. ' Ἐὰν ᾖ δύο μεγέθη δεδομένα, καὶ ἀφαιρεθῇ ἀπ' αὐτῶν μεγέθη πρὸς ἄλληλα λόγον ἔχοντα δεδομένον, τὰ λοιπὰ πρὸς ἄλληλα ἤτοι λόγον ἕξει δεδομένον, ἢ τὸ ἕτερον τοῦ ἑτέρου δοθέντι μεῖζόν ἐστιν ἢ ἐν λόγῳ.

ἔστω δύο μεγέθη δεδομένα τὰ ΑΒ, ΓΔ, καὶ ἀπὸ τῶν ΑΒ, ΓΔ ἀφῃρήσθω μεγέθη τὰ ΑΕ, ΓΖ λόγον ἔχοντα πρὸς ἄλληλα δεδομένον· λέγω, ὅτι τὰ ΕΒ, ΖΔ πρὸς ἄλληλα ἤτοι λόγον ἔχει δεδομένον, ἢ τὸ ἕτερον τοῦ ἑτέρου δοθέντι μεῖζόν ἐστιν ἢ ἐν λόγῳ.
ἐπεὶ γὰρ δοθέν ἐστιν ἑκάτερον τῶν ΑΒ, ΓΔ, λόγος ἄρα τοῦ ΑΒ πρὸς ΓΔ δοθείς. καὶ εἰ μὲν ὁ αὐτός ἐστι τῷ τοῦ ΑΕ πρὸς ΓΖ, ἔσται καὶ λοιποῦ τοῦ ΕΒ πρὸς λοιπὸν τὸ ΖΔ λόγος δοθείς. μὴ ἔστω δὴ ὁ αὐτός, καὶ πεποιήσθω ὡς τὸ ΕΑ πρὸς ΓΖ, οὕτως τὸ ΑΗ πρὸς ΓΔ. λόγος δὲ τοῦ ΑΕ πρὸς ΓΖ δοθείς· λόγος ἄρα καὶ τοῦ ΑΗ πρὸς ΓΔ δοθείς. δοθὲν δὲ τὸ ΓΔ· δοθὲν ἄρα καὶ τὸ ΑΗ. ἔστι δὲ καὶ τὸ ΑΒ δοθέν· καὶ λοιπὸν ἄρα τὸ ΗΒ δοθέν ἐστιν. καὶ ἐπεί ἐστιν ὡς τὸ ΑΕ πρὸς ΓΖ, οὕτως τὸ ΑΗ πρὸς τὸ ΓΔ, καὶ λοιποῦ τοῦ ΗΕ πρὸς λοιπὸν τὸ ΖΔ λόγος ἐστὶ δοθείς· δοθὲν δὲ τὸ ΗΒ· τὸ ΕΒ ἄρα τοῦ ΖΔ δοθέντι μεῖζόν ἐστιν ἢ ἐν λόγῳ.

Dt 20. *If there be two given magnitudes, and magnitudes having a given ratio to one another be subtracted from them, the remainders either will have a given ratio to one another, or one of them is by a given greater than in ratio to the other.*

Let there be two given magnitudes AB,CD, and from AB,CD let the magnitudes AE,CZ which have a given ratio to one another, be subtracted. I say that EB,ZD either have a given ratio to one another, or the one is by a given greater than in ratio to the other.

```
A         E      H                    B
C                Z          D
```

For, since each of AB and CD are given, the ratio AB:CD is given [Dt 1]. And if it is the same as AE:CZ, the ratio of the remainder EB to the remainder ZD will be given [V.19, Def. 2].

But then let it not be the same, and let have been made as EA to CZ, so AH to CD. The ratio AE:CZ is given; therefore the ratio AH:CD is given [Def. 2]. And CD is given; therefore AH is given. And AB is given; therefore the remainder HB is given [Dt 4].

And since AE:CZ :: AH:CD, the ratio of the remainder HE to the remainder ZD is given [V.19, Def. 2]. And HB is given. Therefore EB is by a given greater than in ratio to ZD [Def. 11]. ∎

Remarks: gvn.c AND gvn.d AND gvn.x:y =>
gvn.$(c-x)$:$(d-y)$ OR $c-x$ δ>: $d-y$ [OR $d-y$ δ>: $c-x$].

Dt 20 must replace Dt 15 if the given magnitudes c and d are greater than x and y respectively.

◆

Θε 21. Ἐὰν ᾖ δύο μεγέθη δεδομένα, καὶ προστεθῇ αὐτοῖς μεγέθη πρὸς ἄλληλα λόγον ἔχοντα δεδομένον, τὰ ὅλα πρὸς ἄλληλα ἤτοι λόγον ἕξει δεδομένον ἢ τὸ ἕτερον τοῦ ἑτέρου δοθέντι μεῖζόν ἐστιν ἢ ἐν λόγῳ.

ἔστω δύο μεγέθη δεδομένα τὰ ΑΒ, ΓΔ, καὶ προσκείσθω αὐτοῖς μεγέθη τὰ ΑΕ, ΓΖ λόγον ἔχοντα πρὸς ἄλληλα δεδομένον· λέγω, ὅτι τὰ ὅλα τὰ ΕΒ, ΖΔ πρὸς ἄλληλα ἤτοι λόγον ἕξει δεδομένον, ἢ τὸ ἕτερον τοῦ ἑτέρου δοθέντι μεῖζόν ἐστιν ἢ ἐν λόγῳ.

ἐπεὶ γὰρ δοθέν ἐστιν ἑκάτερον τῶν ΑΒ, ΓΔ, λόγος ἄρα τοῦ ΑΒ πρὸς τὸ ΓΔ δοθείς. καὶ εἰ μὲν ὁ αὐτός ἐστι τῷ τοῦ ΕΑ πρὸς τὸ ΓΖ, ἔσται καὶ ὅλου τοῦ ΕΒ πρὸς ὅλον τὸ ΖΔ λόγος δοθείς. ἔστι δὲ καὶ τὸ ΑΒ δοθέν· καὶ λοιπὸν ἄρα τὸ ΗΒ δοθέν ἐστιν. καὶ ἐπεί ἐστιν ὡς τὸ ΕΑ πρὸς ΖΓ, οὕτως τὸ ΑΗ πρὸς τὸ ΓΔ, καὶ ὅλου τοῦ ΕΗ πρὸς ὅλον τὸ ΖΔ λόγος ἐστὶ δοθείς. καὶ δοθὲν τὸ ΗΒ· τὸ ΕΒ ἄρα τοῦ ΖΔ δοθέντι μεῖζόν ἐστιν ἢ ἐν λόγῳ. εἰ δὲ οὔ, πεποιήσθω ὡς τὸ ΑΕ πρὸς ΓΖ, οὕτως τὸ ΗΑ πρὸς τὸ ΓΔ· λόγος ἄρα τοῦ ΗΑ πρὸς τὸ ΓΔ δοθείς. δοθὲν δὲ τὸ ΓΔ· δοθὲν ἄρα καὶ τὸ ΗΑ. ἔστι δὲ καὶ τὸ ΑΒ δοθέν· καὶ λοιπὸν ἄρα τὸ ΗΒ δοθέν ἐστιν. καὶ ἐπεί ἐστιν ὡς τὸ ΕΑ πρὸς τὸ ΖΓ, οὕτως τὸ ΑΗ πρὸς τὸ ΓΑ, καὶ ὅλου τοῦ ΕΗ πρὸς ὅλον τὸ ΖΔ λόγος ἐστὶ δοθείς. καί δοθὲν τὸ ΗΒ· τὸ ΕΒ ἄρα τοῦ ΖΔ δοθέντι μεῖζόν ἐστιν ἢ ἐν λόγῳ.

Dt 21. *If there be two given magnitudes, and magnitudes having a given ratio to one another be added to them, the wholes will either have a given ratio to one another, or one of them is by a given greater than in ratio to the other.*

Let there be two given magnitudes AB,CD, and let the magnitudes AE,CZ which have a given ratio to one another be added to them. I say that the wholes EB,ZD either will have a given ratio to one another, or the one is by a given greater than in ratio to the other.

```
B_____H_____A_____E
D_____C_____Z
```

For, since each of AB and CD are given, the ratio AB:CD is given [Dt 1]. And if it is the same as EA:CZ, the ratio of the whole EB to the whole ZD is given [V.12, Def. 2].

But if not, let have become as HA:CD :: AE:CZ. Then the ratio HA:CD is given. And CD is given; therefore HA is given[Dt 2]. And AB is given; therefore the remainder HB is given. And since EA:ZC :: AH:CD, the ratio of the whole EH to the whole ZD is given [V.12, Def. 2]. And HB is given; therefore EB is by a given greater than in ratio to ZD [Def. 11]. ∎

Remarks: gvn.*c* AND gvn.*d* AND gvn.*x*:*y* =>
gvn.(*c*+*x*):(*d*+*y*) OR *c*+*x* δ>: *d*+*y* [OR *c*+*y* δ>: *d*+*x*].

Our notation shows that Dt 21 proves the same truths as Dt 14 (since addition is commutative). Clemens Thaer claims it to be spurious 'although all the Greek manuscripts bring it', since it is absent from the Arabic tradition. I think he is going too far to 'save' Euclid: the *Data* is no thoroughly worked out system (as the *Elements, e.g.*), and the Arabs may well have tried to polish it by doing away with such repetitions. Probably we shall never know.

◆

Chapter 4. Magnitudes and Ratio II
Dt 22–24

These propositions seem to be misplaced (like Dt 12): Dt 22 and 23 could have been proved immediately after Dt 8; they are the first to use propositions from Book VI of the *Elements*, and do not belong in the series about 'Greater than in Ratio' (see the survey of deductive structure on p. 92), nor have they much to do with what follows, about position and form. However, a faint connexion with Dt 14 and 21 can be seen, because addition is operating on two different ratios. Dt 24, on the other hand, could be postponed till after the theory of triangles (Dt 39–48).

Θε 22. ᾿Εὰν δύο μεγέθη πρός τι μέγεθος λόγον ἔχῃ δεδομένον, καὶ τὸ συναμφότερον πρὸς τὸ αὐτὸ λόγον ἕξει δεδομένον.

δύο γὰρ μεγέθη τὰ ΑΒ, ΒΓ πρός τι μέγεθος τὸ Δ λόγον ἐχέτω δεδομένον· λέγω, ὅτι καὶ τὸ συναμφότερον τὸ ΑΓ πρὸς τὸ αὐτὸ τὸ Δ λόγον ἔχει δεδομένον.
ἐπεὶ γὰρ ἑκάτερον τῶν ΑΒ, ΒΓ πρὸς τὸ Δ λόγον ἔχει δεδομένον, λόγος ἄρα καὶ τοῦ ΑΒ πρὸς τὸ ΒΓ δοθείς· καὶ συνθέντι τοῦ ΑΓ πρὸς τὸ ΒΓ λόγος ἐστὶ δοθείς. τοῦ δὲ ΒΓ πρὸς Δ λόγος ἐστὶ δοθείς· καὶ τοῦ ΑΓ ἄρα πρὸς τὸ Δ λόγος ἐστὶ δοθείς.

Dt 22.   *If two magnitudes have a given ratio to a third, their sum will have a given ratio to the same.*

Let two magnitudes AB,BC have a given ratio to a magnitude D. I say that their sum AC has a given ratio to the same D.

```
A                    B         C
_____
D
_____
```

For, since each of AB,BC has a given ratio to D, the ratio AB:BC is also given [Dt 8]. And *synthenti* [(AB+BC):BC, *i.e.*] AC:BC is given [Dt 6]. The ratio BC:D is given; therefore the ratio AC:D is given [Dt 8].   ∎

Remarks: gvn.*x*:*z* AND gvn.*y*:*z* => gvn.(*x*+*y*):*z*.

Dt 22 is an extension of Dt 6, by means of which (and the *di 'isou* metamorphosis, Dt 8) it is easily proved. It could be proved by V.24, of which it is an exact counterpart. V.24 can be paraphrased as follows: *If two ratios have a common consequent and the* antecedents be added together while the consequent remains, then the new ratio will be independent of the choice of representatives. In shorthand:

V.24 $x:z :: a:c$ & $y:z :: b:c$ => $(x+y):z :: (a+b):c$.

This *could* have been used to define 'addition of ratios', but never was within Greek theory of magnitude or number (cf. Vitrac, vol 2:125). The idea of a 'common denominator' does not enter Euclidean geometry. The proof of V.24 involves the *di 'isou* theorem (V.22), which likewise can (but should not) be construed as 'multiplication of ratios',– as if that operation were prior to addition. The proof of V.24 can be sketched as follows:

Since	$x:z :: a:c$ AND $y:z :: b:c$,	(*)
anapalin	$x:z :: a:c$ AND $z:y :: c:b$.	
di 'isou, V.22	$x:y :: a:b$,	
synthenti, V.18	$(x+y):y :: (a+b):b$,	
(*), *di 'isou*, V.22	$(x+y):z :: (a+b):c$.	

The proof of Dt 22 runs similarly:

Hypotheses	gvn.$x:z$ & gvn.$y:z$	gvn.AB:D & gvn.BC:D
Dt 8	gvn.$x:y$	gvn.AB:BC
Dt 6	gvn.$(x+y):y$	gvn.AC:BC
Dt 8	gvn.$(x+y):z$.	gvn.AC:D

◆

Θε 23. ' Ἐὰν ὅλον πρὸς ὅλον λόγον ἔχῃ δεδομένον, ἔχῃ δὲ καὶ τὰ μέρη πρὸς τὰ μέρη λόγους δεδομένους, μὴ τοὺς αὐτοὺς δέ, καὶ πάντα πρὸς πάντα λόγους ἕξει δεδομένους.

ἐχέτω γὰρ ὅλον τὸ ΑΒ πρὸς ὅλον τὸ ΓΔ λόγον δεδομένον, ἐχέτω δὲ καὶ τὰ ΑΕ, ΕΒ μέρη πρὸς τὰ ΓΖ, ΖΔ μέρη λόγους δεδομένους, μὴ τοὺς αὐτοὺς δέ· λέγω, ὅτι καὶ πάντα πρὸς πάντα λόγους ἕξει δεδομένους.
ἐπεὶ γὰρ λόγος ἐστὶ τοῦ ΑΕ πρὸς ΓΖ δοθείς, ὁ αὐτὸς αὐτῷ γεγονέτω ὁ τοῦ ΑΒ πρὸς ΓΗ· λόγος ἄρα καὶ τοῦ ΑΒ πρὸς ΓΗ δοθείς. ἔσται καὶ λοιποῦ τοῦ ΕΒ πρὸς λοιπὸν τὸ ΖΗ λόγος δοθείς. τοῦ δὲ ΕΒ πρὸς τὸ ΖΔ λόγος ἐστὶ δοθείς· καὶ τοῦ ΖΔ ἄρα πρὸς ΖΗ λόγος ἐστὶ δοθείς· καὶ ἀναστρέψαντι τοῦ ΖΔ πρὸς ΔΗ λόγος ἐστὶ δοθείς. καὶ ἐπεὶ λόγος ἐστὶ τοῦ ΑΒ πρὸς ἑκάτερον τῶν ΔΓ, ΓΗ, καὶ τοῦ ΔΓ ἄρα πρὸς τὸ ΓΗ λόγος ἐστὶ δοθείς· ἀναστρέψαντι καὶ τοῦ ΓΔ πρὸς ΔΗ λόγος ἐστὶ δοθείς. ἀλλὰ τοῦ ΗΔ πρὸς ΔΖ λόγος ἐστὶ δοθείς· καὶ τοῦ ΓΔ ἄρα πρὸς ΔΖ λόγος ἐστὶ δοθείς· ὥστε καὶ τοῦ ΓΖ πρὸς τὸ ΖΔ λόγος ἐστὶ δοθείς. ἀλλὰ τοῦ μὲν ΓΖ πρὸς ΑΕ λόγος ἐστὶ δοθείς, τοῦ δὲ ΖΔ πρὸς ΒΕ λόγος ἐστὶ δοθείς· ὥστε πάντων πρὸς πάντα λόγος ἐστὶ δοθείς.

Dt 23. *If a whole have to a whole a given ratio, and the parts have to the parts given ratios, but not the same, all* [the magnitudes] *will have given ratios to all.*

For, let the whole AB have to the whole CD a given ratio, and let the parts AE,EB have to the parts CZ,ZD given ratios, but not the same. I say that all [the magnitudes] will have given ratios to all.

```
A                E                       B
C      Z                 H      D
```

For, since the ratio AE:CZ is given, let the same as that have become AB:CH. Then the ratio AB:CH is given [Def. 2]. But the ratio of the remainder EB to the remainder ZH will also be given [V.19]. And the ratio EB:ZD is given; therefore the ratio ZD:ZH is also given [Dt 8].

And *anastrepsanti* the ratio [ZD:(ZD–ZH), *i.e.*] ZD:DH is given [Dt 5]. And since the ratio of AB to each of DC,CH [is given], therefore the ratio DC:CH is given [Dt 8]. Anastrepsanti the ratio CD:DH is given [Dt 5].

But the ratio HD:DZ is given; therefore the ratio CD:DZ is given [Dt 8]. So the ratio CZ:ZD is given. But the ratio CZ:AE is given, and the ratio ZD:BE is given; so that the ratio of all to all [*i.e.* every one to every one] is given. ∎

Remarks: gvn.$(x+y):(u+v)$ AND gvn.$x:u$ AND gvn.$y:v$ => gvn.$u:v$.

This is also a proposition about 'adding ratios', in the specific sense that antecedent is added to antecedent and consequent to consequent. It says that if the two ratios operated on are given and the outcoming ratio is also given, then any one of the (four) magnitudes involved will have a given ratio to any one of the other three.

The conclusion is stated verbally as '*all will have given ratios to all*'; but since this condition is fulfilled (Dt 8) if $u:v$ is a given ratio, Euclid stops after having proved that. A magnificent example of dexterous juggling with ratios, but its mathematical uses are not apparent.

Hypotheses	gvn.AB:CD	gvn.$(x+y):(u+v)$	0
	gvn.AE:CZ	gvn.$x:u$	1
	gvn.EB:ZD	gvn.$y:v$	2

Assertion 'all have a given ratio to all', that is (since the rest follows from that):

gvn.CZ:ZD   gvn.$u:v$

Proof

make	AB:CH :: AE:CZ	get p,[76] such that $(x+y):(u+p) :: x:u$	3
1, Def. 2*	gvn.AB:CH	gvn.$(x+y):(u+p)$	4
V.19	gvn.EB:ZH	[$y:p :: x:u$], gvn.$y:p$	5
2, Dt 8	gvn.ZD:ZH	gvn.$v:p$	6
Dt 5	gvn.ZD:DH	gvn.$v:(v-p)$ [77] *anastrepsanti*	7
0, 3, Dt 8	gvn.DC:CH	gvn.$(u+v):(u+p)$	8
Dt 5	gvn.CD:DH	gvn.$(u+v):(v-p)$ *anastrepsanti*	9
7, Dt 8	gvn.CD:DZ	gvn.$(u+v):v$	10
Dt 6	gvn.CZ:ZD	gvn.$u:v$ *dielonti*	11
1, 2, Dt 8	gvn.(all to all)	i.e. gvn.$x:y$, gvn.$x:v$, gvn.$y:u$.	

No (relative) values of the ratios are found; so Clemens Thaer's comment, if $kx + ly = m(x+y)$, then $x = ((l-m)/(m-k))y$, is his own concoction. You will look in vain in the Greek text for such expressions.

Θε 24. ᾽Εὰν τρεῖς εὐθεῖαι ἀνάλογον ὦσιν, ἡ δὲ πρώτη πρὸς τὴν τρίτην λόγον ἔχῃ δεδομένον, καὶ πρὸς τὴν δευτέραν λόγον ἕξει δεδομένον.

ἔστωσαν τρεῖς εὐθεῖαι ἀνάλογον αἱ Α, Β, Γ, ὡς ἡ Α πρὸς τὴν Β, οὕτως ἡ Β πρὸς τὴν Γ, ἡ δὲ Α πρὸς τὴν Γ λόγον ἐχέτω δεδομένον· λέγω, ὅτι καὶ πρὸς τὴν Β λόγον ἕξει δεδομένον.

ἐκκείσθω γὰρ δοθεῖσα ἡ Δ. καὶ ἐπεὶ λόγος ἐστὶ τῆς Α πρὸς τὴν Γ δοθείς, ὁ αὐτὸς αὐτῷ γεγονέτω ὁ τῆς Δ πρὸς τὴν Ζ· λόγος ἄρα καὶ τῆς Δ πρὸς τὴν Ζ δοθείς· δοθεῖσα δὲ ἡ Δ· δοθεῖσα ἄρα καὶ ἡ Ζ. εἰλήφθω τῶν Δ, Ζ μέση ἀνάλογον ἡ Ε· τὸ ἄρα ὑπὸ τῶν Δ, Ζ ἴσον ἐστὶ τῷ ἀπὸ τῆς Ε. δοθὲν δὲ τὸ ὑπὸ τῶν Δ, Ζ· δοθεῖσα γὰρ ἑκατέρα αὐτῶν· δοθὲν ἄρα καὶ τὸ ἀπὸ Ε· δοθεῖσα ἄρα ἐστὶν ἡ Ε. ἔστι δὲ καὶ ἡ Δ δοθεῖσα· λόγος ἄρα ἐστὶ τῆς Δ πρὸς τὴν Ε δοθείς. καὶ ἐπεί ἐστιν ὡς ἡ Α πρὸς τὴν Γ, οὕτως ἡ Δ πρὸς τὴν Ζ, ἀλλ᾽ ὡς μὲν ἡ Α πρὸς τὴν Γ, οὕτως τὸ ἀπὸ τῆς Α πρὸς τὸ ὑπὸ τῶν Α, Γ, ὡς δὲ ἡ Δ πρὸς τὴν Ζ, οὕτως τὸ ἀπὸ τῆς Δ πρὸς τὸ ὑπὸ τῶν Δ, Ζ, ὡς ἄρα τὸ ἀπὸ τῆς Α πρὸς τὸ ὑπὸ τῶν Α, Γ, οὕτως τὸ ἀπὸ τῆς Δ πρὸς τὸ ὑπὸ τῶν Δ, Ζ. ἀλλὰ τῷ μὲν ὑπὸ τῶν Α, Γ ἴσον ἐστὶ τὸ ἀπὸ τῆς Β· αἱ γὰρ Α, Β, Γ ἀνάλογόν εἰσιν· τῷ δὲ ὑπὸ τῶν Δ, Ζ ἴσον ἐστὶ τὸ ἀπὸ τῆς Ε· ὡς ἄρα τὸ ἀπὸ τῆς Α πρὸς τὸ ἀπὸ τῆς Β, οὕτως τὸ ἀπὸ τῆς Δ πρὸς τὸ ἀπὸ τῆς Ε· καὶ ὡς ἄρα ἡ Α πρὸς τὴν Β, οὕτως ἡ Δ πρὸς τὴν Ε. λόγος δὲ τῆς Δ πρὸς τὴν Ε δοθείς· λόγος ἄρα καὶ τῆς Α πρὸς τὴν Β δοθείς.

---

[76] This is the bright idea: to get another magnitude to which y has a given ratio (by V.19). The sequel is quite straightforward manipulations with ratios.

[77] Tacit assumption $(x+y):(u+v) < x:u$, whence $(x+y):(u+v) < (x+y):(u+p)$], and therefore $p < v$. If not, exchange the lines AB and CD.

Dt 24. *If three straight lines be proportional, and the first have to the third a given ratio, it will have to the second a given ratio.*

Let three line segments A, B, C be proportional such that A:B :: B:C; and let A have a given ratio to C. I say that A will have a given ratio to B.
Let a given [line] D have been set out [Axiom 0*].

```
A _____    D _____
B _____          E _____
C _____              Z _____
```

And since the ratio A:C is given, let the same ratio have become D:Z [VI.12]; then the ratio D:Z is given [Def. 2*]. And D is given; therefore Z is given [Dt 2].
Let the mean proportional E to D and Z have been taken [VI.13]; then ⊏⊐D,Z = ▫E [VI.17]. ⊏⊐D,Z is given because both D and Z are given [Dt 24*A]; therefore ▫E is given; therefore E is given [Dt 24*B]. And D is given; therefore the ratio D:E is given [Dt 1].
And since A:C :: D:Z, but A:C :: ▫A:⊏⊐A,C, and D:Z :: ▫D:⊏⊐D,Z [VI.1], therefore ▫A:⊏⊐A,C :: ▫D.:⊏⊐D,Z [V.11]. But ⊏⊐A,C = ▫B, because A, B, C are proportional [VI.17]. And ⊏⊐D,Z = ▫E [by construction]; therefore ▫A:▫B :: ▫D:▫E; therefore A:B :: D:E [Dt 24*C].
But the ratio D:E is given; therefore the ratio A:B is given.   ∎

Remarks: If A:B :: B:C and gvn.A:C then gvn.A:B.

With this vital proposition about line segments in geometrical progression Euclid leaves the theory of magnitudes in general. In the initial propositions the line segments were meant to illustrate magnitudes of any kind, and the geometrical operations were limited to adding, subtracting or splitting line segments. In Dt 24 rectangles and squares enter, implying propositions from Book VI.
From *Elements* VI.20 we infer that if A:B :: B:C then A:C :: ▫A:▫B; therefore, if gvn.A:C then gvn.▫A:▫B, and so the theorem may be restated as

> **Dt 24*A** *If two squares have a given ratio to one another, their sides will also have a given ratio to one another.*

Anachronistically, we could 'translate' ratio into a real number and state the theorem in a modern idiom as: *If a positive real number is given, its positive square root is also given.* Or: *Extraction of the positive square root is unique.*

Dt 24 is out of place; it must, however, be proved before Dt 54, where it is used, being a special case of that theorem. It is well-known [78] that Dt 24 is very weakly proved, though of course true all the same; the crucial statements (proved nowhere in the *Data*) are

Dt 24*B *If two line segments are given in magnitude, their rectangle is also given in magnitude.* (a special case of Dt 52).

Dt 24*C *If a square is given in magnitude, its side is also given in magnitude* (a special case of Dt 55).

Dt 24*D *If* □A:□B :: □D:□E *then* A:B :: D:E.

The missing proofs are remarkable *lacunas* and confirm the impression of an unfinished and inhomogeneous work. After all, these statements are as fundamental as *e.g.* Dt 3 and 4. But then, intuitively obvious as they are, they are also missing in VI.22, see footnote 78.

Analysis: Before suggesting the missing proofs let us analyse Dt 24. On the hypothesis that A:B :: B:C and that A:C is given, assume that A:B is also a given ratio. That means [Def. 2] that there are given line segments D and E such that A:B :: D:E.

VI.1   □A:⊏⊐A,B :: □D:⊏⊐D,E
and   ⊏⊐A,B:□B :: ⊏⊐D,E: □E
V.22   □A:□B :: □D:□E.   [*di 'isou*, ex aequali]

From the hypothesis (A:B :: B:C) follows [VI.17] □B = ⊏⊐A,C, a rectangle with height A; therefore, if we make □E a rectangle with height D, that is: if we construct Z as the third proportional to D, E, so that D:E :: E:Z, then □E = ⊏⊐D,Z, and the last line of the analysis becomes   □A:⊏⊐A,C :: □D:⊏⊐D,Z; whence [VI.1] A:C :: D:Z.

The analysis shows that if we take given line segments D and Z in the given ratio A:C, and if their mean proportional E is given, then A:B, being the same ratio as D:E, will be given. Now Euclid *assumes* that if D and Z be given, E is also given, though this assumption is equivalent to what he is about to prove, *viz.* Dt 24. It follows from the three unproven propositions, Dt 24*B, *C, and *D, mentioned above.

---

[78] Cf. (Heath 1956, II:246 f.) with a history of emendations of *Elements* VI.22.

A proof of 24*B, imitating Dt 48 or 52, could be: *If a rectangle and a square be described on the same line segment, they will have to one another a given ratio.* That will work if it is true that *if a line segment is given, its square is given in magnitude.* I suggest that we take the latter to be an axiom. Areas are not covered by Def. 1.

Let AB and BC be given line segments. The ratio of the rectangle ABC to the square on AB is the same as the ratio of BC to AB. Since that ratio is given because the lines are given [Dt 1], ⊏⊐ ABC is given in magnitude if □ AB is given in magnitude [Dt 2]. And □ AB *is* given in magnitude by our axiom.

24*C can be 'proved' along the same lines as Dt 39, by involving position and Dt 27 and 29, but that is no better than taking it to be an axiom. It is a special case of Dt 55, where it is used by way of Dt 54.

To prove 24*D we shall involve two straight lines A and B and the statement *If* □A = □B then $A = B$. That theorem is proved by Proclus (Friedlein 424–25, Morrow 337, see below).

Dt 24*D is the 'cheap' version of VI.22 (about squares only). This is less complicated than the general VI.22 (cf. Heath's five pages of comments, see footnote 78); as in the *Elements* we prove it by first proving the converse (cf. our analysis above):

If	A:B :: D:E
then	□A:⊏⊐A,B :: □D:⊏⊐D,E
and	⊏⊐A,B:□B :: ⊏⊐D,E:□E.
therefore *di 'isou*	□A:□B :: □D:□E.

Proof of 24*D: Suppose that the fourth proportional to A, B, and D is H:

If	A:B :: D:H	
then	□A:□B :: □D:□H	[proved above]
but	□A:□B :: □D:□E,	[by hypothesis]
therefore	□E = □H	[V.9]
and	E = H	[Proclus]

Proclus' proof (figure 24.1): Hypothesis: □(BD) = □(BE). Assertion: AB = BC.

∠AZB = ∠CHB = ½ right angle ∴
AZ ∥ CH [I.27].
△ABZ = △CBH = ½ □ [I.34] ∴
△AZC = △HCZ [C.N.2] ∴
AH ∥ CZ [VI.1] ∴
AZ = CH [I.34] ∴
△ABZ ≅ △ CBH [I.26] ∴
AB = BC.

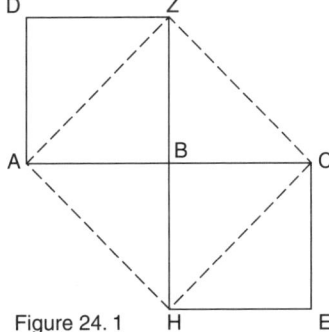

Figure 24. 1

EUCLID'S *DATA*

Deductive Structure of Dt 10-24

		Dt 10	11	12	13	14	15	16	17	18	19	20	21	22	23	24
Data																
Def.	1*	.	.	+	.	.	.	.	.	.	.	.	.	.	.	.
	2*	.	+	.	+	+	+	+	.	+	+	.	+	.	+	+
	9	.	.	+	.	.	.	.	.	.	.	.	.	.	.	.
	11	+	+	.	+	.	+	+	+	+	+	+	.	.	.	.
Dt	1	.	.	.	.	+	+	.	.	.	.	+	+	.	.	+
	2	.	+	.	+	+	+	+	.	+	+	+	.	.	.	+
	3	.	.	.	.	.	.	+	.	.	+	.	.	.	.	.
	4	.	+	+	.	+	+	.	.	.	.	+	+	.	.	.
	5	+	+	.	.	.	.	.	.	.	.	.	.	.	+	.
	6	+	+	.	.	.	.	.	.	.	.	.	.	+	+	.
	8	.	.	.	+	.	.	.	+	+	+	.	.	.	+	.
	14	.	.	.	.	.	.	.	+	.	.	.	.	.	.	.
	15	.	.	.	.	.	.	.	.	+	.	.	.	.	.	.
Elem																
C.N.	3	.	.	+	.	.	.	.	.	.	.	.	.	.	.	.
V.	11	.	.	.	.	.	.	.	.	.	.	.	.	.	.	+
	12	.	+	.	.	+	.	.	.	+	.	.	+	.	.	.
	19	.	+	.	+	.	+	+	.	.	+	+	.	.	.	.
VI.	1	.	.	.	.	.	.	.	.	.	.	.	.	.	.	+
	13	.	.	.	.	.	.	.	.	.	.	.	.	.	.	+
	17	.	.	.	.	.	.	.	.	.	.	.	.	.	.	+
	22	.	.	.	.	.	.	.	.	.	.	.	.	.	.	+
		Dt 10	11	12	13	14	15	16	17	18	19	20	21	22	23	24

Dt 10, 11, 13-21 form a group, being the only ones to use Def 11 (apart from Dt 86) and *Elements* V.19 (whereas V.12 is used again in Dt 45, 46, and 93). They illustrate the importance of Def 2 (2*) and the 'fundamental' theorems Dt 1-6 and 8, which support most propositions apart from the position-theorems Dt 25-30. Dt 14 and 15 serve only to prove Dt 17 and 18. Dt 22 is an extension of Dt 6. Dt 24 is a lone wolf, inportant but weakly proved.

Chapter 5. Position
Distance, Direction, Parallels. Dt 25–38

Θε 25. ᾽Εὰν δύο γραμμαὶ τῇ θέσει δεδομέναι τέμνωσιν ἀλλήλας, δέδοται τὸ σημεῖον, καθ᾽ ὃ τέμνουσιν ἀλλήλας, τῇ θέσει.

δύο γὰρ γραμμαὶ τῇ θέσει δεδομέναι αἱ ΑΒ, ΓΔ τεμνέτωσαν ἀλλήλας κατὰ τὸ Ε σημεῖον. λέγω, ὅτι δοθέν ἐστι τὸ Ε σημεῖον.

εἰ γὰρ μή, μεταπεσεῖται τὸ Ε σημεῖον. μεταπεσεῖται ἄρα καὶ μιᾶς τῶν ΑΒ, ΓΔ ἡ θέσις. οὐ μεταπίπτει δέ. δοθὲν ἄρα ἐστὶ τὸ Ε σημεῖον.

Dt 25. *If two lines given in position cut one another, their point of section is given in position.*

Let two lines given in position AB, CD cut one another at the point E. I say that the point E is given.

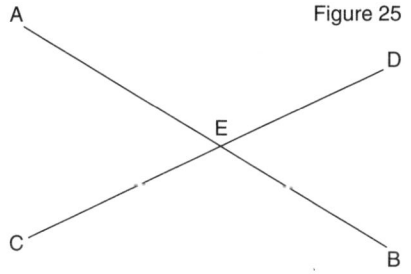

Figure 25

For if not, the point E will change position. But then the position of one of the lines AB, CD will change, too. But it does not change. Therefore the point E is given [Def. 4]. ∎

Remarks: Dt 25–38 are independent of the previous propositions and deal with the concept *given in position*, which was 'defined' in Def. 4: *Given in position* is said of points and lines and angles which *always hold the same place*.

What does it mean to hold the same place, *topos*? And to hold it always? In the proofs to come (all of them *demonstrationes ad absurdum*) this idiom is not used but substituted by what must be considered a negative equivalent of it:

> **Def. 4*A** *Given in position* is said of points and lines and angles which do not *metapipt*.

The etymology of the keyword *metapipt*, μεταπίπτειν, is 'fall beside the mark', 'hit the wrong place', or perhaps 'hop away'; according to Liddell and Scott it is used as passive to μεταβάλλω, throw away, make change. It may be translated 'move', but it is not synonymous with κινεῖν. The verbal root 'pt' has the connotation of 'flying' or 'falling' and signifies that the geometric object is coming from outside the Plane to be put there, but for some reason or other hits the wrong place. It does not denote continuous movement as *e.g.* the moving point in Archimedes' helix; that point is 'carried' (φέρεσθαι or ἐνέγκειν). Let us for the moment borrow the Greek word, *metapipt*, – a nice and handy term. In the theorems we are going to treat, the translation 'change position' will do in most (but not all) places, and contrasts nicely with 'given in position'.[79]

The *Data*'s treatment of positions is as provocative and disturbing to our notions of normal Euclidean behaviour as *Elements* I.4, the first congruence theorem, and may be understood or *saved* within the same context, as follows: The place to be 'always' held is in the Plane (cf. Mueller's Single Plane Law, p. 19), and the locomotion in *Elements* I.4 and in the *Data*'s propositions about position goes *from outside* the Plane into it. If in I.4 it is possible to place one triangle on top of another, it is because it is not in the Plane before being placed that way. The Geometer may hold up his hand and ask, 'Why not, instead of putting this triangle beside that one, put it on top of it' – which he then does.

Curiously, the same feature has not roused similar uneasiness when it appears in Book VI at several places, *e.g.* in VI.14. The *ekthesis* begins (in Heath's translation): *Let AB, BC be equal and equiangular parallelograms having the angles at B equal.* And continues: *and let DB, BE be placed in a straight line.* That is to say: At the opening, the parallelograms are not yet placed, *only thought of* and named; then (without any suggestion about how to do it, but we trust *The Helping Hand*) they are put into the Plane to make a useful *gnomon*.[80]

In the *Data*, if a point or a line is to be given in position, it is put into the Plane from outside. Tacitly the Plane is considered as a continuum of potential positions, τόποι; and from the doings in the *Data* we infer that

(Axiom 0*) *Any point may be (taken and) appointed given*

that is: put in a certain *topos* and meant to stay there. It cannot, at some stage of the theorem, occupy another *topos* and still claim to be the same point. It cannot *metapipt*, hop into another *topos*. In modern words: it is *unique* and recognizable. And it must

---

[79] Suggested by Len Berggren. Notice, however, that in the proof it is the position, *thesis*, that changes, – and in Dt 26 it is the length of the line as well.

[80] To quote (Mueller 1981, 153) '... in book VI Euclid follows the procedure ... of placing figures where he can use them'.

stay in its position *always*, that is to say till the curtain falls at the end of the actual proposition.

When commenting on Def. 1-3 we discovered 'latent coactors': A magnitude is given if there is a (latent) magnitude equal to it. A ratio is given if there is a (latent) ratio equal to it. A form is given if there is a (latent) form similar to it. (We shall arrive at forms in Dt 39). Can this approach be used for positions, too? It can, a little modified: A point is given in position if it is *assigned* to one of the potential positions in the Plane, waiting for points or lines to be fixed so as not to be moved, not to be able to hop away. Axiom 0* had better be called *Axiom of Assignment*. We can rewrite the idea of being *given in position* as follows:

> **Def. 4*B** A point P is said to be given in position if and only if it is *assigned* to a point in the Plane so that no other point in the Plane can claim the name P.
> A line L is said to be given in position if and only if it is *assigned* to a line in the Plane so that no other line in the Plane can claim the name L or do the same job.
> An angle is given in position if its sides are given in position.

(*E.g.*, if L is a line through the given point P parallel to the given line M, then no other line in the Plane through P will be parallel to the line M, – as is proved in Dt 28).

These points and lines share a concept which in due course was to be recognized as *uniqueness*, one of the important twins of mathematics, its companion being *existence*. About uniqueness in Euclid Mueller says (Mueller 1981, 55, note 33)

> it is not possible for me to discuss the many attempts to show that Euclid was concerned with questions of uniqueness.

With the risk of starting that discussion I will assert that Euclid's *Data* is one such concern; nay, one of the main issues of the Data is to show how uniqueness plays an important role in geometrical analysis by proving how that property is inherited from one individual to another.

The *Data* not unexpectedly shows many embryologic features, being the first known approach to a theory of the concept 'unique'. The very word used for it, *viz. dedoménon*, given, is not a self-evident choice; the term has expanded its trivial meaning 'handed over' and is used idiomatically, both of the 'input' to a problem and of the 'output' – and we shall see it used time and again meaning 'taken', 'chosen', that is: in almost the opposite sense of 'offered' or 'granted'.

An importunate question must be asked if not answered: Where are geometric objects when they are not in the Plane? (The triangles in *Elements* I.4, *e.g.*) That question is not answered anywhere in Greek mathematics as far as I know. Philosophers have spent some time and words on the genesis and whereabouts of *mathematicals*, τὰ μαθηματικά; read Proclus' introduction to his *Commentary on the First Book of Euclid's Elements* (Friedlein 3–17, Morrow 3–14). Without going into an ontological discussion about *mathematicals* we may use the following picture to give a fair impression of what is going on in the *Data*:

The Plane [81] is a scene into which the Geometer may put geometric objects, particularly points and lines, which in turn will produce figures. Some of these points and lines are labelled *given in position* and may not hop around: they are supposed to be identified from now on. Other lines and figures do have position, but are 'only' given in *magnitude* so that any object *equal* to a given may serve as well. Once figures have begun to emerge from the points and lines, shape or *form* becomes relevant, and is in fact the dominating category from Dt 39 onwards. – The *Data*'s theorems about position prove how other points, lines, and figures inherit or take part in the givenness. The fundamentals of a point's givenness in position seem to be three: A point is given

1. if it is taken for given, assigned or 'appointed' as it were (Axiom 0*), εἰλήφθω δεδομένον, 'let a given (point) have been taken';
2. or if it inherits its givenness from the givenness of two lines that produce the point (Dt 25);
3. or is at a given distance and direction from a point given in position (Dt 27).

A line is given in position

1. if it is taken for given (Axiom 0*);
2. or if it inherits its givenness from the givenness of two points that produce the line (Dt 26);
3. or is at a given distance and direction from a line given in position (Dt 28). [82]

An angle is given in position if (Dt 25) and only if (Dt 29) its sides are given in position.

I suppose that what offends most in the statement 'P is given in position' is the

---

[81] The Plane has a double status: being a *mathematical* it is one of many planes in space, such as occur in *Elements* XI –XIII; but in the *Data* it plays a dominant role as a unique board in which to fix individual *mathematicals*, turning them into fixed geometric objects.

[82] A rather generalized form of Dt 28 about the unique parallel (to a given line) through a given point.

absolute and universal notion. We are prone to think of coordinates in an appropriately chosen coordinate system, when we are told that some point is given in position and always occupies the same place *relatively to a fixed origin*. But Greek mathematics never got the idea of referring a point to a coordinate system; what would be gained by increasing the number of lines by 'two straight lines, which normally have nothing at all to do with the actual problem', to quote (Julius Petersen, 4)? Any point may serve as its own *origin*, any point or line or several may be taken as *given in position*. By appropriate reasoning others are proved to be as given as those taken.[83] To a modern mathematician there is too little information in these statements, no measures, no coordinates, no slopes or directions. But our coordinate systems camouflages the problem.

When we make a coordinate system, we choose a point, the origin, which is supposed to be the same 'all the time', that is till the very end of the proposition. And when we draw the axes, we choose directions and distances measured by units of our own choice. Distance and direction are undefined and vaguely used concepts in Greek geometry. And the relation between points and lines is quite opaque; we are left with intuition – but the history of mathematics shows that intuition is about the richest, if not the safest, place from which to watch geometry.

That a point or line or angle *is given in position if it does not change position* means that it is unique if it is said or proved to be unique. As in any other mathematical discipline there must be axiomatic statements on which to build the rest, and the propositions depending on the concept *metapipt* must be understood as axioms, if we are not willing to admit undefined movements in the game. The said propositions are Dt 25–27 and 29, whereas Dt 28 and 30 (although they use *metapipt*) are quite sound deductions from the parallel postulate and I.16. To my view the theorems Dt 25–27 are improvable axioms, the 'proofs' of which can be no more than illustrations, albeit useful ones:

Axiom 25* *If two lines given in position cut one another, their point of section is given in position.*

Axiom 26*A *If the extremities of a straight line be given in position, the line is given in position and in magnitude.*

---

[83] A similar relative feature is known from *Elements* Book X (Taisbak 1982, 28): A line segment R is chosen as *the* rational line, to be called ἡ ῥητή, the expressible. The square on R is also said to be expressible. Now, any other line segment is expressible if and only if its square is commensurable with the square on R; the others are irrational, ἄλογοι, with a certain *primus inter pares* whose square is a *meson*, a mean proportional between two expressible squares whose sides are not commensurable.

Axiom 27* *If one extremity of a straight line given in position and in magnitude be given, the other will also be given.*

Axioms 26*A and 27* are important connections between position and magnitude. It is a tenable point of view that position is the more fundamental of the two, although this is not evident from the definitions. That would account for the axiomatic character of most statements about position.

Dt 25 can be no more than an illustration of Axiom 25*; the immovability of the point is as axiomatic as that of the lines, and is in fact one of the fundamental properties of the Euclidean Plane as seen in the *Elements*: *Movement is forbidden.*[84] Once placed, the geometric objects do not hop about. Does Dt 25 *prove* that two straight lines can have only one point in common? I do not think so; that the point of section between two straight lines is unique was as axiomatic to Euclid as it was to Hilbert (cf. Common Notion 9,[85] *two straight lines cannot contain a space*). Dt 25 shows that the point of section is as given as the lines that produce it; if the lines are 'known', *i.e.* uniquely determined, the point is also 'known' and uniquely determined; its identity follows from its being the only common point of the lines. The proof is meant as a *reductio ad absurdum*, but it does not work if *metapipt* is a mere negation of 'being given in position'; to say that a point is given in position if it is not NOT given in position, is a tautology. We are in need of a useful definition of *metapipt*, and furthermore we would like to limit the axiom to points and straight lines only; but it turns out to be meant for circles, too, and for conic sections in Apollonius – nay the most important uses of it are about circles cut by straight lines (*e.g.* Dt 31), which causes some uneasiness as to which of two points is meant. We will leave that discussion to Dt 31.

◆

---

[84] In space things are different: a Euclidean cone is generated by rotating a right triangle about one of the shorter sides (which is itself unmoved). A sphere is generated by rotating a semicircle about the diameter (which is itself unmoved).

[85] There is, nowadays, consensus that common notion 9 is an interpolation, an idea that was accepted, though much debated, by the ancient geometers. See *e.g.* (Mueller 1981, 31).

Θε 26. ’Εὰν εὐθείας γραμμῆς τὰ πέρατα ᾖ δεδομένα τῇ θέσει, δέδοται ἡ εὐθεῖα τῇ θέσει καὶ τῷ μεγέθει.

εὐθείας γὰρ γραμμῆς τὰ πέρατα τὰ Α, Β δεδομένα ἔστω τῇ θέσει. λέγω, ὅτι δέδοται ἡ ΑΒ τῇ θέσει καὶ τῷ μεγέθει.
 εἰ γὰρ μένοντος τοῦ Α μεταπεσεῖται τῆς ΑΒ εὐθείας ἤτοι ἡ θέσις ἢ τὸ μέγεθος, μεταπεσεῖται καὶ τὸ Β σημεῖον. οὐ μεταπίπτει δέ. δέδοται ἄρα ἡ ΑΒ εὐθεῖα τῇ θέσει καὶ τῷ μεγέθει.

Dt 26. *If the extremities of a straight line be given in position, the line is given in position and in magnitude.*

Let the extremities A, B of a straight line be given in position. I say that the [straight line] [86] AB is given in position and in magnitude.

A———————————————B

For if, while A remains fixed, either the position or the magnitude of the straight line AB will *metapipt*, the point B will also *metapipt*. But it does not *metapipt* [Def. 4]. Therefore the straight line AB is given in position and in magnitude. ∎

Remarks: This is a very nice example of a convincing argument that is not a logical proof. That the extremities of a line segment determine the segment as to position and magnitude, must be an axiom; it deserves an extension:

> **Axiom 26*B** If two points are given, [the infinite straight line that passes through them is given in position; and] the segment whose endpoints are the two points is given in position and in magnitude; [and the circle with one of the points as centre and the segment as radius is given in position and in magnitude.]

The last supplement is a consequence of Def. 6: *A circle is said to be given in position and in magnitude, if its center is given in position and its radius is given in magnitude.* Dt 26 seems to be used or presupposed in *Elements* I.2 (cf. p. 14).

◆

---

[86] The definite article ἡ (AB) is sufficient to determine the object, feminine meaning 'the line'.

Θε 27. Ἐὰν εὐθείας γραμμῆς τῇ θέσει καὶ τῷ μεγέθει δεδομένης τὸ ἓν πέρας δοθὲν ᾖ, καὶ τὸ ἕτερον δοθήσεται.

εὐθείας γὰρ γραμμῆς τῇ θέσει καὶ τῷ μεγέθει δεδομένης τῆς ΑΒ τὸ ἓν πέρας τὸ Α δοθὲν ἔστω. λέγω, ὅτι καὶ τὸ Β δοθέν ἐστιν.
εἰ γὰρ μένοντος τοῦ Α σημείου μεταπεσεῖται τὸ Β σημεῖον, μεταπεσεῖται ἄρα καὶ τῆς ΑΒ εὐθείας ἤτοι ἡ θέσις ἢ τὸ μέγεθος. οὐ μεταπίπτει δέ. δοθὲν ἄρα ἐστὶ τὸ Β σημεῖον.

Dt 27. *If the one extremity of a straight line given in position and in magnitude be given, the other will also be given.*

For, let the one extremity A of the straight line AB given in position and in magnitude be given. I say that B is also given.

A   B

For if, while the point A remains fixed the point B will *metapipt*, then either the position or the magnitude of the straight line AB will also *metapipt*. But it does not *metapipt*. Therefore the point B is given [Def. 4]. ∎

Remarks: In Dt 26 and 27 not only the position, but also the magnitude of the line segment are said to *metapipt*, hop around; of magnitudes an understandable translation would be 'grow or shrink', 'wax or wane'. But whatever it means, Dt 26 and 27 should be taken as axioms.

◆

Θε 28. Ἐὰν διὰ δεδομένου σημείου παρὰ θέσει δεδομένην εὐθεῖαν εὐθεῖα γραμμὴ ἀχθῇ, δέδοται ἡ ἀχθεῖσα τῇ θέσει.

διὰ γὰρ δεδομένου σημείου τοῦ Α παρὰ θέσει δεδομένην εὐθεῖαν τὴν ΒΓ εὐθεῖα γραμμὴ ἤχθω ἡ ΔΑΕ. λέγω, ὅτι δέδοται ἡ ΔΑΕ τῇ θέσει.
εἰ γὰρ μή, μένοντος τοῦ Α σημείου μεταπεσεῖται τῆς ΔΑΕ ἡ θέσις. διαμενούσης τῆς ΒΓ παραλλήλου μεταπιπτέτω καὶ ἔστω ἡ ΖΑΗ. παράλληλος ἄρα ἐστὶν ἡ ΓΒ τῇ ΖΑΗ. ἀλλὰ ἡ ΒΓ τῇ ΔΑΕ ἐστι παράλληλος. καὶ ἡ ΔΑΕ ἄρα τῇ ΗΑΖ παράλληλός ἐστιν. ἀλλὰ καὶ συμπίπτει· ὅπερ ἐστὶν ἄτοπον. οὐκ ἄρα μεταπεσεῖται τῆς ΔΑΕ ἡ θέσις. θέσει ἄρα ἐστὶν ἡ ΔΑΕ.

Dt 28. *If through a given point a straight line be drawn parallel to*[87] *a straight line given in position, the straight line drawn is given in position.*

---

[87] παρὰ is idiomatic and means 'parallel to' and not merely 'alongside', as McDowell/Sokolik translate it. The short form is frequent in Apollonius, too.

Through the given point A let the straight line DAE have be drawn parallel to the straight line given in position BC. I say that DAE is given in position.

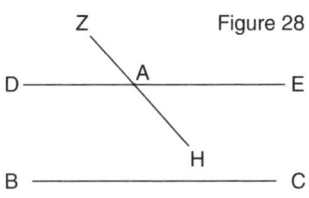

Figure 28

For if not, the position of DAE will *metapipt* while the point A remains where it is. Let its position *metapipt* while the parallel BC all the time remains where it is, and let it be ZAH. Then CB is parallel to ZAH. But BC is parallel to DAE. Therefore DAE is also parallel to HAZ [I.30]; but it also meets it, which is absurd.

Therefore the position of DAE will not *metapipt*. Therefore DAE is given in position [Def. 4]. ∎

Remarks: The proof hinges on the fact that *there cannot be two parallels to one and the same line through a point*. That statement is not explicitly made in the *Elements*, (and that deficiency gave John Playfair a chance of making it his own axiom in 1795, seeing that it was and is a valid substitute of the Fifth Postulate in Book I); however, the truth of it follows from I.29 (*i.e.* from postulate 5) and was well known to the Ancients, as appears from Proclus' note to I.31 (Friedlein 376.8). Data 28 can be understood as proving exactly that – or *can* it? Is Dt 28 using or proving the fact?

We may think of the theorem as two pictures showing two different positions of a parallel through A, DAE and ZAH; we are to overlay one picture over the other, seeing the two positions in one picture. But that amounts to drawing – in one and the same picture – two different parallels to the same line. Which is impossible because of I.30 (a consequence of the parallel postulate), that *two lines that are parallel to the same, are parallel to one another*; but DAE meets ZAH in A.

We infer from Dt 28 that the phrase '(the position of) L will *metapipt*' means 'two different items will deserve the name L'; which (we learn) is impossible if L is given in position. By this interpretation 'L is given in position' is synonymous with 'L is unique'. If L is a point or a line segment, we know where it is. The latter is also given in magnitude (Dt 26).

◆

Θε **29**. Ἐὰν πρὸς θέσει δεδομένῃ εὐθείᾳ καὶ τῷ πρὸς αὐτῇ σημείῳ δεδομένῳ εὐθεῖα γραμμὴ ἀχθῇ δεδομένην ποιοῦσα γωνίαν, δέδοται ἡ ἀχθεῖσα τῇ θέσει.

πρὸς θέσει γὰρ δεδομένῃ εὐθείᾳ τῇ ΑΒ καὶ τῷ πρὸς αὐτῇ σημείῳ δεδομένῳ τῷ Γ εὐθεῖα ἤχθω ἡ ΓΔ δεδομένην ποιοῦσα γωνίαν τὴν ὑπὸ ΒΓΔ. λέγω, ὅτι θέσει ἐστὶν ἡ ΓΔ.
εἰ γὰρ μή, μένοντος τοῦ Γ σημείου μεταπεσεῖται τῆς ΓΔ ἡ θέσις διατηροῦσα τῆς

ὑπὸ τῶν ΒΓΔ γωνίας τὸ μέγεθος. μεταπιπτέτω καὶ ἔστω ἡ ΓΕ. ἴση ἄρα ἐστὶν ἡ ὑπὸ τῶν ΔΓΒ γωνία τῇ ὑπὸ ΕΓΒ, ἡ μείζων τῇ ἐλάσσονι· ὅπερ ἄτοπον. οὐκ ἄρα μεταπεσεῖται τῆς ΔΓ ἡ θέσις. θέσει ἄρα ἐστὶν ἡ ΓΔ.

Dt 29. *If at a straight line given in position and at a given point on it a straight line be drawn making a given angle, the straight line drawn is given in position.*

For, at the straight line AB given in position, and at the given point C on it, let the straight line CD have been drawn making the given angle BCD. I say that CD is given in position. For if not, with the point C remaining [where it is], the position of CD will *metapipt* while preserving the magnitude of ∠BCD.

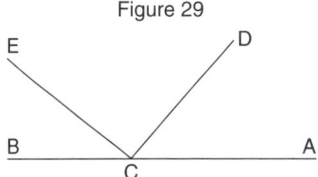

Figure 29

Let its position *metapipt* and be CE. Then ∠DCB is equal to ∠ECB, the greater to the less; which is absurd. Therefore the position of CD will not *metapipt*; therefore CD is given in position [Def. 4]. ∎

Remarks: Dt 29 is as axiomatic as the preceding Dt 25–27: If a given angle has its vertex and one side given in position, the other side is uniquely positioned, that is: given in position. As this is the first proposition to speak of angles, it is appropriate to introduce a vocabulary by which to speak unequivocally about them. I will speak of the vertex of the angle and of the right and left side, determined as if one is looking into the angle from the vertex. No such orientation is known from ancient sources, but would be helpful if the phrase *similarly positioned* (VI.22 and *passim*) should be given a precise meaning. See also (Mueller 1981, 158).

The question whether or not angles count as magnitudes, has been raised in the remarks on Dt 2; they can be *given in magnitude*, – but since they cannot (in the frame of Greek geometry) outgrow two right angles, they must be used in proportions with some care. A famous fallacy in using Dt 2 is the 'proof' that *it is possible to trisect any angle* (cf. footnote 37):

Let ABC be a given angle with ∠ABD as one-third of it. Since the ratio (∠ABC):(∠ABD) is given (it is 3:1), and ∠ABC is given, ∠ABD is given. Then, since AB is given in position and ∠ABD is given, BD is given in position (Dt 29).

But that is not what Dt 29 ensures: it says that

If (to a straight line given in position, and to a given point on it) *a straight line be drawn making a given angle, then* the straight line drawn is given in position.

So we have to make sure that a straight line *can* be drawn making the given angle, and that is not possible by geometric methods, as we know and Ptolemy purports to know, cf. p. 13. There is no doubt, however, that there is (exists) a unique angle which is one third of a given, even if it cannot be constructed by Euclidean means. (Cp.what Marinus says about squaring the circle on p. 245.)

To make Dt 29 un-equivocal, we should rewrite it as an axiom:

> **Axiom 29*** *If a given angle has its vertex and its one side given in position, the other side is also given in position.*

To 'prove' that, we must prove that no other line deserves the name of left side to that angle; any other line will make an angle that is greater or less than the given one, which of course is untenable. Again, what had better be an axiom is 'proved' by appeal to geometrical intution.

◆

Θε 30. Ἐὰν ἀπὸ δεδομένου σημείου ἐπὶ θέσει δεδομένην εὐθεῖαν εὐθεῖα γραμμὴ ἀχθῇ δεδομένην ποιοῦσα γωνίαν, δέδοται ἡ ἀχθεῖσα τῇ θέσει.

ἀπὸ γὰρ δεδομένου σημείου τοῦ Α ἐπὶ θέσει δεδομένην εὐθεῖαν τὴν ΒΓ εὐθεῖα γραμμὴ ἤχθω ἡ ΑΔ δεδομένην ποιοῦσα γωνίαν τὴν ὑπὸ τῶν ΑΔΓ. λέγω, ὅτι θέσει ἐστὶν ἡ ΑΔ.
    εἰ γὰρ μή, μένοντος τοῦ Α σημείου μεταπεσεῖται τῆς ΑΔ ἡ θέσις διατηροῦσα τῆς ὑπὸ ΑΛΓ γωνίας τὸ μέγεθος. μεταπιπτέτω καὶ ἔστω ἡ ΑΖ. ἴση ἄρα ἐστὶν ἡ ὑπὸ τῶν ΑΔΓ γωνία τῇ ὑπὸ τῶν ΑΖΓ, ἡ μείζων τῇ ἐλάττονι· ὅπερ ἐστὶν ἀδύνατον. οὐκ ἄρα μεταπεσεῖται τῆς ΑΔ ἡ θέσις. θέσει ἄρα ἐστὶν ἡ ΑΔ.

Dt 30. *If from a given point to a straight line given in position a straight line be drawn making a given angle, the line drawn is given in position.*

For, from the given point A to the straight line BC given in position let the straight line AD have been drawn, making the given angle ADC. I say that AD is given in position.

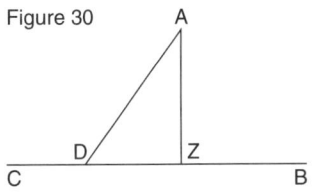

Figure 30

For if not, then with the point A remaining [where it is] the position of AD will *metapipt* while preserving the magnitude of △ADC.
    Let its position *metapipt*, and let it be AZ. Then △ADC is equal to △AZC, the greater to the less; which is impossible [I.16]. Therefore the position of AD will not *metapipt*; therefore AD is given in position [Def. 4]. ∎

Remarks: Dt 30 is a sound deduction from *Elements* I.16. Suppose that two lines through A, AD and AZ would be able to do the job; then a triangle would emerge in conflict with I.16, since the exterior angle would be equal to an internal and opposite one. Tacitly it is assumed that the given angle has AD as its right side; otherwise there may be another angle in the direction of B which meets the conditions, and the proof would be invalid. Again we are confirmed in our interpretation of 'the position of AD does not *metapipt*': it means 'no other line deserves the name AD by doing its job'.

In the appendix Menge prints no fewer than three alternative proofs of Dt 30, numbers 4–6 in his count of alternative proofs. Since all of them uses Dt 29 (which I take as an axiom), we can consider them weaker than Dt 30. They build on the parallel postulate, no. 4 and 5 by way of Dt 28, no. 6 by the sum of angles in a triangle (I.32).

Θε 31. ' Ἐὰν ἀπὸ δεδομένου σημείου ἐπὶ θέσει δεδομένην εὐθεῖαν εὐθεῖα γραμμὴ προσβληθῇ δεδομένη τῷ μεγέθει, δέδοται καὶ τῇ θέσει.

ἀπὸ γὰρ δεδομένου σημείου τοῦ Α ἐπὶ θέσει δεδομένην εὐθεῖαν τὴν ΒΓ εὐθεῖα γραμμὴ ἤχθω δεδομένη τῷ μεγέθει. λέγω, ὅτι καὶ τῇ θέσει δέδοται.

κέντρῳ γὰρ τῷ Α, διαστήματι δὲ τῷ ΑΔ κύκλος γεγράφθω ὁ ΕΔΖ. θέσει ἄρα ἐστὶν ὁ ΕΔΖ κύκλος· δέδοται γὰρ αὐτοῦ τὸ Α κέντρον τῇ θέσει καὶ ἡ ἐκ τοῦ κέντρου ἡ ΑΔ τῷ μεγέθει. θέσει δὲ καὶ ἡ ΒΓ εὐθεῖα. ἐὰν δὲ δύο γραμμαὶ τῇ θέσει δεδομέναι τέμνωσιν ἀλλήλας, δέδοται τὸ σημεῖον, καθ' ὃ τέμνουσιν ἀλλήλας· δοθὲν ἄρα ἐστὶ τὸ Δ. ἔστι δὲ καὶ τὸ Α δοθέν. θέσει ἄρα ἐστὶν ἡ ΑΔ.

Dt 31. *If from a given point a straight line given in magnitude be drawn to meet a straight line given in position, the line drawn is also given in position.*

For, from the given point A to the straight line BC given in position let a straight line [DA] given in magnitude have been drawn. I say that it is also given in position.

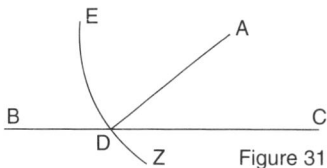

Figure 31

For, with centre A and radius AD let the circle EDZ have be described. Then the circle EDZ is [given] in position, for its centre A is given in position and its radius AD in magnitude [Def. 6].

The straight line BC is also [given] in position; and if two lines given in position cut one another, their point of section is given in position. Therefore D is given [Dt 25]; and A is also given. Therefore AD is [given] in position [Dt 26].

Remarks: Either the proposition is false or 'given in position' does not mean 'unique' – unless we extend the property 'given in position' to cover *symmetrical* positions (as Marinus Neapolitanus does, see p. 244). Consider the following problem: *From a given point A to draw a straight line AD given in magnitude to a straight line BC given in position.*

Analysis: With A as centre and AD as radius [88] let the circle DEZ be described. Let AH be the perpendicular from A on BC. Now there are three possibilities:

1 if AD is less than AH, the problem has no solution.
2 if AD is equal to AH, the problem has one solution, *viz.* AH.
3 if AD is greater than AH, the problem has two solutions because the circle will meet BC in two points D and D*. The lines AD and AD* are symmetrically positioned about AH.

The text illustrates case 3 without caring about the alternative. Dt 31 is an example of what we might call 'repeated construction': The *ekthesis* demands that the straight line DA [89] has been drawn from A to the line BE, without dwelling on how that is done – *The Helping Hand* was there, to be sure. But then it is quite surprising that the *katascheue*, the construction, shows in detail how to do that, by means of the circle with centre A and the given magnitude as radius. [90] However, Euclid needs the circle to make Dt 25 / Axiom 25*A 'prove' the proposition.

What about the double positions? Did he not see the ambiguity, or is his concept of 'given in position' the symmetrical one? The use of προσβάλλω instead of the more frequent ἐκβάλλω is striking; the former denotes the drawing of the line with respect to the goal, the latter to the starting point. Thus προσβληθῇ could be meant to stress the fact that the goal has been reached: 'if the line has been drawn', that is: one of the possible lines chosen. Dt 31 is not applied anywhere in the *Data*, so we cannot even hope to learn something from its use there.

◆

---

[88] Greek has two expressions for 'radius', διάστημα (meaning perhaps 'opening' of the compass) and ἡ ἐκ τοῦ κέντρου (the line from the centre); cf. (Fowler & Taisbak 1999).

[89] I take the reading of ms a 'ἡ ΔΑ' (Menge 52, 20) to be correct.

[90] A similar repetition is seen in Dt 43.

Θε 32. Ἐὰν εἰς παραλλήλους τῇ θέσει δεδομένας εὐθείας εὐθεῖα γραμμὴ ἀχθῇ δεδομένας ποιοῦσα γωνίας, δέδοται ἡ ἀχθεῖσα τῷ μεγέθει.

εἰς γὰρ παραλλήλους τῇ θέσει δεδομένας εὐθείας τὰς ΑΒ, ΓΔ εὐθεῖα γραμμὴ ἤχθω ἡ ΕΖ δεδομένας ποιοῦσα γωνίας τὰς ὑπὸ ΒΕΖ, ΕΖΔ. λέγω, ὅτι δέδοται ἡ ΕΖ τῷ μεγέθει.

εἰλήφθω γὰρ ἐπὶ τῆς ΓΔ δοθὲν σημεῖον τὸ Η, καὶ διὰ τοῦ Η τῇ ΕΖ παράλληλος ἤχθω ἡ ΗΘ. ἐπεὶ παράλληλός ἐστιν ἡ ΗΘ τῇ ΕΖ καὶ εἰς αὐτὰς εὐθεῖα ἐμπέπτωκεν ἡ ΓΔ, ἴση ἄρα ἐστὶν ἡ ὑπὸ ΕΖΔ τῇ ὑπὸ ΘΗΔ. δοθεῖσα δὲ ἡ ὑπὸ τῶν ΕΖΔ· δοθεῖσα ἄρα καὶ ἡ ὑπὸ ΘΗΔ.

ἐπεὶ οὖν πρὸς θέσει δεδομένῃ εὐθείᾳ τῇ ΓΔ καὶ τῷ πρὸς αὐτῇ σημείῳ δεδομένῳ τῷ Η εὐθεῖα γραμμὴ ἦκται ἡ ΗΘ δεδομένην ποιοῦσα γωνίαν τὴν ὑπὸ ΘΗΖ, θέσει ἄρα ἐστὶν ἡ ΗΘ. θέσει δὲ καὶ ἡ ΑΒ. δοθὲν ἄρα ἐστὶ τὸ Θ σημεῖον. ἔστι δὲ καὶ τὸ Η δοθέν. δοθεῖσα ἄρα ἐστὶν ἡ ΗΘ τῷ μεγέθει· καί ἐστιν ἴση τῇ ΕΖ. δοθεῖσα ἄρα ἐστὶ καὶ ἡ ΕΖ τῷ μεγέθει.

Dt 32. *If to parallel straight lines given in position a straight line be drawn making given angles, the straight line drawn is given in magnitude.*

For, to the parallel straight lines given in position AB, CD let the straight line EZ have been drawn making the given angles BEZ, EZD. I say that EZ is given in magnitude.

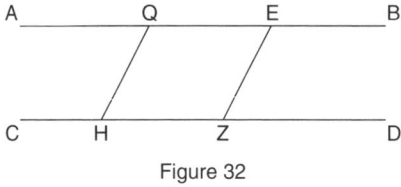

Figure 32

For, on CD let the given point H have been taken, and through H let HQ have been drawn parallel to EZ [I.31].

Since HQ is parallel to EZ and the straight line CD has fallen into them, ∠EZD is equal to ∠QHD [I.29].

And ∠EZD is given; therefore ∠QHD is given [Def. 1].

Since then, to the straight line CD given in position, and to the given point H on it the straight line HQ has been drawn making the given angle QHZ, therefore HQ is given in position [Dt 29]. And AB is given in position; therefore the point Q is given [Dt 25]. And H is given; therefore HQ is given in magnitude [Dt 26]; and it is equal to EZ [I.34]; therefore EZ is given in magnitude [Def. 1]. ∎

Remarks: Dt 32 and 33 follow Dt 28 in dealing with two parallel straight lines, AB through the given point A, and CD given in position. They use the axiomatic Dt 25, 26, and 29: the essence of the three propositions can be summed up as follows (figure 32.1):

If A and CD are given in position
Dt 28 then =I.30=> AB is given in position
 and   if ∠ACD is a given angle
Dt 32 then AC is given in magnitude
Dt 33 and *vice versa.*

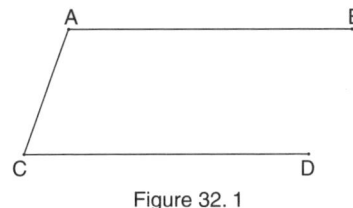

Figure 32. 1

Trigonometrically, the parallels define a constant AH, the distance between them. Since sin ∠ACD = AH:AC, if one of them is known, the other is also known.

◆

Θε 33. 'Εὰν εἰς παραλλήλους τῇ θέσει δεδομένας εὐθείας εὐθεῖα γραμμὴ ἀχθῇ δεδομένη τῷ μεγέθει, δεδομένας ποιήσει γωνίας.

εἰς γὰρ παραλλήλους τῇ θέσει δεδομένας εὐθείας τὰς ΑΒ, ΓΔ εὐθεῖα γραμμὴ ἤχθω ἡ ΕΖ, δεδομένη τῷ μεγέθει. λέγω, ὅτι δεδομένας ποιήσει γωνίας τὰς ὑπὸ τῶν ΒΕΖ, ΕΖΔ.
 εἰλήφθω γὰρ ἐπὶ τῆς ΑΒ δοθὲν σημεῖον τὸ Η καὶ διὰ τοῦ Η τῇ ΕΖ παράλληλος ἤχθω ἡ ΗΘ. ἴση ἄρα ἐστὶν ἡ ΖΕ τῇ ΗΘ. δοθεῖσα δὲ ἡ ΕΖ· δοθεῖσα ἄρα καὶ ἡ ΗΘ. καί ἐστι τὸ Η δοθέν. ὁ ἄρα κέντρῳ μὲν τῷ Η, διαστήματι δὲ τῷ ΗΘ κύκλος γραφόμενος ἔσται τῇ θέσει. γεγράφθω καὶ ἔστω ὁ ΚΘΛ. θέσει ἄρα ἐστὶν ὁ ΚΘΛ. θέσει δὲ καὶ ἡ ΓΔ. δοθὲν ἄρα ἐστὶ τὸ Θ σημεῖον. ἔστι δὲ καὶ τὸ Η δοθέν· θέσει ἄρα ἐστὶν ἡ ΗΘ. θέσει δὲ καὶ ἡ ΓΔ. δοθεῖσα ἄρα ἐστὶν ἡ ὑπὸ τῶν ΗΘΔ γωνία. καί ἐστι τῇ ὑπὸ τῶν ΕΖΔ ἴση. δοθεῖσα ἄρα καὶ ἡ ὑπὸ τῶν ΕΖΔ. καὶ λοιπὴ ἄρα ἡ ὑπὸ τῶν ΖΕΒ δοθεῖσά ἐστιν.

Dt 33. *If on parallel straight lines given in position a straight line given in magnitude be drawn, it will make given angles.*

For, on the parallel straight lines given in position AB, CD let the straight line EZ given in magnitude have been drawn. I say that it will make the given angles BEZ, EZD.

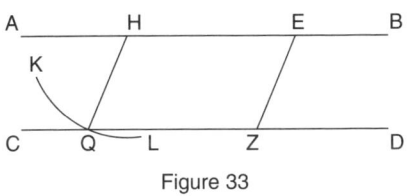

Figure 33

For, on AB let the given point H have been taken, and through H let the straight line HQ have been drawn parallel to EZ. Then ZE = HQ [I.34]. But EZ is given [in magnitude]; therefore HQ is given [in magnitude] [Def. 1].

And H is given; therefore the circle described with H as centre and HQ as radius will be [given] in position [Def. 6].
 Let it have been described, and let it be the [circle] KQL. Then KQL is [given] in position. And CD is [given] in position; therefore the point Q is given [Dt 25]. And H is given; therefore HQ is [given] in position [Dt 26].
 And CD is [given] in position; therefore ∠HQD is given. And it is equal to EZD: therefore ∠EZD is given. Therefore the remaining ∠ZEB is given, too [I.29]. ∎

Remarks: Read the remarks on Dt 32. In the diagram in Menge's edition (and in the mss) CD is tangent to the circle KQL, so that ∠EZD is right, thus concealing the problem that there might be two solutions, making ∠EZD either acute or obtuse.

Θε 34. Ἐὰν εἰς παραλλήλους τῇ θέσει δεδομένας εὐθείας ἀπὸ δεδομένου σημείου εὐθεῖα γραμμὴ ἀχθῇ, εἰς δεδομένον λόγον τμηθήσεται.

εἰς γὰρ παραλλήλους τῇ θέσει δεδομένας εὐθείας τὰς ΑΒ, ΓΔ ἀπὸ δεδομένου σημείου τοῦ Ε εὐθεῖα γραμμὴ ἤχθω ἡ ΕΖΗ. λέγω, ὅτι λόγος ἐστὶ τῆς ΕΖ πρὸς ΖΗ δοθείς.

ἤχθω γὰρ ἀπὸ τοῦ Ε σημείου ἐπὶ τὴν ΓΔ κάθετος ἡ ΕΚΘ. ἐπεὶ ἀπὸ δεδομένου σημείου τοῦ Ε ἐπὶ θέσει δεδομένην εὐθεῖαν τὴν ΓΔ εὐθεῖα γραμμὴ ἦκται ἡ ΕΘ δεδομένην ποιοῦσα γωνίαν τὴν ὑπὸ τῶν ΕΘΗ, θέσει ἄρα ἐστὶν ἡ ΕΘ· θέσει δὲ καὶ ἑκατέρα τῶν ΑΒ, ΓΔ· δοθὲν ἄρα ἐστὶν ἑκάτερον τῶν Κ, Θ. ἔστι δὲ καὶ τὸ Ε δοθέν· δοθεῖσα ἄρα ἐστὶν ἑκατέρα τῶν ΕΚ, ΚΘ. λόγος ἄρα τῆς ΕΚ πρὸς τὴν ΚΘ δοθείς. καί ἐστιν ὡς ἡ ΕΚ πρὸς τὴν ΚΘ, οὕτως ἡ ΕΖ πρὸς τὴν ΖΗ. λόγος ἄρα καὶ τῆς ΕΖ πρὸς τὴν ΖΗ δοθείς.

Dt 34. *If a straight line be drawn from a given point to parallel straight lines given in position, it will be cut in a given ratio.*

For, on the parallel straight lines given in position AB, CD let the straight line EZH have been drawn from the given point E. I say that the ratio EZ:ZH is given.

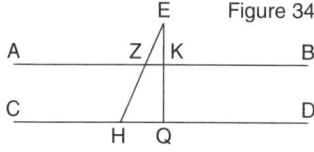

Figure 34

For, from the point E let EKQ have been drawn perpendicular to CD. Since from the given point E to the straight line given in position CD the straight line EQ have been drawn, making the given angle EQH, therefore EQ is [given] in position [Dt 30]. And each of AB, CD is [given] in position. Therefore each of K, Q is given [Dt 25]. And E is given; therefore each of the lines EK, KQ is given [Dt 26]; therefore the ratio EK:KQ is given [Dt 1]. And EK:KQ :: EZ:ZH [VI.2; therefore the ratio EZ:ZH is given [Def. 2*]. ∎

Remarks: Dt 34–38 constitute a theory of parallels and ratio which can be summed up as follows: Four lines are involved, one of which (EZ) is given in position, two (AB, CD) are parallel to that one, and a fourth (HQ) cuts them in three points: H on AB, K on CD, Q on EZ (figure 34.1). Six statements are combined in the five propositions.

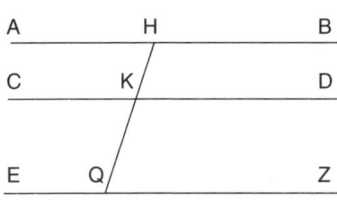

Figure 34. 1

0) pos.EZ, AB ∥ CD ∥ EZ
1) H is given in position
2) K is given in position
3) AB is given in position
4) CD is given in position
5) the ratio HK:KQ is given

Dt 34 proves 1 & 4 => 5: a transversal HQ from a given point H across parallels CD and EZ given in position is cut in a given ratio HK:KQ.

Dt 35 proves 1 & 5 => 4, the partial converse of 34.

Dt 36 proves 2 & 5 => 3. The proof is a variant of 35. Together the two prove that if a transversal HQ from a given point H (or K) to a given line EZ be cut internally (at K) or externally (at H) in a given ratio, the parallel line AB (or CD) through the point of division is given in position.

Dt 34–36 are counterparts of *Elements* VI.2.

Dt 37 proves 3 & 5 => 4.

Dt 38 proves 4 & 5 => 3, the partial converse of 37. Together the two prove that if the interval between two parallels be divided internally in a given ratio and a parallel be drawn through the point of division, the line drawn will be given in position.

The proofs are (of course) very similar, basing themselves on Dt 25 and its sequels: Dt 26, uniqueness of straight line joining two given points; Dt 27, unique endpoint of straight line given in position and magnitude; Dt 28, unique parallel through a given point; Dt 30, uniqueness of perpendicular dropped on a given line.

In Dt 35 and 37 the *synthenti* (*componendo*) operation of Dt 6 is used. From the *Elements* Book VI propositions 2 and 4 are invoked. The expositions are unnecessarily complicated by the Greek tendency to use different names of points in different propositions even if the diagrams are substantially the same, according to Netz (1999) because diagrams were alphabetized in the order in which the points or lines appear in the proof. I have tried to avoid that by using one and the same diagram for my synopsis.

Bernard Vitrac (Vitrac II:161) underlines the importance of propositions VI.2 & 4, which establish a direct connexion between proportionality (and therefore similarity of figures) and parallelism.

◆

Θε 35. ' Ἐὰν ἀπὸ δεδομένου σημείου ἐπὶ θέσει δεδομένην εὐθεῖαν εὐθεῖα γραμμὴ ἀχθῇ καὶ τμηθῇ εἰς δεδομένον λόγον, διὰ δὲ τῆς τομῆς παρὰ τὴν θέσει δεδομένην εὐθεῖαν εὐθεῖα γραμμὴ ἀχθῇ, δέδοται ἡ ἀχθεῖσα τῇ θέσει.

ἀπὸ γὰρ δεδομένου σημείου τοῦ Α ἐπὶ θέσει δεδομένην εὐθεῖαν τὴν ΓΒ εὐθεῖα γραμμὴ ἤχθω ἡ ΑΔ καὶ τετμήσθω εἰς δεδομένον λόγον τὸν τῆς ΔΕ πρὸς ΕΑ, καὶ ἤχθω διὰ τοῦ Ε τῇ ΒΓ παράλληλος ἡ ΖΕΗ. λέγω, ὅτι θέσει ἐστὶν ἡ ΖΕΗ.

ἤχθω γὰρ ἀπὸ τοῦ Α ἐπὶ τὴν ΒΓ κάθετος ἡ ΑΘ. ἐπεὶ ἀπὸ δεδομένου σημείου τοῦ Α ἐπὶ θέσει δεδομένην εὐθεῖαν τὴν ΒΓ εὐθεῖα γραμμὴ ἧκται ἡ ΑΘ δεδομένην ποιοῦσα γωνίαν τὴν ὑπὸ τῶν ΑΘΔ, θέσει ἄρα ἐστὶν ἡ ΑΘ. θέσει δὲ καὶ ἡ ΒΓ· δοθὲν ἄρα τὸ Θ σημεῖον. ἔστι δὲ καὶ τὸ Α δοθέν. δοθεῖσα ἄρα ἐστὶν ἡ ΑΘ. καὶ ἐπεὶ λόγος τῆς ΔΕ πρὸς τὴν ΕΑ δοθείς, ὡς δὲ ἡ ΔΕ πρὸς τὴν ΕΑ, οὕτως ἡ ΘΚ πρὸς τὴν ΚΑ, λόγος ἄρα καὶ ὁ τῆς ΘΚ πρὸς τὴν ΚΑ δοθείς. συνθέντι ἄρα λόγος ἐστὶ τῆς ΘΑ πρὸς ΑΚ δοθείς. δοθεῖσα δὲ ἡ ΘΑ· δοθεῖσα ἄρα καὶ ἡ ΑΚ. ἀλλὰ καὶ τῇ θέσει. καί ἐστι τὸ Α δοθέν· δοθὲν ἄρα καὶ τὸ Κ. ἐπεὶ οὖν διὰ δεδομένου σημείου τοῦ Κ παρὰ θέσει δεδομένην εὐθεῖαν τὴν ΒΓ εὐθεῖα γραμμὴ ἧκται ἡ ΖΗ, θέσει ἄρα ἐστὶν ἡ ΖΗ.

Dt 35. *If from a given point to a straight line given in position, a straight line be drawn and cut in a given ratio, and* [if] *a straight line be drawn through the point of section parallel to the line given in position, the latter straight line drawn will be given in position.*

For, from the given point A to the straight line given in position CB let the straight line AD have been drawn and cut in the given ratio DE:EA, and through the point E let the straight line ZEH have been drawn parallel to BC; I say that ZEH is [given] in position.

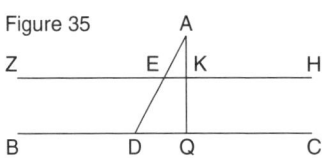

Figure 35

For from the point A let AQ have been drawn perpendicular to BC [I.12].
Since from the given point A to the straight line given in position BC the straight line AQ has been drawn, making the given angle AQD, therefore AQ is [given] in position [Dt 30].

And BC is also [given] in position; therefore the point Q is given [Dt 25]. And A is also given; therefore the straight line AQ is given [Dt 26].

And since the ratio DE:EA is given, but DE:EA :: QK:KA [VI.2], therefore the ratio QK:KA is given [Def. 2]. *Synthenti* therefore the ratio [(QK+KA):AK, *i.e.*] QA:AK is given [Dt 6]. And QA is given; therefore AK is given. But in position, too. And A is given; therefore K is given [Dt 27].

Since then, through the given point K the straight line ZH has be drawn parallel to BC, which is given in position, therefore ZH is [given] in position [Dt 28]. ∎

Θε 36. ’Εὰν ἀπὸ δεδομένου σημείου ἐπὶ θέσει δεδομένην εὐθεῖαν εὐθεῖα γραμμὴ ἀχθῇ καὶ προστεθῇ τις αὐτῇ εὐθεῖα λόγον ἔχουσα πρὸς αὐτὴν δεδομένον, διὰ δὲ τοῦ πέρατος τῆς προστεθείσης παρὰ τὴν τῇ θέσει δεδομένην εὐθεῖαν εὐθεῖα γραμμὴ ἀχθῇ, δέδοται ἡ ἀχθεῖσα τῇ θέσει.

ἀπὸ γὰρ δεδομένου σημείου τοῦ Α ἐπὶ θέσει δεδομένην εὐθεῖαν τὴν ΒΓ εὐθεῖα γραμμὴ ἤχθω ἡ ΑΔ, καὶ προσκείσθω τῇ ΑΔ ἡ ΑΕ λόγον ἔχουσα πρὸς τὴν ΑΔ δεδομένον, διὰ δὲ τοῦ Ε τῇ ΒΓ παράλληλος ἤχθω ἡ ΖΚ. λέγω, ὅτι θέσει ἐστὶν ἡ ΖΚ.

ἤχθω γὰρ ἀπὸ τοῦ Α ἐπὶ τὴν ΒΓ κάθετος ἡ ΑΘ καὶ διήχθω ἐπὶ τὸ Η. ἐπεὶ ἀπὸ δεδομένου σημείου τοῦ Α ἐπὶ θέσει δεδομένην εὐθεῖαν τὴν ΒΓ εὐθεῖα γραμμὴ ἦκται ἡ ΑΘ δεδομένην ποιοῦσα γωνίαν τὴν ὑπὸ ΑΘΓ, θέσει ἄρα ἐστὶν ἡ ΘΑΗ. θέσει δὲ καὶ ἡ ΒΓ· δοθὲν ἄρα ἐστὶ τὸ Θ σημεῖον. ἔστι δὲ καὶ τὸ Α δοθέν· δοθεῖσα ἄρα ἐστὶν ἡ ΑΘ. καὶ ἐπεὶ λόγος ἐστὶ τῆς ΔΑ πρὸς τὴν ΑΕ δοθείς, ὡς δὲ ἡ ΔΑ πρὸς τὴν ΑΕ, οὕτως ἡ ΘΑ πρὸς τὴν ΑΗ, λόγος ἄρα καὶ τῆς ΘΑ πρὸς τὴν ΑΗ δοθείς. δοθεῖσα δὲ ἡ ΘΑ· δοθεῖσα ἄρα καὶ ἡ ΑΗ. ἀλλὰ καὶ τῇ θέσει. καί ἐστι δοθὲν τὸ Α· δοθὲν ἄρα καὶ τὸ Η. ἐπεὶ οὖν διὰ δεδομένου σημείου τοῦ Η παρὰ θέσει δεδομένην εὐθεῖαν τὴν ΒΓ εὐθεῖα γραμμὴ ἦκται ἡ ΖΗΚ, θέσει ἄρα ἐστὶν ἡ ΖΗΚ.

Dt 36. *If from a given point to a straight line given in position a straight line be drawn and a straight line having a given ratio to it be added to it, and* [if] *a straight line be drawn through the extremity of the added line parallel to the line given in position, the* [latter] *straight line drawn is given in position.*

For, from the given point A to the straight line given in position BC let the straight line AD have been drawn, and to AD let AE have been added having a given ratio to AD, and through E let ZK have been drawn parallel to BC. I say that ZK is [given] in position.

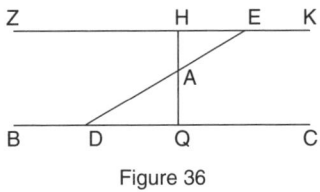

Figure 36

For, from the point A let AQ have been drawn perpendicular to BC and produced to H. Since from the given point A to the straight line given in position BC the straight line AQ has been drawn making the given angle AQC, therefore QAH is [given] in position [Dt 30].

And BC is [given] in position; therefore the point Q is given [Dt 25]. And A is given; therefore AQ is given [Dt 26].

And since the ratio DA:AE is given, and DA:AE :: QA:AH [VI.4], therefore the ratio QA:AH is given. And QA is given; therefore AH is given [Dt 2]. But also in position. And A is given; therefore H is given [Dt 27].

Since then, the straight line ZHK has be drawn through the given point H parallel to BC, which is given in position, ZHK is [given] in position [Dt 28]. ∎

Θε 37. ' Ἐὰν εἰς παραλλήλους τῇ θέσει δεδομένας εὐθείας εὐθεῖα γραμμὴ ἀχθῇ καὶ τμηθῇ εἰς δεδομένον λόγον, διὰ δὲ τῆς τομῆς παρὰ τὰς τῇ θέσει δεδομένας εὐθείας εὐθεῖα γραμμὴ ἀχθῇ, δέδοται ἡ ἀχθεῖσα τῇ θέσει.

εἰς γὰρ παραλλήλους τῇ θέσει δεδομένας εὐθείας τὰς ΑΒ, ΓΔ εὐθεῖα γραμμὴ ἤχθω ἡ ΕΖ καὶ τετμήσθω εἰς δεδομένον λόγον τὸν τῆς ΖΗ πρὸς τὴν ΗΕ, καὶ διήχθω διὰ τοῦ Η ὁποτέρᾳ τῶν ΑΒ, ΓΔ παράλληλος ἡ ΘΚ. λέγω, ὅτι θέσει ἐστὶν ἡ ΘΚ.

εἰλήφθω γὰρ ἐπὶ τῆς ΑΒ δοθὲν σημεῖον τὸ Λ, καὶ κατήχθω ἀπὸ τοῦ Λ ἐπὶ τὴν ΓΔ κάθετος ἡ ΛΝ. ἐπεὶ ἀπὸ δεδομένου σημείου τοῦ Λ ἐπὶ θέσει δεδομένην εὐθεῖαν τὴν ΓΔ εὐθεῖα γραμμὴ ἦκται ἡ ΛΝ, δεδομένην ποιοῦσα γωνίαν τὴν ὑπὸ τῶν ΛΝΔ, θέσει ἄρα ἐστὶν ἡ ΛΝ. θέσει δὲ καὶ ἡ ΓΔ· δοθὲν ἄρα τὸ Ν σημεῖον. ἔστι δὲ καὶ τὸ Λ δοθέν· δοθεῖσα ἄρα ἐστὶν ἡ ΛΝ. καὶ ἐπεὶ λόγος ἐστὶ τῆς ΖΗ πρὸς τὴν ΗΕ δοθείς, ὡς δὲ ἡ ΖΗ πρὸς τὴν ΗΕ, οὕτως ἡ ΝΜ πρὸς τὴν ΜΛ, λόγος ἄρα καὶ τῆς ΝΜ πρὸς τὴν ΜΛ δοθείς· ὥστε καὶ τῆς ΝΛ πρὸς τὴν ΜΛ ἐστι δοθεὶς λόγος. δοθεῖσα δὲ ἡ ΝΛ. δοθεῖσα ἄρα καὶ ἡ ΛΜ. ἀλλὰ καὶ τῇ θέσει. καί ἐστι δοθὲν τὸ Λ· δοθὲν ἄρα καὶ τὸ Μ. ἐπεὶ οὖν διὰ δεδομένου σημείου τοῦ Μ παρὰ θέσει δεδομένην εὐθεῖαν τὴν ΓΔ εὐθεῖα γραμμὴ ἦκται ἡ ΘΚ, θέσει ἄρα ἐστὶν ἡ ΘΚ.

Dt 37. *If on parallel lines given in position a straight line be drawn and cut in a given ratio, and through the point of section a straight line be drawn parallel to the lines given in position, the* [latter] *line drawn is given in position.*

For, on the parallel lines given in position AB, CD let the straight line EZ have been drawn and cut in the given ratio ZH:HE, and through the point H let the QK have been drawn parallel to either of AB, CD. I say that QK is [given] in position.

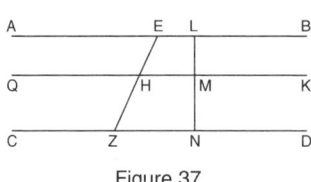

Figure 37

On AB let the given point L have been taken [Axiom 0*], and from L let LN have been drawn perpendicular to CD [I.12].
Since from the given point L to the straight line given in position CD the straight line LN has been drawn making the given angle LND, therefore LN is [given] in position [Dt 30].

And CD is [given] in position; therefore the point N is given [Dt 25]. And L is given; therefore LN is given [Dt 26].

And since the ratio ZH:HE is given, but ZH:HE :: NM:ML [VI.2], therefore the ratio NM:ML is given; so that NL:ML is also given [Dt 6]. And NL is given; therefore LM is given [Dt 2]. But also in position. And L is given; therefore M is given [Dt 27].

Since then, through the given point M the straight line QK has be drawn parallel to CD, which is given in position, therefore QK is [given] in position [Dt 28]. ∎

Θε 38. ' Ἐὰν εἰς παραλλήλους τῇ θέσει δεδομένας εὐθείας εὐθεῖα γραμμὴ ἀχθῇ καὶ προστεθῇ τις αὐτῇ εὐθεῖα λόγον ἔχουσα πρὸς αὐτὴν δεδομένον, διὰ δὲ τοῦ πέρατος παρὰ τὰς τῇ θέσει δεδομένας παράλληλος εὐθεῖα γραμμὴ ἀχθῇ, δέδοται ἡ ἀχθεῖσα τῇ θέσει.

εἰς γὰρ παραλλήλους τῇ θέσει δεδομένας εὐθείας τὰς ΑΒ, ΓΔ εὐθεῖα γραμμὴ ἤχθω ἡ ΕΖ, καὶ προσκείσθω τις αὐτῇ εὐθεῖα ἡ ΕΗ λόγον ἔχουσα πρὸς τὴν ΕΖ δεδομένον, διὰ δὲ τοῦ Η ὁποτέρᾳ τῶν ΑΒ, ΓΔ εὐθειῶν παράλληλος εὐθεῖα γραμμὴ ἤχθω ἡ ΘΚ. λέγω, ὅτι θέσει ἐστὶν ἡ ΘΚ.

εἰλήφθω γὰρ ἐπὶ τῆς ΑΒ δοθὲν σημεῖον τὸ Ν, καὶ ἤχθω ἀπὸ τοῦ Ν ἐπὶ τὴν ΓΔ κάθετος εὐθεῖα γραμμὴ ἡ ΝΜ καὶ διήχθω ἐπὶ τὸ Λ. ἐπεὶ ἀπὸ δεδομένου σημείου τοῦ Ν ἐπὶ θέσει δεδομένην εὐθεῖαν τὴν ΓΔ εὐθεῖα γραμμὴ ἦκται ἡ ΝΜ δεδομένην ποιοῦσα γωνίαν τὴν ὑπὸ ΝΜΔ, θέσει ἄρα ἐστὶν ἡ ΛΝΜ. θέσει δὲ καὶ ἡ ΓΔ. δοθὲν ἄρα ἐστὶ τὸ Μ σημεῖον. ἔστι δὲ καὶ τὸ Ν δοθέν. δοθεῖσα ἄρα ἐστὶν ἡ ΝΜ. καὶ ἐπεὶ λόγος ἐστὶ τῆς ΖΕ πρὸς τὴν ΕΗ δοθείς, ὡς δὲ ἡ ΖΕ πρὸς τὴν ΗΕ, οὕτως ἡ ΝΜ πρὸς τὴν ΝΛ, λόγος ἄρα καὶ τῆς ΜΝ πρὸς τὴν ΝΛ δοθείς. δοθεῖσα δὲ ἡ ΝΜ· δοθεῖσα ἄρα καὶ ἡ ΝΛ. ἀλλὰ καὶ τῇ θέσει. καί ἐστι τὸ Ν δοθέν· δοθὲν ἄρα καὶ τὸ Λ. ἐπεὶ οὖν διὰ δεδομένου σημείου τοῦ Λ παρὰ θέσει δεδομένην εὐθεῖαν τὴν ΑΒ εὐθεῖα γραμμὴ ἦκται ἡ ΘΚ, θέσει ἄρα ἐστὶν ἡ ΘΚ.

Dt 38. *If on parallels given in position a straight line be drawn and a straight line having a given ratio to it be added to it, and if through the extremity a parallel straight line be drawn parallel to the lines given in position, the [latter] line drawn is given in position.*

For, on the parallel lines given in position AB, CD let the straight line EZ have been drawn and let a straight line EH have been added to it having a given ratio to EZ, and through H let the straight line QK have been drawn parallel to either of AB, CD. I say that QK is [given] in position.

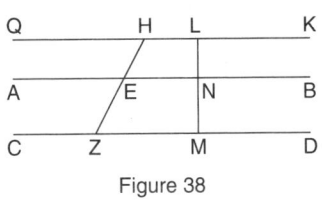
Figure 38

On AB, let the given point N have been taken [Axiom 0*], and from N let NM have been drawn perpendicular to CD and produced to L. Since from the given point N to the straight line CD, given in position, the straight line NM has been drawn, making the given angle NMD, therefore LNM is [given] in position [Dt 30].

And CD is [given] in position; therefore the point M is given [Dt 25]. And N is given; therefore NM is given [Dt 26]. And since the ratio ZE:EH is given, but ZE:EH :: NM:NL [VI.2], therefore the ratio NM:NL is given. And NM is given; therefore NL is given. But also in position. And N is given; therefore L is given [Dt 27].

Since then, through the given point L the straight line QK has be drawn parallel to AB, which is given in position, therefore QK is [given] in position [Dt 28]. ∎

Deductive Structure of Dt 25–38

	Dt	25	26	27	28	29	30	31	32	33	34	35	36	37	38
Data															
Def.	1*	.	.	.	.	.	.	+	+	.	.	.	.	.	.
	2*	.	.	.	.	.	.	.	.	.	+	+	+	+	+
	4*A	+	+	+	+	+	+	.	.	.	.	.	.	.	.
	6	.	.	.	.	.	.	+	.	+	.	.	.	.	.
Dt	1	.	.	.	.	.	.	.	.	.	+	.	.	.	.
	2	.	.	.	.	.	.	.	.	.	.	+	+	+	+
	4	.	.	.	.	.	.	.	.	+	.	.	.	.	.
	6	.	.	.	.	.	.	.	.	.	+	.	+	.	.
	25	.	.	.	.	.	.	+	+	+	+	+	+	+	+
	26	.	.	.	.	.	.	+	+	+	+	+	+	+	+
	27	.	.	.	.	.	.	.	.	.	.	+	+	+	+
	28	.	.	.	.	.	.	.	.	.	.	+	+	+	+
	29	.	.	.	.	.	.	.	+	.	.	.	.	.	.
	30	.	.	.	.	.	.	.	.	.	+	+	+	+	+
Elem															
Post	2	.	.	.	.	.	.	.	.	.	.	+	.	+	.
I.	12	.	.	.	.	.	.	.	.	+	.	+	.	.	.
	16	.	.	.	.	+	.	.	.	.	.	.	.	.	.
	29	.	.	.	.	.	.	+	+	.	.	.	.	.	.
	30	.	.	.	+	.	.	.	.	.	.	.	.	.	.
	34	.	.	.	.	.	.	+	+	.	.	.	.	.	.
VI.	2	.	.	.	.	.	.	.	.	.	+	+	.	+	+
	4	.	.	.	.	.	.	.	.	.	.	.	+	.	.
Data	Dt	25	26	27	28	29	30	31	32	33	34	35	36	37	38

Dt 25–27 and 29 are axiomatic in so far as they use Def 4. Dt 28 and 30 involve I.16 and 30 resp. to make sound reductions *ad absurdum*. They are put to work in Dt 31–38, which are there for their own sake and not used again in the *Data*.

Chapter 6. Form
Triangles and Polygons. Dt 39–55

Definition 3

Intuitively, most people know what it means that a plane figure has a fixed shape of its own, different from some and, perhaps, similar to others; something to do with angles and proportion. So the *Data*'s definition of being 'given in form' is hardly surprising. Def. 3 is tailored after the pattern of *Elements* VI, def. 1:

Data, Def. 3	*Elements* VI, def. 1
Εὐθύγραμμα σχήματα τῷ εἴδει δεδόσθαι λέγεται, ὧν αἵ τε γωνίαι δεδομέναι εἰσὶ κατὰ μίαν καὶ οἱ λόγοι τῶν πλευρῶν πρὸς ἀλλήλας δεδομένοι.	῞Ομοια σχήματα εὐθύγραμμά ἐστιν, ὅσα τάς τε γωνίας ἴσας ἔχει κατὰ μίαν καὶ τὰς περὶ τὰς ἴσας γωνίας πλευρὰς ἀνάλογον.
Rectilineal figures are said to be *given in form* if the angles are given one by one and the ratios of the sides to one another are given.	*Similar* rectilineal figures are such as have their angles equal one by one and the sides around the equal angles proportional.

From this parallelism follows a lemma or definition (cf. p. 33):

> **Def. 3*B** *A rectilineal figure is given in form if it is similar to one that is given in form.*

That reminds us of our discussion about Def. 1, where we concluded that *a magnitude is given if it is equal to one that is given*, Def. 1*; and this was saved from circularity by Axiom *0, *Any point or line segment or angle may be (taken and) appointed given.* Thus a magnitude is given if it is equal to one that is already present and fixed in the Plane. The same situation holds for Def. 3:

> **Def. 3*A** *A rectilineal figure is given in form if its vertices are given in position,*

and consequently its sides are given in position and in magnitude (Axiom 26*A). Both definitions satisfy Def. 3, since in either case each angle is given and the ratios between the sides are given. They are often used in the proofs to establish a situation where Def. 3 is active, though never quoted. It looks as if Euclid realized that *position* of figures implies *form* of figures, but did not put that insight into the definitions, because givenness in form is a 'wider' concept than givenness in position, the latter being included in the former. The following reasoning is absent from his proofs, but must be understood, so to speak (*e.g.* in Dt 40):[91]

> If the vertices of a figure are given, the sides are given in position and in magnitude (Axiom 26*A); therefore the angles are given in position and in magnitude, and the ratios of the sides are given (Dt 1).

We need an axiom about given angles, but no such axiom or any ingredient for it is forthcoming in the *Data*:

> **Def. 1*A** *An angle is given in position and in magnitude if its vertex and one point on each side is given in position.*

When a triangle or polygon is *known* to be given in form (first in Dt 48), Def. 3 is used in the 'only-if-case', *If a rectilineal figure is (said to be) given in form, then each angle is given and the ratios of the sides to one another are given.*

Dt 55 The Main Theorem?

On p. 90 I supplied Dt 24*C: *If a square is given in magnitude, its side is also given in magnitude,* a special case of the *Data*'s most general assertion about givenness in form and in magnitude, namely

> Dt 55 *If a rectilineal figure is given in form and in magnitude, its sides will also be given in magnitude.*

To assess that theorem, we may remark heuristically: If a rectilineal figure P is given in form, its angles are given, and its sides have given ratios to each other (Def. 3). If the figure grows without changing its shape, the sides grow simultaneously, keeping pace with each other. This suggests that (in an anachronistic idiom) the length of one side (and therefore of all sides) is a function of its area, and *vice versa*.

I now give an analysis for Dt 55 (figure 55.1): P is a rectilineal figure, given in form (as defined in Def. 3), and a is any side of P. It suffices to prove that one side of P is given in magnitude, since, if one is given, they all are (Def. 3). Thus we want to prove that *if mag.P then mag.a*.

---

[91] Since no angles are greater than two right angles, all rectilineal figures (but gnomons) in the *Data* (and the *Elements*) are supposed to be convex, although nothing is ever said to that effect.

Now, suppose the conclusion to be true: an arbitrary side *a* of P is given in magnitude. Let *b* be a line segment, taken (and therefore given) in position and therefore also in magnitude (Axiom 26*A).

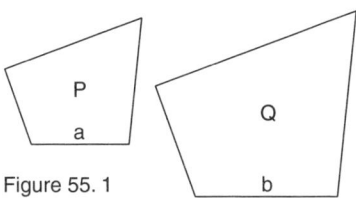

Figure 55. 1

Then *a:b* is a given ratio (Dt 1). Let Q be constructed on the line segment *b*, similar and similarly situated to P, (with *a* and *b* homologous sides, although Dt 53 ensures that the proof is valid even for arbitrary sides); then Q will be given in form (Def. 3*B).

The ratio P:Q will be the 'double' (διπλασίων, we would say the 'square') of *a:b*, by VI.20 coroll. Therefore, since the 'double' of a given ratio is itself given (Dt 8, *di 'isou*), we have gvn.P:Q. And then, since P was given in magnitude, so is Q (Dt 2). We can count the steps in the analysis as follows:

Hypotheses: mag.P, frm.P, Q sim P, mag.*b*. Assume mag.*a*.
1) mag.*a* & mag.*b* => gvn.*a:b* [Dt 1]
2) P:Q = dbl(*a:b*) => gvn.P:Q [VI.20, Dt 8]
3) mag.P => mag.Q [Dt 2]

Dt 55 reverses the argument; a few lemmas are needed (first and foremost the proof of its converse), which are provided in the sequence 48–54:

Hypotheses: mag.P, frm.P, Q sim P, mag.*b*. Assertion: mag.*a*.
3) mag.*b*, frm.Q => mag.Q [Dt 52, conv. of Dt 55]
2) mag.P => gvn.P:Q [Dt 1]
1) P:Q = dbl(*a:b*) => gvn.*a:b* [Dt 54, Dt 24]
0) mag.*b* => mag.*a* [Dt 2]

The stemma of the deductive structure (on p. 144, 145) will show how the sequence of Dt 49–54 leads up to Dt 55; Dt 8, the *di 'isou* metamorphosis, assists most of them. The problems raised by Dt 24, (which is the 'square case' of Dt 54) were discussed on p. 89 ff.

Dt 50 serves to establish Dt 51 (the converse of Dt 54), which is not used in the proof of Dt 55, since the figures P and Q are not arbitrary polygons given in form, but similar. Now follows a brief survey of the sequence, *a* and *b* being homologous sides:

Dt 49 If P and Q are drawn on the same line b, then gvn.P:Q.
Dt 50 If gvn.*a:b* & P sim Q, then gvn.P:Q.

Dt 51 If gvn.*a:b*, then gvn.P:Q.
Dt 52 If mag.*a*, then mag.P.
Dt 53 If gvn.*a:b* for one pair, then gvn.*c:d* for any pair *c*, *d*.
Dt 54 If gvn.P:Q, then gvn.*a:b*   (51 converse).
Dt 55 If mag.P, then mag.*a*   (52 converse).

I shall comment on the members of this sequence in due course, after a brief treatment of their foundations, namely

Triangles
Dt 49 is a generalisation of Dt 48, which deals with triangles. Def. 3 speaks of rectilineal figures, *i.e.* polygons; but polygons given in form are resolved (Dt 47) into triangles given in form. Therefore the proof of Dt 55 profits from the fact that triangles are simpler figures than polygons: two given angles, or one angle and the ratio between any two sides, or the ratio between all three sides will ensure givenness in form. Dt 40–46 are meant to establish these relations – and a couple more, as can be seen from the following diagram. The close connection with *Elements* VI.4–7 is hinted at in the utmost right column:

∡A	∡B	*a:b*	*b:c*	(*a+b*):*c*	(*b+c*):*a*	Data	Elem
gvn	gvn					40	VI.4
gvn			gvn			41	VI.6
right		gvn				43	
gvn		gvn				44	VI.7
		gvn	gvn			42	VI.5
gvn					gvn	45	
gvn				gvn		46	

In all the theorems, except Dt 42, one angle is given and something more:

In Dt 40 it is one of the other angles; it follows immediately, from the sum of angles in a triangle (I.32), that the third angle is also given; however, the enunciation of Dt 40 takes the three of them to be given.

In Dt 41 the ratio of the sides containing the given angle is given, in Dt 43 and 44 the ratio of the sides containing another angle, the latter being dubious, though it could be saved by a precondition as in VI.7.[92] Finally (and somewhat unexpectedly) in Dt 45 and 46 the ratio of the sum of two sides to the third side is given – a relation that turns out more useful than one would expect.

For argument's sake I now present Dt 39 and 42 followed by Dt 40 and 41.

◆

Θε 39. Ἐὰν τριγώνου ἑκάστη τῶν πλευρῶν δεδομένη ᾖ τῷ μεγέθει, δέδοται τὸ τρίγωνον τῷ εἴδει.

τριγώνου γὰρ τοῦ ΑΒΓ ἑκάστη τῶν πλευρῶν δεδομένη ἔστω τῷ μεγέθει· λέγω, ὅτι τὸ ΑΒΓ τρίγωνον δέδοται τῷ εἴδει.
ἐκκείσθω γὰρ εὐθεῖα τῇ θέσει δεδομένη ἡ ΔΜ, πεπερατωμένη μὲν κατὰ τὸ Δ, ἄπειρος δὲ κατὰ τὸ λοιπόν, καὶ κείσθω τῇ μὲν ΑΒ ἴση ἡ ΔΕ· δοθεῖσα δὲ ἡ ΑΒ· δοθεῖσα ἄρα καὶ ἡ ΔΕ· ἀλλὰ καὶ τῇ θέσει· καί ἐστι δοθὲν τὸ Δ· δοθὲν ἄρα καὶ τὸ Ε· τῇ δὲ ΒΓ ἴση ἡ ΕΖ· δοθεῖσα δὲ ἡ ΒΓ· δοθεῖσα ἄρα καὶ ἡ ΕΖ· ἀλλὰ καὶ τῇ θέσει· καί ἐστι δοθὲν τὸ Ε· δοθὲν ἄρα καὶ τὸ Ζ· τῇ δὲ ΑΓ ἴση ἡ ΖΗ. δοθεῖσα δὲ ἡ ΑΓ· δοθεῖσα ἄρα καὶ ἡ ΖΗ. ἀλλὰ καὶ τῇ θέσει. καί ἐστι δοθὲν τὸ Ζ· δοθὲν ἄρα καὶ τὸ Η. καὶ κέντρῳ μὲν τῷ Ε, διαστήματι δὲ τῷ ΕΔ κύκλος γεγράφθω ὁ ΔΚΘ· θέσει ἄρα ἐστὶν ὁ ΔΚΘ. πάλιν κέντρῳ μὲν τῷ Ζ, διαστήματι δὲ τῷ ΖΗ κύκλος γεγράφθω ὁ ΗΚΛ· θέσει ἄρα ἐστὶν ὁ ΗΚΛ· θέσει δὲ καὶ ὁ ΔΘΚ κύκλος· δοθὲν ἄρα ἐστὶ καὶ τὸ Κ σημεῖον. ἔστι δὲ καὶ ἑκάτερον τῶν Ε, Ζ δοθέν· δοθεῖσα ἄρα ἐστὶν ἑκάστη τῶν ΚΕ, ΕΖ, ΖΚ τῇ θέσει καὶ τῷ μεγέθει· δέδοται ἄρα τὸ ΚΕΖ τρίγωνον τῷ εἴδει. καί ἐστιν ἴσον τε καὶ ὅμοιον τῷ ΑΒΓ· δέδοται ἄρα τὸ ΑΒΓ τρίγωνον τῷ εἴδει.

Dt 39. *If each of the sides of a triangle be given in magnitude, the triangle is given in form.*

For, let each of the sides of △ABC be given in magnitude; I say that △ABC is given in form.

For, let a straight line given in position DM have been set out [Axiom 0*], terminated at D, infinite in the other direction, and let DE lie equal to AB; and AB is given [in magnitude]; therefore DE is also given [Def. 1], but in position, too [for DM is given in position]; and D is given, therefore E is also given [Dt 27].

And [let] EZ [lie] equal to BC; and BC is given, therefore EZ is also given, but in position, too; and E is given, therefore Z is also given.

And [let] ZH [lie] equal to AC; and AC is given, therefore ZH is also given, but in position too; and Z is given, therefore H is also given.

---

[92] If the antecedent in a ratio is greater than the consequent (which is most often the case in figures in the mss, and may have been an ancient convention), no ambiguity arises in Dt 44.

And with centre E and radius ED let the circle DKG have been described; then the [circle] DKG is given in position [Def. 6].

Again with centre Z and radius ZH let the circle HKL have been described; then the [circle] HKL is given in position. And the circle DGK is given in position, too; therefore the point K is given [Dt 25].

And each of the points E and Z is also given; therefore each of KE, EZ, ZK is given in position and in magnitude [Dt 26]; therefore △KEZ is given in form [Def. 3*A]. And it is equal and similar to △ABC [I.8, I.4, VI.Def. 1]; therefore △ABC is given in form [Def. 3*B]. ∎

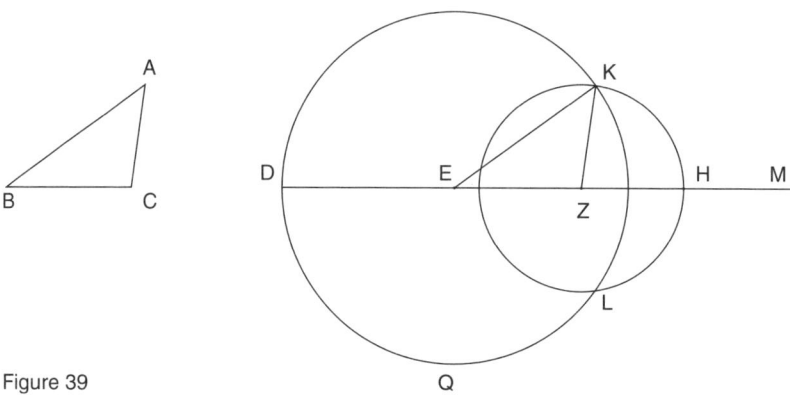

Figure 39

Note that Dt 40 is postponed till after Dt 42.

◆

Θε 42. ' Ἐὰν τριγώνου αἱ πλευραὶ πρὸς ἀλλήλας λόγον ἔχωσι δεδομένον, δέδοται τὸ τρίγωνον τῷ εἴδει.

τριγώνου γὰρ τοῦ ΑΒΓ αἱ πλευραὶ πρὸς ἀλλήλας λόγον ἐχέτωσαν δεδομένον· λέγω, ὅτι τὸ ΑΒΓ τρίγωνον δέδοται τῷ εἴδει.

ἐκκείσθω γὰρ δεδομένη τῷ μεγέθει εὐθεῖα ἡ Δ. καὶ ἐπεὶ λόγος ἐστὶ τῆς ΑΒ πρὸς ΒΓ δοθείς, ὁ αὐτὸς αὐτῷ γεγονέτω ὁ τῆς Δ πρὸς τὴν Ε. δοθεῖσα δὲ ἡ Δ· δοθεῖσα ἄρα καὶ ἡ Ε. πάλιν ἐπεὶ λόγος ἐστὶ τῆς ΒΓ πρὸς τὴν ΑΓ δοθείς, ὁ αὐτὸς αὐτῷ γεγονέτω ὁ τῆς Ε πρὸς τὴν Ζ. δοθεῖσα δὲ ἡ Ε· δοθεῖσα ἄρα καὶ ἡ Ζ.

καὶ ἐκ τριῶν εὐθειῶν, αἵ εἰσιν ἴσαι τρισὶ ταῖς δοθείσαις ταῖς Δ, Ε, Ζ, ὧν αἱ δύο τῆς λοιπῆς μείζονές εἰσι πάντη μεταλαμβανόμεναι, τρίγωνον συνεστάτω τὸ ΗΘΚ· ὥστε ἴσην εἶναι τὴν μὲν Δ τῇ ΗΘ, τὴν δὲ Ε τῇ ΘΚ, τὴν δὲ Ζ τῇ ΗΚ. δοθεῖσα δὲ ἑκάστη τῶν Δ, Ε, Ζ· δοθεῖσα ἄρα καὶ ἑκάστη τῶν ΗΘ, ΘΚ, ΚΗ τῷ μεγέθει· δέδοται ἄρα τὸ ΗΘΚ τρίγωνον τῷ εἴδει.

καὶ ἐπεί ἐστιν ὡς ἡ ΑΒ πρὸς τὴν ΒΓ, οὕτως ἡ Δ πρὸς τὴν Ε, ἴση δὲ ἡ μὲν Δ τῇ ΗΘ, ἡ δὲ Ε τῇ ΘΚ, ἔστιν ἄρα ὡς ἡ ΑΒ πρὸς τὴν ΒΓ, οὕτως ἡ ΗΘ πρὸς τὴν ΘΚ. πάλιν ἐπεί ἐστιν ὡς ἡ ΒΓ πρὸς τὴν ΓΑ, οὕτως ἡ Ε πρὸς τὴν Ζ, ἴση δὲ ἡ μὲν Ε τῇ ΘΚ, ἡ δὲ Ζ τῇ ΗΚ, ἐστιν ἄρα ὡς ἡ ΒΓ πρὸς τὴν ΓΑ, οὕτως ἡ ΘΚ πρὸς τὴν ΚΗ. ἐδείχθη δὲ καὶ ὡς ἡ ΑΒ πρὸς τὴν ΒΓ, οὕτως ἡ ΘΗ πρὸς τὴν ΘΚ· δι' ἴσου ἄρα ἐστὶν ὡς ἡ ΑΒ πρὸς τὴν ΑΓ, οὕτως ἡ ΘΗ πρὸς τὴν ΗΚ. ὅμοιον ἄρα ἐστὶ τὸ ΑΒΓ τρίγωνον τῷ ΗΘΚ τριγώνῳ. δέδοται δὲ τὸ ΗΘΚ τρίγωνον τῷ εἴδει· δέδοται ἄρα καὶ τὸ ΑΒΓ τρίγωνον τῷ εἴδει.

Dt 42. *If the sides of a triangle have a given ratio to one another, the triangle is given in form.*

For, let the sides of △ABC have a given ratio to one another. I say that the triangle is given in form.

For, let a straight line D have been set out given in magnitude [Axiom 0*]. And since the ratio AB:BC is given, let the same ratio have become D:E [VI.12]; and D is given, therefore E is also given [Dt 2].

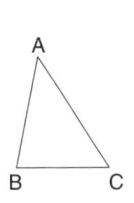
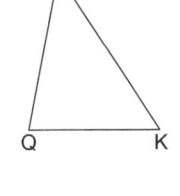

Figure 42

Again since the ratio BC:AC is given, let the same ratio have become E:Z; and E is given, therefore Z is also given.

And out of three straight lines which are equal to the three given straight lines D, E, Z, of which any two taken together in any manner are greater than the remaining one, let △HQK have been constructed; such that D is equal to HQ, E is equal to QK, Z is equal to HK [I.22].

Each of D, E, Z is given; therefore each of HQ, QK, KH is given in magnitude. Therefore △HQK is given in form [Dt 39 or Def. 3*A].

And since AB:BC :: D:E, and D = HQ, E = QK, therefore AB:BC :: HQ:QK.
And since BC:CA :: E:Z, and E = QK, Z = HK, therefore BC:CA :: QK:KH.
And it was shown that AB:BC :: HQ:QK; therefore *di' isou* AB:AC :: QH:HK [V.22]. Therefore △ABC is similar to △HQK [VI.5]; and △HQK is given in form; therefore △ABC is given in form [Def. 3*B]. ∎

Remarks: *Is Dt 39 necessary for Dt 42 ?*

It is instructive to begin with Dt 42, because it does not involve a given angle; it seems to be based on Dt 39, which stands somewhat apart from the others, having sides with given magnitudes, not only given ratios. Dt 39 and 42 'prove' one of the oldest lessons of statics, that triangles are rigid while any other polygon is collapsible. Obviously, if Dt 42 can be proved independently of Dt 39, the latter is an immediate consequence of the former and needs no proposition of its own, but can be seen as a corollary.

Let us study a paraphrase of Dt 39:

A triangle ABC has its sides given in magnitude. Another triangle is constructed, after the fashion of I.22, on a segment, given in position, and with sides equal to the sides of △ABC.

The next passage (emphasized in my Greek text) is remarkable:

> The point K is given; and each of the points E and Z is also given; therefore each of KE, EZ, ZK is given in position and in magnitude; therefore △KEZ is given in form.

Since we knew from the hypothesis that the sides were given in magnitude, the vital property must be their *being given in position*, that is, the new triangle is given in form because its vertices are given in position. Menge refers to Def. 3. But this definition cannot be applied directly, since neither angles nor ratios are mentioned.

The problem is reduced to prove that the first triangle is similar to another one which is given in position *and therefore in form*. Obviously, if the vertices are fixed, the sides are fixed, too, and thus the whole triangle is stiff, and that, of course, is part of the meaning of 'given in form'. – Euclid certainly subscribes to Def. 3*B in his conclusion of Dt 39:[93]

> △KEZ is equal and similar to (*i.e.* congruent with) △ABC, which is therefore given in form.

One question remains, and probably will remain, unanswered: why did Euclid not conclude *en passant* that △ABC is given in magnitude? To believe Def 1, this property of the triangle is actually proved, in so far as another figure, *in casu* another triangle, is provided, equal to the one concerned. But Def. 1 does not cover areas, as we mentioned already in Dt 24. And then, it was not easy to compute its area from its sides. (Archimedes [94] had not yet invented Heron's formula for the area of a triangle.) But normally, to be given in magnitude means to be reproducible by geometric methods, not necessarily to be computable. (Ptolemy seems to distinguish the two meanings, cf. p. 13.) Nor is it easy to compute the angles, although they, too, are 'given in magnitude'. But perhaps the answer is the simplest possible: *There is no*

---

[93] Clemens Thaer remarks laconically that (I translate) ' the proof of Dt 39, which is transmitted differently in the Arabic edition(s), seems to have been recast (überarbeitet)'. I have not seen the Arabic text.

[94] Cf. my paper "An Archimedean Proof of Heron's Formula for the Area of a Triangle, Reconstructed", *Centaurus* 1980, vol 24, 110-116.

need of triangles given in magnitude in the following theorems, which are all about form. If a given magnitude is needed, it is demanded, and other magnitudes are proved to be given by way of ratio.

In fact, there is no need of Dt 39 at all; of the four theorems that seem to have used it (Dt 40–43) only Dt 42 does not repeat the argument from Dt 39 (that three points are given and therefore three line segments are given in position and in magnitude); thus Def. 3 is latently active through Def. 3*A and Def. 3*B. Dt 42 seems to make vital use of Dt 39, but we shall see that it can be dispensed with.

Let us have a look at Dt 42: Consider a triangle ABC with sides having given ratios. Using I.22 another triangle is constructed from line segments, taken and therefore given in magnitude, in the given ratios. The new triangle will be given in form (apparently referring to Dt 39) since its sides are given in magnitude. And it is similar to △ABC (VI.5), which therefore is given in form (Def. 3*B).

As it stands, Dt 42 presupposes Dt 39 about sides given in magnitude; but Euclid could dodge that by constructing △QHK *in position* (as he did with EHK in Dt 39), and concluding that since HQ, QK, KH are given *in position* and in magnitude, exactly as KE, EZ, ZK in Dt 39, *therefore* △QHK is given in form.

Both propositions begin with the imperative ἐκκείσθω, let be set out. In Dt 39 the line set out is explicitly said to be given in position, whereas in Dt 42 it is given in magnitude (only). But nothing prevents the line HQ (equal to D) from being given in position,[95] too, since the construction that is supposed to have been done is the same as the one elaborated in Dt 39, namely I.22. In appendix B (p. 250) Dt 39 and I.22 are viewed synoptically.

Superfluous or not, Dt 39 is a gateway to understanding the *Data* from its doings and not from its definitions, and to fathoming its dependence on and coherence with the *Elements*. It lines up with other suspects, Dt 2 and Dt 25, to mention a couple of the more important ones. Dt 42 could (and in my view should) join Dt 40 in being the foundation of the theory about triangles given in form.

◆

---

[95] If ἐκκείσθω means 'put into position' (cognate with *thesis*), the proof of Dt 42 is almost equivalent to Dt 39, which then becomes superfluous in this context, since it follows immediately from Dt 42. If it does not mean that, I do not know what it means.

Θε 40. Ἐὰν τριγώνου ἑκάστη τῶν γωνιῶν δεδομένη ᾖ τῷ μεγέθει, δέδοται τὸ τρίγωνον τῷ εἴδει.

τριγώνου γὰρ τοῦ ΑΒΓ ἑκάστη τῶν γωνιῶν δεδομένη ἔστω τῷ μεγέθει· λέγω, ὅτι δέδοται τὸ ΑΒΓ τρίγωνον τῷ εἴδει.
ἐκκείσθω γὰρ τῇ θέσει καὶ τῷ μεγέθει δεδομένη εὐθεῖα ἡ ΔΕ, καὶ συνεστάτω πρὸς τῇ ΔΕ καὶ τοῖς πρὸς αὐτῇ σημείοις τοῖς Δ, Ε τῇ μὲν ὑπὸ ΓΒΑ γωνίᾳ ἴση γωνία εὐθύγραμμος ἡ ὑπὸ ΕΔΖ, τῇ δὲ ὑπὸ τῶν ΑΓΒ ἴση ἡ ὑπὸ τῶν ΔΕΖ· λοιπὴ ἄρα ἡ ὑπὸ τῶν ΒΑΓ λοιπῇ ἴση τῇ ὑπὸ τῶν ΔΖΕ ἐστιν. δοθεῖσα δὲ ἑκάστη τῶν πρὸς τοῖς Α, Β, Γ· δοθεῖσα ἄρα καὶ ἑκάστη τῶν πρὸς τοῖς Δ, Ε, Ζ. ἐπεὶ οὖν πρὸς θέσει δεδομένῃ εὐθείᾳ τῇ ΔΕ καὶ τῷ πρὸς αὐτῇ σημείῳ δεδομένῳ τῷ Δ εὐθεῖα γραμμὴ ἦκται ἡ ΔΖ δεδομένην ποιοῦσα γωνίαν τὴν πρὸς τῷ Δ, θέσει ἄρα ἐστὶν ἡ ΔΖ. διὰ τὰ αὐτὰ δὴ καὶ ἡ ΕΖ θέσει ἐστίν· δοθὲν ἄρα ἐστὶ τὸ Ζ σημεῖον. ἔστι δὲ καὶ ἑκάτερον τῶν Δ, Ε δοθέν· δοθεῖσα ἄρα ἐστὶν ἑκάστη τῶν ΔΖ, ΔΕ, ΕΖ τῇ θέσει καὶ τῷ μεγέθει· δέδοται ἄρα τὸ ΔΖΕ τρίγωνον τῷ εἴδει. καί ἐστιν ὅμοιον τῷ ΑΒΓ τριγώνῳ· δέδοται ἄρα καὶ τὸ ΑΒΓ τρίγωνον τῷ εἴδει.

Dt 40. *If each of the angles of a triangle be given in magnitude, the triangle is given in form.*

For, let each of the angles of △ABC be given in magnitude; I say that △ABC is given in form.

For, let a straight line DE have been set out given in position and in magnitude [Axiom 0*]; and on DE at the points D and E on it let the rectilineal angle EDZ have been constructed equal to ∠CBA, and ∠DEZ equal to ∠ACB [I.23]; then the remaining ∠BAC is equal to the remaining ∠DZE [I.32].[96] And each of the angles at A, B, C is given; therefore each of the angles at D, E, Z is given [Def. 1].

Since then, at the straight line DE given in position and at the given point D on it the straight line DZ has been drawn making the given angle at D, therefore DZ is given in position [Dt 29]. For the same reason EZ is also given in position.

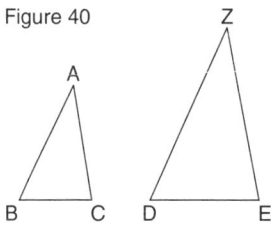

Figure 40

Therefore the point Z is given [Dt 25]. And each of [the points] D, E is given; therefore each of [the lines] DZ, DE, EZ is given in position and in magnitude [Dt 26]; therefore △DZE is given in form [Def. 3*A]. And it is similar to △ABC [VI.4]; therefore △ABC is given in form [Def. 3*B]. ∎

---

[96] In VI.4 care is taken that the angle exists, *i.e.* that the lines meet. Was the author of the *Elements* more meticulous?

Remarks: Given the angles, we expect the proof to establish the given ratios of the sides of △ABC, but the word ratio, λόγος, does not even occur in the text. Another triangle DEZ, equiangular with △ABC, is constructed on a line segment given in position and in magnitude. Using the axiomatic theorems Dt 25, 26, and 29, Euclid concludes that the vertices of DEZ are given in position, and therefore its sides given in position and magnitude. Again Menge refers to Dt 39 at that point, but without warrant: Dt 40 reaches its conclusion in exactly the same wording as Dt 39, and thus independently of it. Let us compare the passage:

Dt 39
δοθὲν ἄρα ἐστὶ καὶ τὸ Κ σημεῖον. ἔστι δὲ καὶ ἑκάτερον τῶν Ε, Ζ δοθέν· δοθεῖσα ἄρα ἐστὶν ἑκάστη τῶν ΚΕ, ΕΖ, ΖΚ τῇ θέσει καὶ τῷ μεγέθει· δέδοται ἄρα τὸ ΚΕΖ τρίγωνον τῷ εἴδει. καί ἐστιν ἴσον τε καὶ ὅμοιον τῷ ΑΒΓ· δέδοται ἄρα τὸ ΑΒΓ τρίγωνον τῷ εἴδει.

Dt 40
δοθὲν ἄρα ἐστὶ τὸ Ζ σημεῖον. ἔστι δὲ καὶ ἑκάτερον τῶν Δ, Ε δοθέν· δοθεῖσα ἄρα ἐστὶν ἑκάστη τῶν ΔΖ, ΔΕ, ΕΖ τῇ θέσει καὶ τῷ μεγέθει· δέδοται ἄρα τὸ ΔΖΕ τρίγωνον τῷ εἴδει. καί ἐστιν ὅμοιον τῷ ΑΒΓ τριγώνῳ· δέδοται ἄρα καὶ τὸ ΑΒΓ τρίγωνον τῷ εἴδει.

The only difference, apart from the names of the points, is that △DZE of Dt 40 is not equal but (only) similar to △ABC. Euclid subscribes to Def. 3*A. Now, if he did not want to use position in his definitions before Def. 4, why did he not construe definition 3 analogously to Def. 1: *A figure is given in form if we can provide a figure similar to it* ? – The answer could be that it would not help him, if 'provide' means 'construct in the Plane', *i.e.* in position.

Θε 41. Ἐὰν τρίγωνον μίαν ἔχῃ γωνίαν δεδομένην, περὶ δὲ τὴν δεδομένην γωνίαν αἱ πλευραὶ πρὸς ἀλλήλας λόγον ἔχωσι δεδομένον, δέδοται τὸ τρίγωνον τῷ εἴδει.
 ἐχέτω γὰρ τρίγωνον τὸ ΑΒΓ μίαν γωνίαν δεδομένην τὴν ὑπὸ τῶν ΒΑΓ, περὶ δὲ τὴν ὑπὸ τῶν ΒΑΓ αἱ πλευραὶ αἱ ΒΑ, ΑΓ πρὸς ἀλλήλας λόγον ἐχέτωσαν δεδομένον· λέγω, ὅτι τὸ ΑΒΓ τρίγωνον δέδοται τῷ εἴδει.
 ἐκκείσθω γὰρ τῇ θέσει καὶ τῷ μεγέθει δεδομένη εὐθεῖα ἡ ΔΖ καὶ συνεστάτω πρὸς τῇ ΔΖ εὐθείᾳ καὶ τῷ πρὸς αὐτῇ σημείῳ τῷ Ζ τῇ ὑπὸ τῶν ΒΑΓ γωνίᾳ ἴση ἡ ὑπὸ τῶν ΔΖΕ. δοθεῖσα δὲ ἡ ὑπὸ τῶν ΒΑΓ· δοθεῖσα ἄρα καὶ ἡ ὑπὸ τῶν ΔΖΕ. ἐπεὶ οὖν πρὸς θέσει δεδομένῃ εὐθείᾳ τῇ ΔΖ καὶ τῷ πρὸς αὐτῇ δεδομένῳ σημείῳ τῷ Ζ εὐθεῖα γραμμὴ ἦκται ἡ ΖΕ δεδομένην ποιοῦσα γωνίαν τὴν ὑπὸ τῶν ΔΖΕ, θέσει ἄρα ἐστὶν ἡ ΖΕ. καὶ ἐπεὶ λόγος ἐστὶ τῆς ΒΑ πρὸς τὴν ΑΓ δοθείς, ὁ αὐτὸς αὐτῷ γεγονέτω ὁ τῆς ΔΖ πρὸς τὴν ΖΕ καὶ ἐπεζεύχθω ἡ ΔΕ· λόγος ἄρα καὶ τῆς ΔΖ πρὸς τὴν ΖΕ δοθείς· δοθεῖσα δὲ ἡ ΔΖ· δοθεῖσα ἄρα καὶ ἡ ΖΕ. ἀλλὰ καὶ τῇ θέσει. καί ἐστι τὸ Ζ δοθέν· δοθὲν ἄρα καὶ τὸ Ε. ἔστι δὲ καὶ ἑκάτερον τῶν Δ, Ζ δοθέν· δοθεῖσα ἄρα ἐστὶν ἑκάστη τῶν ΔΖ, ΖΕ, ΔΕ τῇ θέσει καὶ τῷ μεγέθει· δέδοται ἄρα τὸ ΔΖΕ τῷ εἴδει. καὶ ἐπεὶ δύο τρίγωνα τὰ ΑΒΓ, ΔΕΖ μίαν γωνίαν μία γωνία ἴσην ἔχει, τὴν ὑπὸ τῶν ΒΑΓ τῇ ὑπὸ τῶν ΔΖΕ, περὶ δὲ τὰς ὑπὸ τῶν ΒΑΓ, ΔΖΕ γωνίας τὰς πλευρὰς ἀνάλογον, ὅμοιον ἄρα ἐστὶ τὸ ΑΒΓ τρίγωνον τῷ ΔΕΖ τριγώνῳ. δέδοται δὲ τὸ ΔΖΕ τῷ εἴδει· δέδοται ἄρα καὶ τὸ ΑΒΓ τρίγωνον τῷ εἴδει.

Dt 41. *If a triangle have one given angle, and the sides about the given angle have a given ratio to one another, the triangle is given in form.*

For, let △ABC have one given angle BAC; and let the sides BA, AC about ∠BAC have a given ratio to one another; I say that △ABC is given in form (figure 40).

For, let a straight line DZ have been set out given in position and in magnitude [Axiom 0*]; and let ∠DZE, equal to ∠BAC, be constructed on the straight line DZ and at the point Z on it [I.23]. ∠BAC is given; therefore ∠DZE is also given [Def. 1].

Since then, at the straight line DZ given in position and at the given point Z on it the straight line ZE has been drawn making a given angle DZE, therefore EZ is given in position [Dt 29].

And since the ratio BA:AC is given, let DZ:ZE have been made the same, and let DE have been joined; then the ratio DZ:ZE is given [Def. 2]. But DZ is given, and therefore ZE is given [Dt 2]; but in position, too. And [the point] Z is given; therefore [the point] E is also given [Dt 27].

And each of [the points] D and Z is given; therefore each of the lines DZ, ZE, DE is given in position and magnitude; therefore △DZE is given in form [Def. 3*A].

And since the two triangles ABC and DEZ have one angle equal to one angle, ∠BAC = ∠DZE, and the sides about the angles BAC and DZE proportional, therefore △ABC is similar to △DEZ [VI.6].

But DEZ is given in form. Therefore ABC is given in form [Def. 3*B]. ∎

Remarks. Dt 41 calls for comments similar to those for Dt 40: We expect a proof that one more angle is given, or that the ratio of another pair of sides is given. But what happens? A triangle is constructed *in position* with the given angle and the given ratio between the sides about the angle. Therefore the two triangles are similar (VI.6 is quoted at some length), so Euclid concludes that the first triangle is given in form. Again Menge refers to Def. 3, but what is proved is that *it is similar to another triangle whose vertices and sides are given in position.* Nor is it necessary or relevant to refer to Dt 39, as Dt 41 simply repeats the steps of that proposition.

I suppose that something like theorem 41 was in Knorr's mind when he wrote of 'parallel results' (cf. p. 16): In the Elements VI.6 two triangles T1 and T2 are compared, having one angle equal to one angle and the sides about the angle in the same ratio. A triangle T3 is constructed, equiangular with T1 and congruent with T2. Therefore T1 is equiangular with T2. In Data 41, one triangle T1 is set out, having one angle 'given' and the sides about the angle in a 'given' ratio. A triangle T2 is constructed 'in position', with the same properties as T1, and therefore equiangular with it. Since T2 is 'given in position', it is also 'given in form', and therefore T1 is 'given in form'. *The Elements compare triangles. The Data deals with individuals,* and with the 'knowledge' we may have of such individual triangles, within the language of Givens. A superfluous discipline? A useful one in geometric analysis, where individual objects prevail.

◆

Θε 43. ' Ἐὰν τριγώνου ὀρθογωνίου περὶ μίαν τῶν ὀξειῶν γωνιῶν αἱ πλευραὶ πρὸς ἀλλήλας λόγον ἔχωσι δεδομένον, δέδοται τὸ τρίγωνον τῷ εἴδει.

τριγώνου γὰρ ὀρθογωνίου τοῦ ΑΒΓ ὀρθὴν ἔχοντος τὴν ὑπὸ τῶν ΒΑΓ γωνίαν, περὶ μίαν τῶν ὀξειῶν αὐτοῦ γωνιῶν τὴν ὑπὸ ΑΒΓ αἱ πλευραὶ αἱ ΓΒ, ΒΑ πρὸς ἀλλήλας λόγον ἐχέτωσαν δεδομένον· λέγω, ὅτι δέδοται τὸ ΑΒΓ τρίγωνον τῷ εἴδει.

ἐκκείσθω γὰρ τῇ θέσει καὶ τῷ μεγέθει δεδομένη εὐθεῖα ἡ ΔΕ, καὶ γεγράφθω ἐπὶ τῆς ΔΕ ἡμικύκλιον τὸ ΔΗΕ· θέσει ἄρα ἐστὶ τὸ ΔΗΕ ἡμικύκλιον. καὶ ἐπεὶ λόγος ἐστὶ τῆς ΓΒ πρὸς τὴν ΒΑ δοθείς, ὁ αὐτὸς αὐτῷ γεγονέτω ὁ τῆς ΔΕ πρὸς τὴν Ζ· λόγος ἄρα καὶ τῆς ΔΕ πρὸς τὴν Ζ δοθείς. δοθεῖσα δὲ ἡ ΔΕ· δοθεῖσα ἄρα καὶ ἡ Ζ. καί ἐστι μείζων ἡ ΓΒ τῆς ΒΑ· μείζων ἄρα καὶ ἡ ΕΔ τῆς Ζ. ἐνηρμόσθω τῇ Ζ ἴση ἡ ΔΗ, καὶ ἐπεζεύχθω ἡ ΗΕ, καὶ κέντρῳ μὲν τῷ Δ, διαστήματι δὲ τῷ ΔΗ κύκλος γεγράφθω ὁ ΘΗΚ· θέσει ἄρα ἐστὶν ὁ ΘΗΚ κύκλος· δέδοται γὰρ αὐτοῦ τὸ κέντρον τῇ θέσει καὶ ἡ ἐκ τοῦ κέντρου τῷ μεγέθει. θέσει δὲ καὶ τὸ ΔΗΕ ἡμικύκλιον. δοθὲν ἄρα ἐστὶ τὸ Η σημεῖον. ἔστι δὲ καὶ ἑκάτερον τῶν Δ, Ε δοθέν· δοθεῖσα ἄρα ἐστὶν ἑκάστη τῶν ΗΔ, ΔΕ, ΕΗ τῇ θέσει καὶ τῷ μεγέθει· δέδοται ἄρα τὸ ΗΔΕ τρίγωνον τῷ εἴδει.

ἐπεὶ οὖν δύο τρίγωνά ἐστι τὰ ΑΒΓ, ΔΕΗ μίαν γωνίαν μιᾷ γωνίᾳ ἴσην ἔχοντα τὴν ὑπὸ τῶν ΒΑΓ τῇ ὑπὸ τῶν ΔΗΕ, περὶ δὲ ἄλλας γωνίας τὰς ὑπὸ τῶν ΓΒΑ, ΕΔΗ τὰς πλευρὰς ἀνάλογον, τῶν δὲ λοιπῶν τῶν ὑπὸ ΒΓΑ, ΔΕΗ ἑκατέραν ἅμα ἐλάσσονα ὀρθῆς, ὅμοιον ἄρα ἐστὶ τὸ ΑΒΓ τρίγωνον τῷ ΔΕΗ τριγώνῳ. δέδοται δὲ τὸ ΔΕΗ τρίγωνον τῷ εἴδει· δέδοται ἄρα καὶ τὸ ΑΒΓ τρίγωνον τῷ εἴδει.

Dt 43. *If in a right-angled triangle the sides about one of the acute angles have a given ratio to one another, the triangle is given in form.*

For, in the right-angled triangle ABC with the right angle BAC let the sides CB, BA about one of the acute angles ABC have a given ratio to one another; I say that △ABC is given in form.

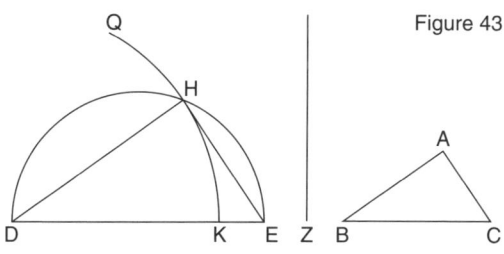

Figure 43

For, let the straight line DE have been set out given in position and in magnitude; and on DE let the semicircle DHE have been described; then the semicircle DHE is given in position [Def. 8].

And since the ratio CB:BA is given, let the same ratio have become DE:Z; then the ratio DE:Z is given. And DE is given; therefore Z is given [Dt 2]. And CB is greater than BA [I.19]; therefore ED is also greater than Z [V.16, 14, or 14a, see p. 41].

Let DH, equal to Z, have been fitted in[to the semicircle] [IV.1], and let HE be joined. And with centre D and radius DH let the circle QHK have been described. Then the circle QHK is given in position [Def. 6]; for its centre is given in position and its radius in magnitude.

And the semicircle DHE is given in position; therefore the point H is given [Dt 25]. And each of D, E is given; therefore each of HD, DE, EH is given in position and in magnitude [Dt 26]; therefore △HDE is given in form [Def. 3*A].

Since then, the two triangles ABC, DEH have one angle BAC equal to one angle DHE [III.31], and the sides about the other angles CBA, EDH proportional, and the remaining angles both less than right angles [I.32], therefore △ABC is similar to △DEH [VI.7]. And △DEH is given in form; therefore △ABC is given in form [Def. 3*B]. ■

Remarks: A right-angled triangle DHE is constructed with hypotenuse given in position and with the ratio between hypotenuse and one of the sides given. Its vertices are proved to be given in position, whence the triangle is given in form (Def. 3*B); and it is similar to ABC (VI.7); therefore the latter is also given in form (Def. 3*A).

The proof includes another example of 'repeated construction' (see on Dt 31 above p. 105): After fitting in the segment DH, Euclid goes on to draw the arc of circle KHQ, which is exactly the one he had to use to fit in DH, according to IV.1. But then, the fitting of DH is just commanded (and he does not care how it is done, since it is admitted by IV.1), whereas the arc KHQ is needed to establish that H is not a moving point (Dt 25).

Since the sine of △B is $a:b$, some moderns will want to interpret this theorem in trigonometric terms, indicating that the sine of an acute angle determines the angle uniquely. That is why I would like to call the whole theory of triangles given in form *Trigonosophy*, Wisdom of the Triangle, a kind of embryo trigonometry. At a time when Ptolemy's trigonometry was still far off, the use of such wisdom was very limited. But wisdom it is. What makes trigono*metry* what it is, is just the feature which is absent from the *Data* and the *Elements*: measuring.

◆

Θε 44. ' Ἐὰν τρίγωνον μίαν ἔχῃ γωνίαν δεδομένην, περὶ δὲ ἄλλην γωνίαν αἱ πλευραὶ πρὸς ἀλλήλας λόγον ἔχωσι δεδομένον, δέδοται τὸ τρίγωνον τῷ εἴδει.

ἔστω τρίγωνον τὸ ΑΒΓ μίαν ἔχον γωνίαν δεδομένην τὴν ὑπὸ τῶν ΒΑΓ, περὶ δὲ ἄλλην γωνίαν τὴν ὑπὸ τῶν ΑΒΓ αἱ πλευραὶ αἱ ΑΒ, ΒΓ λόγον ἐχέτωσαν πρὸς ἀλλήλας δεδομένον· λέγω, ὅτι τὸ ΑΒΓ τρίγωνον δέδοται τῷ εἴδει.

μὴ ἔστω δὴ ἡ ὑπὸ τῶν ΒΑΓ γωνία ὀρθή, ἀλλ' ἔστω πρότερον ὀξεῖα, καὶ ἤχθω ἀπὸ τοῦ Β σημείου ἐπὶ τὴν ΑΓ κάθετος ἡ ΒΔ. ἐπεὶ δοθεῖσά ἐστιν ἡ ὑπὸ ΒΔΑ γωνία, ἔστι δὲ καὶ ἡ ὑπὸ τῶν ΒΑΔ δοθεῖσα, καὶ λοιπὴ ἄρα ἡ ὑπὸ τῶν ΑΒΔ δοθεῖσά ἐστιν· δέδοται ἄρα τὸ ΒΑΔ τρίγωνον τῷ εἴδει· λόγος ἄρα τῆς ΒΑ πρὸς τὴν ΒΔ δοθείς. ἀλλὰ τῆς ΑΒ πρὸς τὴν ΒΓ λόγος ἐστὶ δοθείς· καὶ τῆς ΒΔ ἄρα πρὸς τὴν ΒΓ λόγος ἐστὶ δοθείς. καί ἐστιν ὀρθὴ ἡ ὑπὸ τῶν ΒΔΓ· δέδοται ἄρα τὸ ΒΔΓ τρίγωνον τῷ εἴδει· δοθεῖσα ἄρα ἐστὶν ἡ ὑπὸ τῶν ΒΓΔ γωνία. ἔστι δὲ καὶ ἡ ὑπὸ τῶν ΒΑΓ δοθεῖσα· καὶ λοιπὴ ἄρα ἡ ὑπὸ τῶν ΑΒΓ ἐστι δοθεῖσα· δέδοται ἄρα τὸ ΑΒΓ τρίγωνον τῷ εἴδει.

ἀλλὰ δὴ ἔστω ἡ ὑπὸ τῶν ΒΑΓ γωνία ἀμβλεῖα, καὶ ἐκβεβλήσθω ἡ ΓΑ ἐπὶ τὸ Ε, καὶ ἤχθω ἀπὸ τοῦ Β σημείου ἐπὶ τὴν ΑΕ κάθετος ἡ ΒΕ. ἐπεὶ δοθεῖσά ἐστιν ἡ ὑπὸ τῶν ΒΑΓ, καὶ ἡ ἐφεξῆς ἄρα ἡ ὑπὸ τῶν ΒΑΕ δοθεῖσά ἐστι. ἔστι δὲ καὶ ἡ ὑπὸ τῶν ΒΕΑ δοθεῖσα· καὶ λοιπὴ ἄρα ἡ ὑπὸ τῶν ΕΒΑ δοθεῖσά ἐστιν· δέδοται ἄρα τὸ ΕΒΑ τρίγωνον τῷ εἴδει· λόγος ἄρα τῆς ΕΒ πρὸς τὴν ΒΑ δοθείς. τῆς δὲ ΑΒ πρὸς τὴν ΒΓ λόγος ἐστὶ δοθείς· καὶ τῆς ΕΒ ἄρα πρὸς τὴν ΒΓ λόγος ἐστὶ δοθείς. καί ἐστιν ὀρθὴ ἡ ὑπὸ τῶν ΒΕΓ γωνία· δέδοται ἄρα τὸ ΕΒΓ τρίγωνον τῷ εἴδει· δοθεῖσα ἄρα ἐστὶν ἡ ὑπὸ ΒΓΕ. ἔστι δὲ καὶ ἡ ὑπὸ ΒΑΓ γωνία δοθεῖσα· καὶ λοιπὴ ἄρα ἡ ὑπὸ ΑΒΓ γωνία δοθεῖσά ἐστιν· δέδοται ἄρα τὸ ΑΒΓ τρίγωνον τῷ εἴδει.

Dt 44. *If a triangle have one given angle, and the sides about another angle have a given ratio to one another, the triangle is given in form.*

Let ABC be a triangle having one given angle BAC, and let the sides AB, BC about another angle ABC have a given ratio to one another; I say that △ABC is given in form.

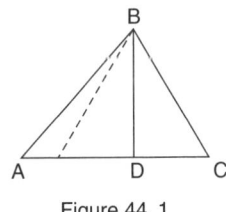
Figure 44. 1

[1] Let ∠BAC not be right, but let it first be acute, and let BD have been drawn from the point B perpendicular to AC. Since ∠BDA is given, and ∠BAD is given, therefore the remaining ∠ABD is given [I.32, Dt 3, 4]; therefore △BAD is given in form [Dt 40]. Therefore the ratio BA:BD is given [Def. 3].

But the ratio AB:BC is given; therefore the ratio BD:BC is given [Dt 8]. And ∠BDC is right; therefore △BDC is given in form [Dt 43]; therefore ∠BCD is given.

And ∠BAC is also given; therefore the remaining ∠ABC is given; therefore △ABC is given in form [Dt 40].

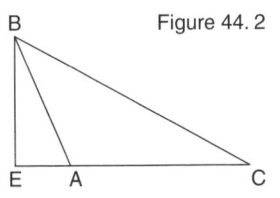
Figure 44. 2

[2] But then let ∠BAC be obtuse, and let CA have been produced to E, and let BE have been drawn from the point B perpendicular to AE.

Since ∠BAC is given, the adjacent ∠BAE is given [I.13, Dt 4]. And BEA is given; therefore the remaining angle EBA is given; therefore △EBA is given in form [Dt 40]. Therefore the ratio EB:BA is given [Def. 3]. But the ratio AB:BC is given; therefore the ratio EB:BC is given [Dt 8].

And ∠BEC is right; therefore ∆EBC is given in form [Dt 43]; therefore ∠BCE is given.

And ∠BAC is given; therefore the third ∠ABC is given; therefore ∆ABC is given in form [Dt 40]. ∎

Remarks: Roughly speaking, a figure with certain given properties should be given in form, if it is similar to a figure with the same given properties. On this assumption, one can easily verify that Dt 44 is ambiguous: if the given angle is A, and the given ratio is a:c, the theorem is true only if $a > c$, or if ∠C is a right angle (i.e.: if $a:c = \sin A$). Otherwise, two very different triangles will both have the relevant properties (figure 44.1). But then there are perhaps ideas about ambiguity and being given in form that I do not understand.[97]

Within the *Data*, Dt 44 is use only to prove Dt 45, where the two 'different' triangles happen to be congruent (as shown below). This is the first instance of Dt 40 at work with Dt 8, an effective method used in 16 propositions about triangles given in form.

◆

Θε 45. ' Ἐὰν τρίγωνον μίαν ἔχῃ γωνίαν δεδομένην, αἱ δὲ περὶ τὴν δεδομένην γωνίαν πλευραὶ συναμφότεραι ὡς μία πρὸς τὴν λοιπὴν λόγον ἔχωσι δεδομένον, δέδοται τὸ τρίγωνον τῷ εἴδει.

ἔστω τρίγωνον τὸ ΑΒΓ μίαν γωνίαν δεδομένην ἔχον τὴν ὑπὸ τῶν ΒΑΓ, περὶ δὲ τὴν ὑπὸ ΒΑΓ γωνίαν αἱ πλευραί, τουτέστι συναμφότερος ἡ ΒΑΓ ὡς μία πρὸς τὴν ΓΒ λόγον ἐχέτω δεδομένον· λέγω, ὅτι τὸ ΑΒΓ τρίγωνον δέδοται τῷ εἴδει.

τετμήσθω γὰρ ἡ ὑπὸ τῶν ΒΑΓ γωνία δίχα τῇ ΑΔ εὐθείᾳ· δοθεῖσα ἄρα ἐστὶν ἡ ὑπὸ τῶν ΒΑΔ γωνία. καὶ ἐπεί ἐστιν ὡς ἡ ΒΑ πρὸς τὴν ΑΓ, οὕτως ἡ ΒΔ πρὸς τὴν ΔΓ, ἐναλλὰξ ὡς ἡ ΑΒ πρὸς τὴν ΒΔ, οὕτως ἡ ΑΓ πρὸς τὴν ΓΔ· καὶ ὡς συναμφότερος ἄρα ἡ ΒΑΓ πρὸς τὴν ΒΓ, οὕτως ἡ ΑΒ πρὸς τὴν ΒΔ. λόγος δὲ συναμφοτέρου τῆς ΒΑΓ πρὸς τὴν ΒΓ δοθείς· λόγος ἄρα καὶ τῆς ΒΑ πρὸς τὴν ΒΔ δοθείς. καί ἐστι δοθεῖσα ἡ ὑπὸ τῶν ΒΑΔ γωνία· δέδοται ἄρα τὸ ΑΒΔ τρίγωνον τῷ εἴδει· δοθεῖσα ἄρα ἐστὶν ἡ ὑπὸ τῶν ΑΒΔ γωνία. ἔστι δὲ καὶ ἡ ὑπὸ τῶν ΒΑΓ γωνία δοθεῖσα· καὶ λοιπὴ ἄρα ἡ ὑπὸ τῶν ΑΓΒ δοθεῖσά ἐστιν· δέδοται ἄρα τὸ ΑΒΓ τρίγωνον τῷ εἴδει.

Dt 45. *If a triangle have one angle given, and the sum [98] of the sides about the given angle has a given ratio to the remaining side, the triangle is given in form.*

---

[97] Len Berggren suggests that we understand 'given in form' to mean 'given within a finite – usually two! – set of possibilities'. Unlike 'given in position', symmetry does not always apply to 'given in form' (cf. p. 105).

[98] συναμφότεραι ὡς μία, literally 'both together as one'.

Let ABC be the triangle having one given angle BAC, and let the sides about ∡BAC, that is both together BA, AC taken as one, have to the side CB a given ratio. I say that △ABC is given in form.

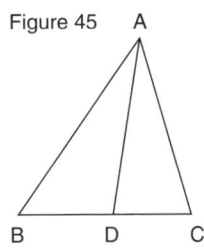

Figure 45

Let ∡BAC have been bisected by the straight line AD [I.9]; then ∡BAD is given [Dt 2]. And since BA:AC :: BD:DC, *enallax* AB:BD::AC:CD [V.16]; therefore (BA+AC):BC :: AB:BD [V.12].

But the ratio (BA+AC):BC is given; therefore the ratio BA:BD is given.

And ∡BAD is given; therefore △ABD is given in form; [Dt 44] therefore ∡ABD is given. And ∡BAC is given; therefore the remaining ∡ACB is given; therefore △ABC is given in form [Dt 40]. ∎

Remarks: The isosceles adjuncts.

For convenience I shall introduce some very useful auxiliary triangles: the lesser and the greater *isosceles adjuncts*. The lesser adjunct is known from I.20 (triangle inequality) and VI.3 (bisector of angle in a triangle):

In a triangle ABC (figure 67, p. 172) let the side BA be produced to D, AD = AC, and let CD be joined. Then △CAD is a lesser isosceles adjunct with respect to ∡A. From the vertex B let BE be drawn parallel to AC, to meet DC produced in E. Then △EBD is the greater isosceles adjunct with respect to ∡A. The adjuncts are similar; their base angles D and E equal half of ∡A. (By interchanging letters B and C, you get two others; while the greater adjuncts are congruent, the lesser are so only if AB = AC).

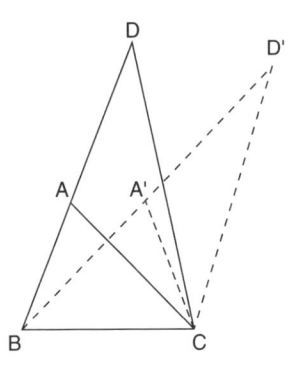

Figure 45.1

Dt 45 and 46 are not elementary, but applications of some of the elements of the Data. Euclid, however, obviously regarded them as belonging among the elementary truths about triangles given in form. Both of them can be (and in the appendix to Menge's edition are) proved by bringing in the lesser isosceles adjunct; then Dt 45 follows directly from 44, Dt 46 from 41.

Since 44 is equivocal if applied to this situation, 45 ought to be so, too; but the two triangles (△BCA and BCA´) which both meet the conditions are congruent symmetrically positioned on BC (Figure 45.1).

As this is the only instance (that I know of) where Dt 44 is used, the author may well have chosen to ignore its ambiguity.

In the proofs which Menge edited as original (because they come first in the manuscripts) the given angle is halved, and VI.3 is applied, followed by some appropriate manipulations with ratios from book V of the *Elements*.
In the end, the propositions follow from Dt 44 and 41 respectively. And the lesser adjunct has done its work in VI.3.

◆

Θε 46. 'Εὰν τρίγωνον μίαν ἔχῃ γωνίαν δεδομένην, περὶ δὲ ἄλλην γωνίαν αἱ πλευραὶ συναμφότεραι ὡς μία πρὸς τὴν λοιπήν λόγον ἔχωσι δεδομένον, δέδοται τὸ τρίγωνον τῷ εἴδει.

ἔστω τρίγωνον τὸ ΑΒΓ μίαν ἔχον γωνίαν δεδομένην τὴν ὑπὸ τῶν ΑΒΓ, περὶ δὲ ἄλλην γωνίαν τὴν ὑπὸ τῶν ΒΑΓ αἱ πλευραί, τουτέστι συναμφότερος ἡ ΒΑΓ πρὸς τὴν ΒΓ λόγον ἐχέτω δεδομένον· λέγω, ὅτι τὸ ΑΒΓ τρίγωνον δέδοται τῷ εἴδει.
τετμήσθω γὰρ ἡ ὑπὸ τῶν ΒΑΓ γωνία δίχα τῇ ΑΔ εὐθείᾳ· ἔστιν ἄρα ὡς συναμφότερος ἡ ΒΑΓ πρὸς τὴν ΓΒ, ἡ ΑΒ πρὸς τὴν ΒΔ. λόγος δὲ τοῦ συναμφοτέρου τῆς ΒΑΓ πρὸς τὴν ΓΒ δοθείς· λόγος ἄρα καὶ τῆς ΑΒ πρὸς τὴν ΒΔ δοθείς. καί ἐστι δοθεῖσα ἡ ὑπὸ τῶν ΑΒΔ γωνία· δέδοται ἄρα τὸ ΑΒΔ τρίγωνον τῷ εἴδει· δοθεῖσα ἄρα ἐστὶν ἡ ὑπὸ τῶν ΒΑΔ γωνία. καί ἐστιν αὐτῆς διπλασίων ἡ ὑπὸ ΒΑΓ· δοθεῖσα ἄρα ἐστὶ καὶ ἡ ὑπὸ τῶν ΒΑΓ. ἔστι δὲ καὶ ἡ ὑπὸ τῶν ΑΒΓ δοθεῖσα· καὶ λοιπὴ ἄρα ἡ ὑπὸ τῶν ΑΓΒ δοθεῖσά ἐστιν· δέδοται ἄρα τὸ ΑΒΓ τρίγωνον τῷ εἴδει.

Dt 46. *If a triangle have one angle given, and the sum of the sides about another angle have a given ratio to the remaining side, the triangle is given in form.*

Let ABC be the triangle having one given angle ABC, and let the sides about another angle BAC, that is, both together BA+AC have to the side BC a given ratio. I say that △ABC is given in form (figure 45).

For, let △BAC have been bisected by the straight line AD; then [99] (BA+AC):CB :: AB:BD. And the ratio (BA+AC):CB is given; therefore the ratio AB:DB is given. And △ABD is given; therefore △ABD is given in form [Dt 41]; therefore △BAD is given [Def. 3]. And △BAC is the double of it; therefore △BAC is given [Dt 2]. And △ABC is given; therefore the remaining △ACB is given; therefore △ABC is given in form [Dt 40]. ■

---

[99] Re-using the argument from Dt 45.

CHAPTER 6. FORM. (39–55)

Θε 47. Τὰ δεδομένα εὐθύγραμμα τῷ εἴδει εἰς δεδομένα τρίγωνα διαιρεῖται τῷ εἴδει.

ἔστω δεδομένον εὐθύγραμμον τῷ εἴδει τὸ ΑΒΓΔΕ· λέγω, ὅτι τὸ ΑΒΓΔΕ εὐθύγραμμον εἰς δεδομένα τρίγωνα διαιρεῖται τῷ εἴδει.
ἐπεζεύχθωσαν γὰρ αἱ ΒΕ, ΕΓ. ἐπεὶ δέδοται τὸ ΑΒΓΔΕ εὐθύγραμμον τῷ εἴδει, δοθεῖσα ἄρα ἐστὶν ἡ ὑπὸ τῶν ΒΑΕ γωνία. καί ἐστι λόγος τῆς ΒΑ πρὸς τὴν ΕΑ δοθείς. ἐπεὶ οὖν δοθεῖσά ἐστιν ἡ ὑπὸ τῶν ΒΑΕ γωνία καί ἐστι λόγος τῆς ΒΑ πρὸς τὴν ΕΑ δοθείς, δέδοται ἄρα τὸ ΒΑΕ τρίγωνον τῷ εἴδει· δοθεῖσα ἄρα ἐστὶν ἡ ὑπὸ τῶν ΑΒΕ γωνία. ἔστι δὲ καὶ ὅλη ἡ ὑπὸ τῶν ΑΒΓ γωνία δοθεῖσα· καὶ λοιπὴ ἄρα ἡ ὑπὸ τῶν ΕΒΓ δοθεῖσά ἐστιν. καί ἐστι λόγος τῆς ΑΒ πρὸς τὴν ΒΕ δοθείς, τῆς δὲ ΑΒ πρὸς τὴν ΒΓ λόγος ἐστὶ δοθείς· καὶ τῆς ΕΒ ἄρα πρὸς τὴν ΒΓ λόγος ἐστὶ δοθείς. καί ἐστι δοθεῖσα ἡ ὑπὸ τῶν ΓΒΕ γωνία· δέδοται ἄρα τὸ ΒΓΕ τρίγωνον τῷ εἴδει. διὰ τὰ αὐτὰ δὴ καὶ τὸ ΓΔΕ τρίγωνον τῷ εἴδει δέδοται· τὰ ἄρα δεδομένα εὐθύγραμμα τῷ εἴδει εἰς δεδομένα τρίγωνα διαιρεῖται τῷ εἴδει.

Dt 47. *Rectilineal figures given in form can be divided into triangles given in form.*

Let ABCDE be the rectilineal figure given in form; I say that the rectilineal figure ABCDE can be divided [100] into triangles given in form.

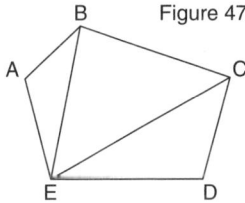

Figure 47

For, let BE, EC have been joined. Since the rectilineal figure ABCDE is given in form, △BAE is given, and the ratio BA:EA is given [Def. 3].
{Since then, △BAE is given and the ratio BA:EA is given,} [101] therefore △BAE is given in form [Dt 41]; therefore △ABE is given [Def. 3].

And the whole △ABC is given; therefore the remaining △EBC is given [Dt 4]. And the ratio AB:BE is given, and the ratio AB:BC is given; therefore the ratio EB:BC is given [Dt 8].
And △CBE is given; therefore △BCE is given in form [Dt 41].
Then for the same reasons, △CDE is given in form; therefore rectilineal figures given in form are divisible into triangles given in form.

Remarks: Dt 47 follows immediately from Dt 41, helped by Dt 8. By means of this proposition problems concerning polygons can be reduced to statements about triangles, beginning with Dt 48 and 49.

◆

---

[100] διαιρεῖται, potential present indicative.

[101] Menge suspects the passage in brackets to be an interpolation.

Θε 48. ' Ἐὰν ἀπὸ τῆς ἀτῆς εὐθείας δύο τρίγωνα ἀναγραφῇ δεδομένα τῷ εἴδει, λόγον ἕξει πρὸς ἄλληλα δεδομένον.

ἀπὸ γὰρ τῆς αὐτῆς εὐθείας τῆς ΑΒ δύο τρίγωνα δεδομένα τῷ εἴδει ἀναγεγράφθω τὰ ΑΒΓ, ΑΔΒ· λέγω, ὅτι λόγος ἐστὶ τοῦ ΑΓΒ πρὸς τὸ ΑΔΒ δοθείς.
ἤχθωσαν ἀπὸ τῶν Α, Β σημείων τῇ ΑΒ εὐθείᾳ πρὸς ὀρθὰς αἱ ΑΕ, ΗΒ καὶ ἐκβεβλήσθωσαν ἐπὶ τὰ Ζ, Θ, καὶ διὰ τῶν Γ, Δ σημείων τῇ ΑΒ εὐθείᾳ παράλληλοι ἤχθωσαν αἱ ΕΓΗ, ΖΔΘ. ἐπεὶ δέδοται τὸ ΑΒΓ τρίγωνον τῷ εἴδει, λόγος ἐστὶ τῆς ΓΑ πρὸς τὴν ΒΑ δοθείς. ἐπεὶ οὖν δοθεῖσά ἐστιν ἡ ὑπὸ τῶν ΓΑΒ γωνία, ἔστι δὲ καὶ ἡ ὑπὸ τῶν ΕΑΒ δοθεῖσα, καὶ λοιπὴ ἄρα ἡ ὑπὸ τῶν ΕΑΓ ἐστι δοθεῖσα. ἔστι δὲ καὶ ἡ ὑπὸ τῶν ΑΕΓ γωνία δοθεῖσα· καὶ λοιπὴ ἄρα ἡ ὑπὸ τῶν ΕΓΑ δοθεῖσά ἐστιν· δέδοται ἄρα τὸ ΑΕΓ τρίγωνον τῷ εἴδει· λόγος ἄρα τῆς ΕΑ πρὸς τὴν ΑΓ δοθείς. τῆς δὲ ΓΑ πρὸς τὴν ΑΒ λόγος ἐστὶ δοθείς· καὶ τῆς ΕΑ ἄρα πρὸς τὴν ΑΒ λόγος ἐστὶ δοθείς. διὰ τὰ αὐτὰ δὴ καὶ τῆς ΖΑ πρὸς τὴν ΑΒ λόγος ἐστὶ δοθείς· ὥστε καὶ τῆς ΕΑ πρὸς τὴν ΑΖ λόγος ἐστὶ δοθείς. καί ἐστιν ὡς ἡ ΑΕ πρὸς τὴν ΑΖ, οὕτως τὸ ΑΗ πρὸς τὸ ΘΑ· ὥστε καὶ τοῦ ΑΗ πρὸς τὸ ΑΘ λόγος ἐστὶ δοθείς. καί ἐστι τοῦ μὲν ΑΗ ἥμισυ τὸ ΑΒΓ, τοῦ δὲ ΑΘ ἥμισυ τὸ ΑΔΒ· καὶ τοῦ ΑΒΓ ἄρα πρὸς τὸ ΑΔΒ λόγος ἐστὶ δοθείς.

Dt 48. *If on the same straight line two triangles be described, given in form, they will have a given ratio to one another.*

For, on the same straight line AB let two triangles given in form ABC, ADB have been described. I say that the ratio △ACB:△ADB is given.

For, let the straight lines AE, HB have been drawn from the points A, B at right angles to the straight line AB, and let them have been produced to Z, Q, and let ECH, ZDQ have been drawn through the points C, D parallel to AB.

Since △ABC is given in form, the ratio CA:BA is given [Def. 3]. Then, since ∠CAB is given, and ∠EAB is given, therefore the remaining ∠EAC is given [Dt 4]. And ∠AEC is given [I.29]; therefore the remaining ∠ECA is given [I.32]; therefore △AEC is given in form [Dt 40]; therefore the ratio EA:AC is given [Def. 3].

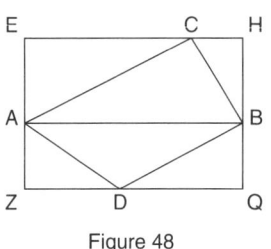

Figure 48

And the ratio CA:AB is given; therefore the ratio EA:AB is also given [102] [Dt 8].

Then, for the same reasons, the ratio ZA:AB is given; so the ratio EA:AZ is given. And AE:AZ :: (AH):(QA) [VI.1]; so the ratio (AH):(AQ) is given. And △ABC is half of [the rectangle] (AH), while ADB is half of (AQ); therefore the ratio ABC:ADB is given [V.15, Def. 2*]. ∎

---

[102] This crucial property is proved again in Dt 76; it could have been put before Dt 47 and used in several proofs in the sequel.

Remarks: Arbitrary polygons given in form described on one and the same line have a given ratio to one another (Dt 49) because the same holds for triangles (Dt 48). The proof of the latter is remarkably complicated, proving on its way what is repeated in Dt 76, that *in a triangle given in form the height has a given ratio to the base*. Using VI.1, that *triangles under the same heights are to each other as their bases*, Dt 48 proves that this ratio is given if the triangles are given in form. A caring editor would have put Dt 76 before 48, but Euclid apparently followed a source or canon which he did not want to tamper with. The word for 'height' (ὕψος), which makes its first appearance in VI.1, does not appear in either of the two, although VI.1 is crucial to Dt 48. A stenographic representation of Dt 48 will show Dt 76 embedded:

Hypotheses:	frm.△ABC, frm.△ABD.	
Assertion:	gvn.△ABC:△ABD.	
Construction:	The triangles are 'circumscribed' by 'their' rectangles.	
Proof:		
[Hyp, Def. 3]	gvn.CA:BA	(1)
[Def. 3]	gvn.∡BAC, gvn.∡EAB,	(2)
[Dt 4]	gvn.∡EAC.	(3)
[i.29, Dt 40]	gvn.∡AEC, therefore frm.△AEC.	(4)
[1, Dt 8]	gvn.EA:AC, gvn.EA:AB.	(5)
similarly	gvn.ZA:AB, therefore [Dt 8] gvn.EA:ZA.	(6)
[VI.1]	⊏⊐(AH):⊏⊐(AQ) :: AE:AZ;	(7)
[Def. 2*]	gvn.⊏⊐(AH):⊏⊐(AQ).	(8)
[V.15, Def. 2*]	gvn.(AH)/2 : (AQ)/2	(9)
[I.41]	gvn.△ABC : △ABD.	

Steps (1) – (5) are seen again *mutatis mutandis* in Dt 76, EA being represented by the perpendicular from the vertex C to the base AB.

Dt 76 would have served not only Dt 48, but also Dt 52, or rather the triangular case of Dt 52 (52T), which would make a perfect analogy to Dt 48, the triangular case of Dt 49. I give a brief summary, T denoting a triangle, h and b its height and base respectively, and P a polygon on base b:

Dt 76	frm.T	=>	gvn.h:b.
Dt 52T	frm.T	=[Dt 76]=>	gvn.h:b
	mag.b	=[Dt 2]=>	gvn.h
		=[Dt 24*B]=>	gvn.⊏⊐b,h
		=[I.41]=>	mag.T.
Dt 52	mag.b & frm.P	=>	mag.P.

As usual, several paths through the mathematical jungle lead to the same tree. But some jungles and paths are better organized than others. Is the *Data* less well organized than the *Elements*? Not less well than Book VI, which suffers from 'the lack of overall coherence' (Mueller 1981, 158), but much behind, say, Books IV and V. More material for the discussion about whether Euclid was compilator or author.

◆

Θε 49. ' Ἐὰν ἀπὸ τῆς αὐτῆς εὐθείας δύο εὐθύγραμμα, ἃ ἔτυχεν, ἀναγραφῇ δεδομένα τῷ εἴδει, λόγον ἕξει πρὸς ἄλληλα δεδομένον.

ἀπὸ γὰρ τῆς αὐτῆς εὐθείας τῆς ΑΒ δύο εὐθύγραμμα, ἃ ἔτυχεν, δεδομένα τῷ εἴδει ἀναγεγράφθω τὰ ΑΕΓΖΒ, ΑΔΒ· λέγω, ὅτι λόγος ἐστὶ τοῦ ΑΕΓΖΒ πρὸς ΑΔΒ δοθείς.

ἐπεζεύχθωσαν γὰρ αἱ ΑΖ, ΖΕ· δέδοται ἄρα ἕκαστον τῶν ΕΓΖ, ΕΖΑ, ΖΑΒ τριγώνων τῷ εἴδει. καὶ ἐπεὶ ἀπὸ τῆς αὐτῆς εὐθείας τῆς ΕΖ δύο τρίγωνα δεδομένα τῷ εἴδει ἀναγέγραπται τὰ ΕΖΓ, ΕΖΑ, λόγος ἄρα ἐστὶ τοῦ ΓΕΖ πρὸς τὸ ΖΕΑ δοθείς· καὶ συνθέντι ἄρα λόγος ἐστὶ τοῦ ΓΕΑΖ πρὸς τὸ ΖΕΑ δοθείς. τοῦ δὲ ΖΕΑ πρὸς τὸ ΖΑΒ λόγος ἐστὶ δοθείς, ἐπειδήπερ ἀπὸ τῆς αὐτῆς εὐθείας τῆς ΑΖ ἀναγέγραπται· καὶ τοῦ ΓΕΑΖ ἄρα πρὸς τὸ ΖΑΒ λόγος ἐστὶ δοθείς· καὶ συνθέντι τοῦ ΓΕΒΖΑ πρὸς τὸ ΖΒΑ λόγος ἐστὶ δοθείς. τοῦ δὲ ΖΑΒ πρὸς τὸ ΑΔΒ λόγος ἐστὶ δοθείς· καὶ τοῦ ΓΕΑΒΖ ἄρα πρὸς τὸ ΑΔΒ λόγος ἐστὶ δοθείς.

Dt 49. *If on the same straight line two arbitrary rectilineal figures given in form be described, they will have a given ratio to one another.*

Let two arbitrary rectilineal figures AECZB, ADB given in form have been described on the same straight line AB; I say that the ratio AECZB:ADB is given.

Let AZ, ZE have been joined; then each of the triangles ECZ, EZA, ZAB is given in form [Dt 47]. And since on the same straight line EZ two triangles two triangles given in form EZC, EZA have been described, the ratio CEZ:ZEA is given [Dt 48].

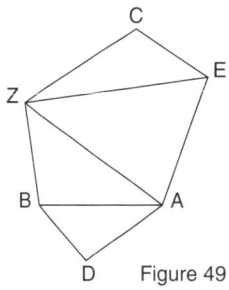

Figure 49

Therefore *synthenti* the ratio CEAZ:ZEA is given [Dt 6]. And the ratio ZEA:ZAB is given, since they have been described on the same straight line AZ [Dt 48]; therefore the ratio CEAZ:ZAB is given [Dt 8].

And *synthenti* CEABZ:ZBA [103] is given [Dt 6]. And the ratio ZAB:ADB is given [Dt 48]; therefore the ratio CEABZ:ADB is given [Dt 48]. ■

Remarks: Dt 49 is a trivial, but very useful, consequence of Dt 47 and 48.

---

[103] I follow ms b; Menge prints CEBZA, which is a rather un-Greek polygon.

Θε 50. 'Εὰν δύο εὐθεῖαι πρὸς ἀλλήλας λόγον ἔχωσι δεδομένον, καὶ τὰ ἀπ' αὐτῶν εὐθύγραμμα ὅμοια καὶ ὁμοίως ἀναγεγραμμένα πρὸς ἄλληλα λόγον ἕξει δεδομένον.

δύο γὰρ εὐθεῖαι αἱ ΑΒ, ΓΔ πρὸς ἀλλήλας λόγον ἐχέτωσαν δεδομένον, καὶ ἀναγεγράφθω ἀπὸ τῶν ΑΒ, ΓΔ ὅμοια καὶ ὁμοίως κείμενα εὐθύγραμμα τὰ Ε, Ζ· λέγω, ὅτι καὶ ὁ πρὸς ἄλληλα αὐτῶν λόγος ἔσται δοθείς.

εἰλήφθω γὰρ τῶν ΑΒ, ΓΔ τρίτη ἀνάλογον ἡ Η· ἔστιν ἄρα ὡς ἡ ΑΒ πρὸς τὴν ΓΔ, ἡ ΓΔ πρὸς τὴν Η· λόγος δὲ ὁ τῆς ΑΒ πρὸς ΓΔ δοθείς· λόγος ἄρα καὶ τῆς ΓΔ πρὸς τὴν Η δοθείς· ὥστε καὶ τῆς ΑΒ πρὸς τὴν Η λόγος ἐστὶ δοθείς. ὡς δὲ ἡ ΑΒ πρὸς τὴν Η, οὕτως τὸ Ε πρὸς τὸ Ζ· λόγος ἄρα τοῦ Ε πρὸς τὸ Ζ δοθείς.

Dt 50. *If two straight lines have a given ratio to one another, the similar and similarly described rectilineal figures on them will have a given ratio to one another.*

Let two straight lines AB, CD have a given ratio to one another, and on AB and CD let the similar and similarly situated rectilineal figures E and Z have been described; I say that also their ratio to one another will be given.

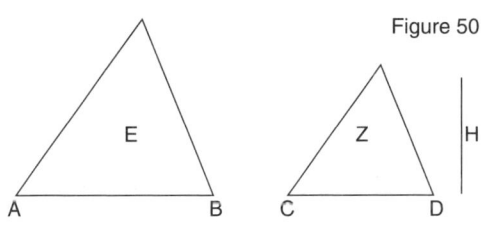

Figure 50

Let the third proportional H to AB, CD have been taken [VI.11]; then AB:CD :: CD:H; and the ratio AB:CD is given; therefore the ratio CD:H is given; so the ratio AB:H is given [Dt 8]. And AB:H :: E:Z [VI.19 corr]; therefore the ratio E:Z is given. ∎

Remarks: While Dt 49 deals with arbitrary polygons [104] given in form, described on *one and the same line segment*, Dt 51 extends this to polygons described on line segments *having a given ratio*. The intermediary is Dt 50, involving the third proportional to the given lines and the corollary to VI.19. The latter gets quite a comment by (Heath 1956, II:234), being misplaced from the following VI.20. The problem is that the corollary speaks of 'figure' (εἶδος) but is as yet proved only for triangles. Dt 50 does something similar, and one might expect the 'error' in the *Elements* to be inherited from the *Data*. That is hardly the case; while the *Data* uses εἶδος in a very idiolectic way in Dt 52–54, meaning both 'form' and 'rectilineal figure', Dt 50 speaks of a rectilineal figure (εὐθύγραμμα) and not of εἶδος.

---

[104] See (Mueller 1981, 158) for a definition (missing in the *Elements*, but almost always self-evident) of *similarly situated*. (Vitrac 1994, 2:198), discusses the idiom and compares it with the modern concept of *homothetic*.

Θε 51. ' Εὰν δύο εὐθεῖαι πρὸς ἀλλήλας λόγον ἔχωσι δεδομένον καὶ ἀπ' αὐτῶν εὐθύγραμμα, ἃ ἔτυχεν, ἀναγραφῇ δεδομένα τῷ εἴδει, λόγον ἕξει πρὸς ἄλληλα δεδομένον.

δύο γὰρ εὐθεῖαι αἱ ΑΒ, ΓΔ πρὸς ἀλλήλας λόγον ἐχέτωσαν δεδομένον, καὶ ἀναγεγράφθω ἀπὸ τῶν ΑΒ, ΓΔ εὐθύγραμμα, ἃ ἔτυχεν, δεδομένα τῷ εἴδει τὰ Ε, Ζ· λέγω, ὅτι τοῦ Ε πρὸς τὸ Ζ λόγος ἐστὶ δοθείς.

ἀναγεγράφθω γὰρ ἀπὸ τῆς ΑΒ τῷ Ζ ὅμοιον καὶ ὁμοίως κείμενον τὸ ΑΗΒ. δέδοται δὲ τὸ Ζ τῷ εἴδει· δέδοται ἄρα καὶ τὸ ΑΗΒ τῷ εἴδει. ἀλλὰ μὴν καὶ τὸ Ε δέδοται τῷ εἴδει καὶ ἀναγέγραπται ἀπὸ τῆς αὐτῆς εὐθείας τῆς ΑΒ· λόγος ἄρα τοῦ Ε πρὸς τὸ ΑΗΒ δοθείς. καὶ ἐπεὶ λόγος ἐστὶ τῆς ΑΒ πρὸς τὴν ΓΔ δοθείς, καὶ ἀναγέγραπται ἀπὸ τῶν ΑΒ, ΓΔ ὅμοια καὶ ὁμοίως κείμενα εὐθύγραμμα τὰ ΑΗΒ, Ζ, λόγος ἄρα τοῦ ΑΗΒ πρὸς τὸ Ζ δοθείς· τοῦ δὲ ΑΗΒ πρὸς τὸ Ε λόγος ἐστὶ δοθείς· καὶ τοῦ Ε ἄρα πρὸς τὸ Ζ λόγος ἐστὶ δοθείς.

Dt 51. *If two straight lines have a given ratio to one another, and arbitrary rectilineal figures given in form be described on them, they will have a given ratio to one another.*

Let two straight lines AB and CD have a given ratio to one another, and let arbitrary rectilineal figures given in form E and Z have been described on AB and CD. I say that the ratio E:Z is given.

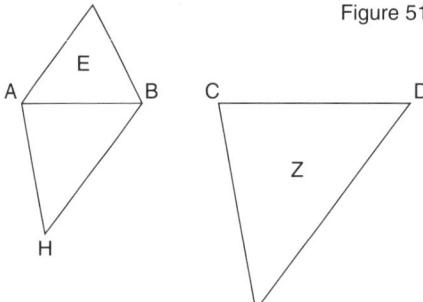

Figure 51

For, on the straight line AB let AHB have been described similar and similarly situated to Z [VI.18]; and Z is given in form; therefore AHB is given in form [Def. 3*A]. But E, too, is given in form and described on the same straight line AB; therefore the ratio E:AHB is given [Dt 49].

And since the ratio AB:CD is given, and on AB, CD similar and similarly situated rectilineal figures AHB, Z have been described, therefore the ratio AHB:Z is given [Dt 50]; and the ratio AHB:E is given; therefore the ratio E:Z is given [Dt 8]. ∎

Remarks: Generalisation of Dt 50. The figures need not be similar if only they are given in form.

◆

CHAPTER 6. FORM. (39–55)

Θε 52. Ἐὰν ἀπὸ δεδομένης εὐθείας τῷ μεγέθει δεδομένον τῷ εἴδει εἶδος ἀναγραφῇ, δέδοται τὸ ἀναγραφὲν τῷ μεγέθει.

ἀπὸ γὰρ δεδομένης εὐθείας τῷ μεγέθει τῆς ΑΒ δεδομένον τῷ εἴδει εἶδος ἀναγεγράφθω τὸ ΑΓΔΕΒ· λέγω, ὅτι τὸ ΑΓΔΕΒ δέδοται τῷ μεγέθει.
ἀναγεγράφθω γὰρ ἀπὸ τῆς ΑΒ τετράγωνον τὸ ΑΖ· δέδοται ἄρα τὸ ΑΖ τῷ εἴδει καὶ τῷ μεγέθει. καὶ ἐπεὶ ἀπὸ τῆς αὐτῆς εὐθείας τῆς ΑΒ δύο εὐθύγραμμα ἀναγέγραπται δεδομένα τῷ εἴδει τὰ ΑΓΔΕΒ, ΑΖ, λόγος ἄρα τοῦ ΑΓΔΕΒ πρὸς τὸ ΑΖ δοθείς· δέδοται δὲ τὸ ΑΖ τῷ μεγέθει· δέδοται ἄρα καὶ τὸ ΑΓΔΕΒ τῷ μεγέθει.

Dt 52. *If on a straight line given in magnitude a form* [105] *given in form be described, the* [form] *described is given in magnitude.*

For let the form ACDEB given in form have been described on the straight line AB given in magnitude; I say that the [form] ACDEB is given in magnitude.

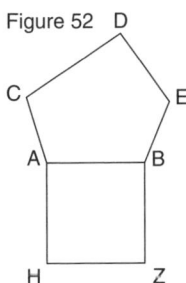

Figure 52

For, let the square (AZ) have been described on the straight line AB [I.46]; then (AZ) is given in form and in magnitude [Def. 3, Def. 1 [106]].

And since on the same straight line AB two rectilineal figures given in form ACDEB, (AZ) have been described, therefore the ratio ACDEB : (AZ) is given [Dt 49]. And (AZ) is given in magnitude; therefore ACDEB is given in magnitude [Dt 2]. ∎

Remarks: Dt 52 is the (partial) converse of the one we are heading for, Dt 55. One may ask why it needs a proof at all – it should be an immediate consequence of Def. 1: If a polygon given in form is described on a straight line given in magnitude, it is possible to *provide, i.e.* construct its equal, since all angles are given and the sides have given ratios. Nevertheless, Euclid invokes 'the square case' of the proposition he is about to prove, namely the unproven *If the side of a square* (which is also a given form) *is given in magnitude, the square is given in magnitude*, and then uses Dt 49 and Dt 2. Obviously, areas are not covered by Def. 1 (cf. the remarks to Dt 24 and 39).

◆

---

[105] Dt 52–54 (and some of the later propositions) have a curious terminology, a 'form (εἶδος) given in form', pointing to an esoteric jargon, like so many other Greek mathematical terms. Below it turns out to be a 'rectilineal figure', but I think it is right to keep the two identical words when they occur in the text; it is no more disturbing in English than in Greek.

[106] See footnote 110.

Θε 53. 'Εὰν δύο εἴδη τῷ εἴδει δεδομένα ᾖ καὶ μία πλευρὰ τοῦ ἑνὸς πρὸς μίαν πλευρὰν τοῦ ἑτέρου λόγον ἔχῃ δεδομένον, καὶ αἱ λοιπαὶ πλευραὶ πρὸς τὰς λοιπὰς πλευρὰς λόγον ἕξουσι δεδομένον.

ἔστω δύο εἴδη τῷ εἴδει δεδομένα τὰ ΑΔ, ΕΘ, καὶ λόγος τῆς ΒΔ πρὸς τὴν ΖΘ δοθείς· λέγω, ὅτι καὶ τῶν λοιπῶν πλευρῶν προς τὰς λοιπὰς πλευρὰς λόγος ἐστὶ δοθείς.

ἐπεὶ γὰρ λόγος ἐστὶ τῆς ΔΒ πρὸς τὴν ΖΘ δοθείς, τῆς δὲ ΔΒ πρὸς τὴν ΒΑ λόγος ἐστὶ δοθείς, καὶ τῆς ΑΒ ἄρα πρὸς τὴν ΖΘ λόγος ἐστὶ δοθείς. τῆς δὲ ΖΘ πρὸς ΕΖ λόγος ἐστὶ δοθείς· καὶ τῆς ΑΒ ἄρα πρὸς τὴν ΕΖ λόγος ἐστὶ δοθείς. διὰ τὰ αὐτὰ δὴ καὶ τῶν λοιπῶν πλευρῶν πρὸς τὰς λοιπὰς λόγος ἐστὶ δοθείς.

Dt 53. *If two forms be given in form, and one side of one form have a given ratio to one side of the other, the other sides will also have a given ratio to the other sides.*

Let AD, EQ be two forms given in form, and let the ratio BD:ZQ be given. I say that the other sides will also have a given ratio to the other sides.

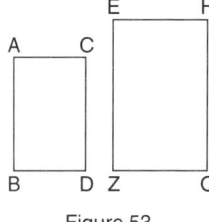

Figure 53

Since the ratio DB:ZQ is given, and the ratio DB:BA is given, therefore the ratio AB:ZQ is given [Dt 8].

And the ratio ZQ:EZ is given [Def. 3]; therefore the ratio AB:EZ is also given [Dt 8].

And for the same reasons the ratio of the other sides to the other sides is also given. ∎

◆

Θε 54. ' Εὰν δύο εἴδη δεδομένα τῷ εἴδει πρὸς ἄλληλα λόγον ἔχῃ δεδομένον, καὶ αἱ πλευραὶ αὐτῶν πρὸς ἀλλήλας λόγον ἕξουσι δεδομένον.

δύο γὰρ εἴδη δεδομένα τῷ εἴδει τὰ Α, Β πρὸς ἄλληλα λόγον ἐχέτω δεδομένον· λέγω, ὅτι καὶ αἱ πλευραὶ αὐτῶν πρὸς ἀλλήλας λόγον ἕξουσι δεδομένον.

τὸ γὰρ Α τῷ Β ἤτοι ὅμοιόν ἐστιν ἢ οὔ. ἔστω πρότερον ὅμοιον, καὶ εἰλήφθω τῶν ΓΔ, ΕΖ τρίτη ἀνάλογον ἡ Η. ἔστιν ἄρα ὡς ἡ ΓΔ πρὸς τὴν Η, οὕτως τὸ Α πρὸς τὸ Β. λόγος δὲ τοῦ Α πρὸς τὸ Β δοθείς· λόγος ἄρα καὶ τῆς ΓΔ πρὸς τὴν Η δοθείς. καί εἰσιν αἱ ΓΔ, ΕΖ, Η ἀνάλογον· καὶ τῆς ΓΔ ἄρα πρὸς τὴν ΕΖ λόγος ἐστὶ δοθείς. καί ἐστιν ὅμοιον τὸ Α τῷ Β· καὶ αἱ λοιπαὶ ἄρα πλευραὶ πρὸς τὰς λοιπὰς πλευρὰς λόγον ἕξουσι δεδομένον.

μὴ ἔστω δὴ ὅμοιον τὸ Α τῷ Β, καὶ ἀναγεγράφθω ἀπὸ τῆς ΕΖ τῷ Α ὅμοιον καὶ ὁμοίως κείμενον τὸ ΕΘ· δέδοται ἄρα καὶ τὸ ΕΘ τῷ εἴδει· δέδοται δὲ καὶ τὸ Β· λόγος ἄρα τοῦ Β πρὸς τὸ ΕΘ δοθείς· τοῦ δὲ Β πρὸς τὸ Α λόγος ἐστὶ δοθείς· καὶ τοῦ Α ἄρα πρὸς τὸ ΕΘ λόγος ἐστὶ δοθείς. καὶ ὅμοιον τὸ Α τῷ ΕΘ· λόγος ἄρα τῆς ΓΔ πρὸς τὴν ΕΖ δοθείς. διὰ τὰ αὐτὰ δὴ καὶ τῶν λοιπῶν πλευρῶν πρὸς τὰς λοιπὰς πλευρὰς λόγος ἐστὶ δοθείς.

Dt 54. *If two forms given in form have a given ratio to one another, their sides will have a given ratio to one another.*

Let the two forms given in form A and B have a given ratio to one another. I say that their sides will also have a given ratio to one another.

The form A is either similar to B or not.

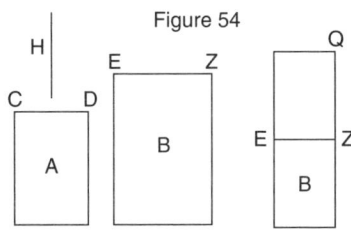

Figure 54

[1] Let it first be similar, and let the third proportional H to CD and EZ have been taken. Then CD:H :: A:B [VI.19 corr].

And the ratio A:B is given; therefore the ratio CD:H is given. And CD, EZ, H are proportional; therefore the ratio CD:EZ is also given [Dt 24].

And A is similar to B; therefore the other sides will have a given ratio to the other sides [Dt 53].

[2] But then let A be not similar to B, and on EZ let EQ have been described similar and similarly situated to A; then EQ is given in form.

And B is given; therefore the ratio B:EQ is given [Dt 49]; and the ratio B:A is given; therefore the ratio A:EQ is also given [Dt 8]. And A is similar to EQ; therefore the ratio CD:EZ is given.

For the same reasons, the ratio of the other sides to the other sides is given. ∎

Remarks: In the 3'rd step of Dt 55 the first, special, case of Dt 54 is used. The proof of Dt 54 goes by two steps, in order that VI.20 coroll. can be used (figure 55.1, p. 117):

1. The form P is similar to the form Q.

Get the third proportional c to the similarly situated sides a and b; then $a:c$ :: P:Q (VI.19 cor), a given ratio. Then (Dt 24) the ratio $a:b$ is also given; since P and Q are given in form, each side in P has a given ratio to each side in Q (Dt 53 and Dt 8).

2. The form P is not similar to the form Q.

On one side b of Q is constructed a rectilineal figure R, similar and similarly situated to P; then R is given in form, and so was Q; therefore (Dt 49) Q:R is a given ratio; and P:Q is given; thus P:R is given (Dt 8). The rest follows from case 1 of the theorem.

In fact, step 2 is not necessary for Dt 55. As it stands, Dt 54 is a much more general proposition than is needed. Dt 53 shows that the sides compared need not be homologous. The assertion follows immediately from Def. 3 and Dt 8. A variant of Dt 54 is proved in Dt 77, applying Dt 49 and a lemma to be derived from Dt 24*D, that *if two squares have a given ratio, their sides will also have a given ratio.* Menge refers to Dt 54, but in fact this special 'square' case is used.

◆

Θε 55. ' Ἐὰν χωρίον τῷ εἴδει καὶ τῷ μεγέθει δεδομένον ᾖ, καὶ αἱ πλευραὶ αὐτοῦ τῷ μεγέθει δεδομέναι ἔσονται.

ἔστω χωρίον τῷ εἴδει καὶ τῷ μεγέθει δεδομένον τὸ Α· λέγω, ὅτι καὶ αἱ πλευραὶ αὐτοῦ δεδομέναι εἰσὶ τῷ μεγέθει.
ἐκκείσθω γὰρ τῇ θέσει καὶ τῷ μεγέθει δεδομένη εὐθεῖα ἡ ΒΓ, καὶ ἀναγεγράφθω ἀπὸ τῆς ΒΓ τῷ Α ὅμοιόν τε καὶ ὁμοίως κείμενον τὸ Δ. δέδοται δὴ τὸ Δ τῷ εἴδει. καὶ ἐπεὶ ἀπὸ δεδομένης εὐθείας τῆς ΒΓ τῷ μεγέθει δεδομένον εἶδος ἀναγέγραπται τὸ Δ, δέδοται ἄρα καὶ τὸ Δ τῷ μεγέθει· δέδοται δὲ καὶ τὸ Α· λόγος ἄρα τοῦ Α πρὸς τὸ Δ δοθείς. καί ἐστιν ὅμοιον τὸ Α τῷ Δ· λόγος ἄρα τῆς ΕΖ πρὸς τὴν ΒΓ δοθείς. δοθεῖσα δὲ ἡ ΒΓ· δοθεῖσα ἄρα καὶ ἡ ΕΖ. καί ἐστι λόγος τῆς ΖΕ πρὸς τὴν ΕΗ δοθείς· δοθεῖσα ἄρα καὶ ἡ ΕΗ. διὰ τὰ αὐτὰ δὴ καὶ ἑκάστη τῶν λοιπῶν δέδοται τῷ μεγέθει.

Dt 55. *If a [rectilineal] figure* [107] *be given in form and in magnitude, its sides will also be given in magnitude.*

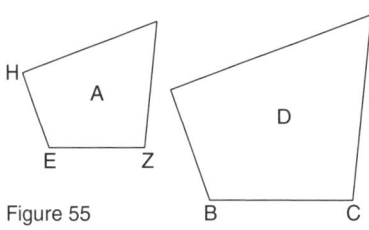

Figure 55

Let the [rectilineal] figure A be given in form and in magnitude; I say that its sides are given in magnitude.

Let the straight line BC have been set out given in position and in magnitude, and on BC let D have been described similar and similarly situated to A [VI.18]. Then D is given in form.

And since on the straight line BC given in magnitude the given form D has been described, therefore D is given in magnitude [Dt 52].

And A is given; therefore the ratio A:D is given [Dt 1]. And A is similar to D;

---

[107] In this crucial theorem the rectilineal figure is called *chorion*, a small place – in fact the normal word in the *Elements* Book X, whereas *euthygrammon* is used in Book VI.

therefore the ratio EZ:BC is given [Dt 54]. And BC is given [Dt 2]; therefore EZ is given [Def. 3]. And the ratio ZE:EH is given; therefore EH is given.

For the same reason each of the other sides is also given in magnitude. ∎

Remarks: The status of Dt 55 as the main theorem was discussed on p. 116 It will play a dominant role in the theory of 'application of areas' (Dt 58 and 59) by way of *Elements* VI.25, which according to Dt 55 has a 'given' solution.

◆

Deductive Structure of Dt 39–55
From the *Data*

		Dt 39	40	41	42	43	44	45	46	47	48	49	50	51	52	53	54	55 use
Data																		
Axiom	0*	+	+	+	+	.	.	.	.	.	.	.	.	.	.	.	.	.
Def.	1*	+	+	+	+	.	.	.	.	.	.	+	.	.	+	.	.	.
	2*	.	.	+	+	+	.	+	.	.	+	.	+	.	.	.	+	.
	3	+	+	+	+	+	+	+	+	+	+	.	.	+	+	+	.	+
	6	+	.	.	.	+	.	.	.	.	.	.	.	.	.	.	.	.
	8	.	.	.	.	+	.	.	.	.	.	.	.	.	.	.	.	.
Dt	1	.	.	.	.	.	.	.	.	.	.	.	.	.	.	.	.	+
	2	.	.	+	+	+	.	+	+	.	.	.	.	.	+	.	.	+
	3	.	.	.	.	.	+	+	+	.	+	.	.	.	.	.	.	.
	4	.	.	.	.	.	+	+	+	+	+	.	.	.	.	.	.	.
	6	.	.	.	.	.	.	.	.	.	.	.	+	.	.	.	.	.
	8	.	.	.	.	.	+	.	.	+	+	+	+	+	.	+	+	.
	24	.	.	.	.	.	.	.	.	.	.	.	.	.	.	.	+	.
	25	+	+	.	.	+	.	.	.	.	.	.	.	.	.	.	.	.
	26	+	+	+	.	+	.	.	.	.	.	.	.	.	.	.	.	.
	27	+	.	+	.	.	.	.	.	.	.	.	.	.	.	.	.	.
	29	.	+	+	.	.	.	.	.	.	.	.	.	.	.	.	.	.
	39	.	?	?	?	?	.	.	.	.	.	.	.	.	.	.	.	.
	40	.	.	.	.	+	+	+	.	+	.	.	.	.	.	.	.	.
	41	.	.	.	.	.	.	+	+	+	+	.	.	.	.	.	.	.
	43	.	.	.	.	.	+	.	.	.	.	.	.	.	.	.	.	.
	44	.	.	.	.	.	.	+	.	.	.	.	.	.	.	.	.	.
	47	.	.	.	.	.	.	.	.	.	.	+	.	.	.	.	.	.
	48	.	.	.	.	.	.	.	.	.	.	+	.	.	.	.	.	.
	49	.	.	.	.	.	.	.	.	.	.	.	.	+	+	.	+	.
	50	.	.	.	.	.	.	.	.	.	.	.	.	+	.	.	.	.
	52	.	.	.	.	.	.	.	.	.	.	.	.	.	.	.	.	+
	53	.	.	.	.	.	.	.	.	.	.	.	.	.	.	.	+	.
	54	.	.	.	.	.	.	.	.	.	.	.	.	.	.	.	.	+

Dt 39 40 41 42 43 44 45 46 47 48 49 50 51 52 53 54 55

Note the 'diagonal theorems': 43 serves 44, 44 serves 45, 47 and 48 serve 49 – and are not used any more (except that 88 is proved by 43, though 44 would do it quicker).

Deductive Structure of Dt 39–55
From the *Elements*

Elem		Dt	39	40	41	42	43	44	45	46	47	48	49	50	51	52	53	54	55	use
Post		2	.	.	.	.	+	.	.	.	.	.	.	.	.	.	.	.	.	.
		3	+	.	.	.	.	.	.	.	.	.	.	.	.	.	.	.	.	.
I.		8	+	.	.	.	.	.	.	.	.	.	.	.	.	.	.	.	.	.
		9	.	.	.	.	.	.	+	+	.	.	.	.	.	.	.	.	.	.
		12	.	.	.	.	+	.	.	.	.	.	.	.	.	.	.	.	.	.
		13	.	.	.	.	+	.	.	.	.	.	.	.	.	.	.	.	.	.
		17	.	.	.	+	.	.	.	.	.	.	.	.	.	.	.	.	.	.
		19	.	.	.	+	.	.	.	.	.	.	.	.	.	.	.	.	.	.
		22	.	.	+	.	.	.	.	.	.	.	.	.	.	.	.	.	.	.
		23	.	+	+	.	.	.	.	.	.	.	.	.	.	.	.	.	.	.
		29	.	.	.	.	.	.	.	.	.	+	.	.	.	.	.	.	.	.
		32	.	+	.	.	.	+	+	+	.	+	.	.	.	.	.	.	.	.
		41	.	.	.	.	.	.	.	.	.	+	.	.	.	.	.	.	.	.
		46	.	.	.	.	.	.	.	.	.	.	.	.	.	.	+	.	.	.
III.		31	.	.	.	+	.	.	.	.	.	.	.	.	.	.	.	.	.	.
V.		7	.	.	+	.	.	.	.	.	.	.	.	.	.	.	.	.	.	.
		12	.	.	.	.	.	.	+	+	.	.	.	.	.	.	.	.	.	.
		14	.	.	.	+	.	.	.	.	.	.	.	.	.	.	.	.	.	.
		15	.	.	.	.	.	.	.	.	.	+	.	.	.	.	.	.	.	.
		16	.	.	.	+	.	+	+	.	.	.	.	.	.	.	.	.	.	.
		22	.	.	+	.	.	.	.	.	.	.	.	.	.	.	.	.	.	.
VI.def.		1	+	.	.	.	.	.	.	.	.	.	.	.	.	.	.	.	.	.
VI.		1	.	.	.	.	.	.	.	.	.	+	.	.	.	.	.	.	.	.
		3	.	.	.	.	.	.	+	+	.	.	.	.	.	.	.	.	.	.
		4	.	+	.	.	.	.	.	.	.	.	.	.	.	.	.	.	.	.
		5	.	.	.	+	.	.	.	.	.	.	.	.	.	.	.	.	.	.
		6	.	.	+	.	.	.	.	.	.	.	.	.	.	.	.	.	.	.
		7	.	.	.	+	.	.	.	.	.	.	.	.	.	.	.	.	.	.
		11	.	.	.	.	.	.	.	.	.	.	.	.	+	.	.	+	.	.
		18	.	.	.	.	.	.	.	.	.	.	.	.	.	+	.	+	+	.
		19cor	.	.	.	.	.	.	.	.	.	.	.	.	+	.	.	+	.	.
		Dt	39	40	41	42	43	44	45	46	47	48	49	50	51	52	53	54	55	

Dt 47, 49, and 53 are only indirectly dependent on the *Elements*. Def. 3 and the fundamental propositions Dt 2, 3, 4, and 8 play vital roles. Dt 39 is not used, but its argument repeated, in Dt 40–43.

Chapter 7. Equiangular parallelograms I
Reciprocal proportion. Dt 56 & (68–75)

Θε 56. ῎Εὰν δύο ἰσογώνια παραλληλόγραμμα πρὸς ἄλληλα λόγον ἔχῃ δεδομένον, ἔσται ὡς ἡ τοῦ πρώτου πλευρὰ πρὸς τὴν τοῦ δευτέρου πλευράν, οὕτως ἡ λοιπὴ τοῦ δευτέρου πλευρὰ πρὸς ἣν ἡ ἑτέρα τοῦ πρώτου λόγον ἔχει δεδομένον, ὃν τὸ παραλληλόγραμμον πρὸς τὸ παραλληλόγραμμον.

δύο γὰρ ἰσογώνια παραλληλόγραμμα τὰ Α, Β πρὸς ἄλληλα λόγον ἐχέτω δεδομένον· λέγω, ὅτι ἐστὶν ὡς ἡ ΓΔ πρὸς τὴν ΕΖ, οὕτως ἡ ΕΗ πρὸς ἣν ἡ ΓΘ λόγον ἔχει δεδομένον, ὃν τὸ Α παραλληλόγραμμον πρὸς τὸ Β παραλληλόγραμμον.

ἐκβεβλήσθω γὰρ ἐπ' εὐθείας τῆς ΓΘ εὐθεῖα ἡ ΓΚ, καὶ πεποιήσθω ὡς ἡ ΓΔ πρὸς τὴν ΕΖ, οὕτως ἡ ΕΗ πρὸς τὴν ΓΚ, καὶ συμπεπληρώσθω τὸ ΓΛ παραλληλόγραμμον. ἐπεὶ οὖν ἐστιν ὡς ἡ ΓΔ πρὸς τὴν ΕΖ, οὕτως ἡ ΕΗ πρὸς τὴν ΓΚ, ἴση δέ ἐστιν ἡ ΓΔ τῇ ΚΛ, ἔστιν ἄρα ὡς ἡ ΚΛ πρὸς τὴν ΕΖ, οὕτως ἡ ΕΗ πρὸς τὴν ΓΚ. καὶ περὶ ἴσας γωνίας τὰς ὑπὸ τῶν ΓΚΛ, ΗΕΖ αἱ πλευραὶ ἀντιπεπόνθασιν· ἴσον ἄρα ἐστὶ τὸ ΚΔ τῷ ΗΖ. καὶ ἐπεὶ λόγος ἐστὶ τοῦ Α πρὸς τὸ Β δοθείς, ἴσον δὲ τὸ Β τῷ ΓΛ, λόγος ἄρα ἐστὶ τοῦ ΘΔ πρὸς τὸ ΓΛ δοθείς. ὡς δὲ τὸ ΘΔ πρὸς τὸ ΓΛ, οὕτως ἡ ΘΓ πρὸς τὴν ΓΚ· καὶ τῆς ΘΓ ἄρα πρὸς τὴν ΓΚ λόγος ἐστὶ δοθείς. καὶ ἐπεί ἐστιν ὡς ἡ ΓΔ πρὸς τὴν ΕΖ, οὕτως ἡ ΕΗ πρὸς τὴν ΓΚ, ἡ δὲ ΓΘ πρὸς τὴν ΓΚ λόγον ἔχει δοθέντα, ὃν τὸ Α χωρίον πρὸς τὸ Β, ἔστιν ἄρα ὡς ἡ ΓΔ πρὸς τὴν ΕΖ, οὕτως ἡ ΕΗ πρὸς ἣν ἡ ΘΓ λόγον ἔχει, ὃν τὸ Α χωρίον πρὸς τὸ Β χωρίον.

Dt 56. *If two equiangular parallelograms have a given ratio to one another, as the side of the first parallelogram is to the side of the second, so the other side of the second will be to that [line] to which the other side of the first has the given ratio which the first parallelogram has to the second.*

For, let two equiangular parallelograms A and B have a given ratio to one another. I say that as CD is to EZ, so is EH to that straight line to which CQ has the given ratio which the parallelogram A has to the parallelogram B.

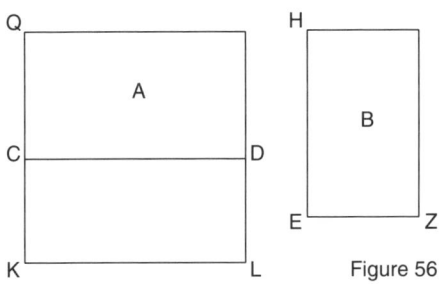

Figure 56

For, let the straight line CK have been produced in a straight line with CQ, and let have been made CD:EZ :: EH:CK, [VI.12] and let the parallelogram (CL) have been completed.

Since then CD:EZ :: EH:CK, and CD = KL [I.34], therefore KL:EZ :: EH:CK.

And the sides about the equal angles CKL, HEZ are reciprocally proportional, therefore (KD) = (HZ) [VI.14].

And since the ratio A:B is given, and B is equal to (CL), therefore the ratio (QD):(CL) is given.

But (QD):(CL) :: QC:CK [VI.1];therefore the ratio QC:CK is given [Def. 2*].

And since CD:EZ :: EH:CK, and CQ has the given ratio to CK which the area A has to B, therefore as CD is to EZ, so is EH to that [line] to which QC has the ratio that the area A has to the area B. ■

Remarks: A couple of anomalies (that might have text-historical significance) spring to mind in this theorem; first, it is quite unusual (although we have seen it a couple of times from Dt 50) to have areas named A and B while their sides have names, too, and take part in the proof. In fact, CD, EZ, EH, and CQ pop up unexpectedly and are undefined in the ekthesis. As always, the argument is hard to follow without the diagram. Normally, A would be DCQ or DC,CQ or DQ and B would be ZEH or ZE,EH or ZH.
– In the manuscripts A and B are drawn as rectangles, but of course any equiangular parallelograms will do. For some reason or other the word *parallelogrammon* was altered to *trigonon*, triangle, throughout this proposition and Dt 68–74. Perhaps some scribe had not learned about equiangular parallelograms, but only about triangles. Menge corrected it, rightly.

Secondly, a suspicion arises that this theorem (which introduces a new series of propositions) belongs rather in the *Elements* than in the context of *Givens*, – because of the unusual information that a certain ratio is not only given, but the same as a third one. Normally that kind of knowledge is suppressed in the *Data*.

Dt 56 seems to be a mixture of two propositions, (Dt 56*A and B below), and thus is a good illustration of the difference between the *Data* and the *Elements*. It belongs in a sequence about (equiangular) parallelograms and triangles, Dt 68 – Dt 75, which say more or less the same couple of things and would profit from proving, once and for all, the following *elementary* proposition, 56*A – which I call elementary because it would fit well into Book VI of the *Elements* and does not use the concept 'given'.

Dt 56*A

Let P and Q be parallelograms with sides a, b and c, d respectively (figure 56.1).

    If   P is equiangular with Q

    and if e is the fourth proportional to a, c, and d (so that $a:c :: d:e$)

    then $b:e :: P:Q$.

Construction (I.44): Let a third parallelogram R, equiangular with Q, be applied to a with width e, so that e is in a straight line with b.

CHAPTER 7. EQUIANGULAR PARALLELOGRAMS I. (56)

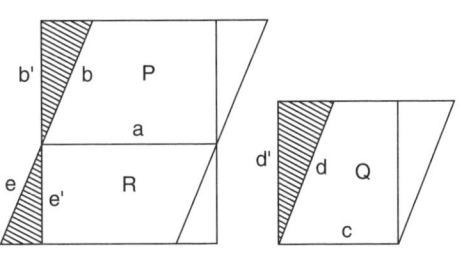

Proof:
Since $a:c :: d:e$,
∴ $R = Q$ [VI.14].
∴ $P:Q :: P:R$ [V.7].
But $P:R :: b:e$ [VI.1],
∴ $P:Q :: b:e$. ∎

Figure 56. 1

Without loss of generality the parallelograms may be rectangles, if a and c are homologous sides [108] (and therefore b and d homologous, too).

Since the ratio $b:e$ is compounded (see note 63) from $b:d$ and $a:c$, because $a:c :: d:e$, what is proved here is that *the ratio between equiangular parallelograms is compounded from the ratios of their sides,* a truth known from VI.23. But the term 'compounded' is not found in the *Data*, which is one more (faint) piece of evidence that the *Data* belongs in a context different from Euclid's *Elements* – albeit it belongs in *some* elements. But then, the treatment of compound ratio is very undeveloped in the *Elements*.

Mixed up in Dt 56 is the corollary **Dt 56*B** If *gvn*.P:Q then *gvn*.b:e [Def. 2*].

Dt 68–Dt 75 can also be stated in terms of Dt 56*A. We will consider them in their proper place; it seems that Dt 56 is misplaced, since it is not used or quoted in Dt 57–67. It is proved once more in Dt 74, which is undoubtedly later (or at least by a more mature hand), since it also proves the more general case of non-equiangular parallelograms with given angles. Perhaps Dt 56 was copied from a different source and put before Dt 57 because the construction of the parallelogram (CL) is a parabolic application of area.

The series Dt 68–75 is more complicated than the mathematical content deserves, and the more so because of the author's reluctance to reuse his figures. So what is CD in Dt 56, is EB in 68, DB in 69, CB in 70 and 73, and so on. And this is rather peculiar, since the wording of the propositions is very repetitive; thus Dt 74, first part, uses almost the same phrases as Dt 56.

◆

---

[108] Say, both of them are 'left' side of equal angles. Euclid took great pains to make sure that the relevant sides are considered, KL instead of CD, – although he forgot to tell us in the first place which angles are equal.

Chapter 8. Application of areas I
Dt 57–62 & (84–85)

Θε 57 ' Ἐὰν δοθὲν παρὰ δοθεῖσαν παραβληθῇ ἐν δεδομένῃ γωνίᾳ, δέδοται τὸ πλάτος τῆς παραβολῆς.

δοθὲν γὰρ τὸ ΑΗ παρὰ δοθεῖσαν τὴν ΒΑ παραβεβλήσθω ἐν δεδομένῃ γωνίᾳ τῇ ὑπὸ τῶν ΓΑΒ· λέγω, ὅτι δοθεῖσά ἐστιν ἡ ΓΑ.
 ἀναγεγράφθω γὰρ ἀπὸ τῆς ΑΒ τετράγωνον τὸ ΕΒ· δοθὲν ἄρα ἐστὶ τὸ ΕΒ. καὶ διήχθωσαν αἱ ΕΑ, ΖΒ, ΓΗ ἐπὶ τὰ Δ, Θ. καὶ ἐπεὶ δοθέν ἐστιν ἑκάτερον τῶν ΕΒ, ΑΗ, λόγος ἄρα τοῦ ΕΒ πρὸς τὸ ΑΗ δοθείς. ἴσον δὲ τὸ ΗΑ τῷ ΑΘ· λόγος ἄρα καὶ τοῦ ΕΒ πρὸς τὸ ΑΘ δοθείς· ὥστε καὶ τῆς ΕΑ πρὸς τὴν ΑΔ λόγος ἐστὶ δοθείς. ἴση δὲ ἡ ΕΑ τῇ ΑΒ· λόγος ἄρα καὶ τῆς ΒΑ πρὸς ΑΔ δοθείς. καὶ ἐπεὶ δοθεῖσά ἐστιν ἡ ὑπὸ τῶν ΓΑΒ, ὧν ἡ ὑπὸ ΔΑΒ δοθεῖσά ἐστιν, λοιπὴ ἄρα ἡ ὑπὸ τῶν ΓΑΔ ἐστι δοθεῖσα. ἔστι δὲ καὶ ἡ ὑπὸ τῶν ΓΔΑ δοθεῖσα· ὀρθὴ γάρ· λοιπὴ ἄρα ἡ ὑπὸ τῶν ΑΓΔ δοθεῖσά ἐστιν· δέδοται ἄρα τὸ ΑΓΔ τρίγωνον τῷ εἴδει· λόγος ἄρα ἐστὶ τῆς ΓΑ πρὸς τὴν ΑΔ δοθείς. τῆς δὲ ΔΑ πρὸς τὴν ΑΒ λόγος ἐστὶ δοθείς· καὶ τῆς ΓΑ ἄρα πρὸς τὴν ΑΒ λόγος ἐστὶ δοθείς. καί ἐστι δοθεῖσα ἡ ΒΑ· δοθεῖσα ἄρα καὶ ἡ ΑΓ. καί ἐστι τὸ πλάτος τοῦ παραβλήματος.

Dt 57. *If a given [area] be applied to a given [straight line]* [109] *in a given angle, the width of the applied [area] is given.*

For let the given [area] (AH) have been applied to the given [straight line] BA in the given angle CAB; I say that CA is given.

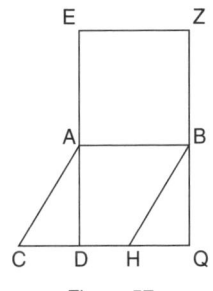
Figure 57

For let the square (EB) have been described on AB; then (EB) is given. [110]
 And let EA, ZB, and CH have been produced to D, Q.
 And since each of (EB) and (AH) is given, therefore the ratio (EB):(AH) is given [Dt 1].
 And [the parallelogram] (HA) is equal to ⊏⊐(AQ) [I.35]; therefore the ratio (EB):(AQ) is given; so that the ratio EA:AD is given [VI.1, Def. 2*].

---

[109] The gender determines the meaning, as always in The Mathematical Lexicon (as Reviel Netz calls it), neuter being χωρίον, feminine εὐθεῖαν.

[110] Menge refers to Dt 52 without noticing that that theorem uses this inference without proof. I have treated this tacit assumption with Dt 24.

And EA = AB; therefore the ratio BA:AD is given [V.7]. And since ∠CAB is given, of which [111] DAB is given, the remaining ∠CAD is given [Dt 4].

And ∠CDA is also given, for it is [a] right [angle]; therefore the remaining [angle of the triangle] ACD is given [I.32]. Therefore △ACD is given in form [Dt 40]. Therefore the ratio CA:AD is given [Def. 3].

And the ratio DA:AB is given. Therefore the ratio CA:AB is also given [Dt 8]. And BA is given; therefore AC is also given [Dt 2], and it is the width of the applied area.[112]   ∎

Remarks: The problem *To a given straight line to apply, in a given angle, a parallelogram equal to a given triangle* is solved in *Elements* I.44, which 'will always remain one of the most impressive in all geometry' (Heath 1956, I.342). It was later to be known as *The parabolic application of area*. 'To apply an area to a [straight] line' always means 'to construct a parallelogram along that line'. The parallelogram may have the line segment as one of its sides (Dt 57), or may exceed (Dt 59) or fall short of (Dt 58) the segment.

The effective lemmas are I.43 (quoted on p. 156) and I.42. Now, what does Dt 57 add to that? I.44 seems to underlie the protasis and ekthesis of Dt 57: When the curtain rises, the parallelogram (AH) is there in the given angle, hanging under the given side AB and being equal to the given area (not seen). Therefore, Dt 57 cannot be understood as a proof of existence; the possibility of applying (AH) is taken for granted. The proposition *can* be interpreted as proving, by means of the apparatus of Givens, that there is a (unique) solution which is worth looking for, even if we do not yet know how to produce it.[113]

So it is tempting to take it as a proof of uniqueness, but such a proof would hardly be needed after I.44; for, although two positions of the parallelogram are thinkable, they must turn out to be congruent, so that the width is given in magnitude in either case.

Since there is not the slightest reference to I.44, one could pretend not to know how to construct AC, as if I.44 was not yet invented? Euclid, however, obviously believes in the possibility of constructing it; or else, he would say 'if it be possible, let the given

---

[111] Plural, considered as two angles CAD + DAB.

[112] παραβλήματος; the protasis has παραβολῆς, which normally denotes the *operation* 'to apply'.

[113] Cf. Ptolemy's introduction to his table of chords, (Heiberg 37-38, Ptolemy-Toomer 51) where he demonstrates the givenness of certain chords before calculating them.

[area] AH ...' etc.[114] And Dt 58 and 59 immediately spoil the pretension that we do not know how to get the Givens constructed: in fact, both those theorems use the constructions of *Elements* VI.28-29 as a matter of course. – Let us watch the individual steps:[115]

1 □AB is constructed [I.46] and seen to be given (since AB is given).
2 ⊏⊐(AQ) is constructed and seen to be equal to (AH), by I.35.
3 Since the ratio (EB):(AQ) is given [Dt 1, because both are given areas], and EA:AD is the same ratio [VI.1], therefore EA:AD is also given, that is AB:AD.
4 The right triangle ACD is given in form because of the given angle [Dt 40]. Therefore AD:AC is given [Def. 3].
5 Gvn.AB:AD & gvn.AD:AC =[Dt 8]=> AB:AC ∴ gvn.AC [Dt 2].

The theorem first transforms the hypotheses into a ratio between given rectangles, namely gvn.⊏⊐(AQ):□AB, and then (by VI.1) represents this as a ratio between line segments gvn.AD:AB. Obviously, the crux is to apply ⊏⊐(AQ) to AB equal to the given area. So we are back at I.44.

The property *given* is handed on from one object to the next, using ratios as intermediaries which are generated partly by theorems from the *Elements*, partly by appeal to the notion 'given in form, and finally *the side AC is* as given as *the side AB* (which was given at the outset). That ought to mean that if we know (?) the area, and the length of AB, and one of the angles of the parallelogram, we also know (?) the length of AC. But what sort of knowledge is that? As usual, nothing is calculated and nothing can be calculated without trigonometry. If we have that, and if we allow ourselves to put a length to AB, say a, and an area to (AH), say Q, and if the given angle is v, then the theorem anachronistically can be said to lead up to the formula $x = Q / a \sin v$, where the width AC is the 'unknown' x. But that is of little consequence so long as trigonometry is not invented, let alone developed.

We are not informed of the values of the intermediate ratios, and so it will be difficult, if not impossible, to set up a formula relating the first side AB with the second

---

[114] Whenever the existence is in doubt, Euclidlid uses the formula εἰ γὰρ δυνατόν, *if it be possible*; in fact only in propositions proving that something *is* impossible. E.g. III.10, that two circles cannot have more than two points in common. The formula is never met in the *Data*.

[115] The vocabulary points remarkably to an esoteric idiolect, *The Lexicon of Application of Areas* (as Reviel Netz calls it). The word *platos*, meaning 'width', is used here for 'the other side' of the parallelogram, not the 'height' of it. (Cf. remarks on the plural πλάτη in Dt 58).

AC by means of the given angle and area.

The single piece of information established is that AC is given [in magnitude, and in position because of the angle]. I confess that I am at a loss as to the meaning of the proposition. Are we taught anything that we did not learn in I.44? On the other hand, I am convinced that the key to understanding the concept of *given* must lie in correctly interpreting Dt 57 and the following theorems Dt 58-59.

Clemens Thaer from his algebraic universe reluctantly admits that the *Data* might be geometry: 'If $ax = C$, then $x = C / a$; geometrically (*sic*) the width can also be oblique.' True, but hardly what Dt 57 says.

◆

Θε 58 'Εὰν δοθὲν παρὰ δοθεῖσαν παραβληθῇ ἐλλεῖπον εἴδει δεδομένῳ τῷ εἴδει, δέδοται τὰ πλάτη τοῦ ἐλλείματος.

δοθὲν γὰρ τὸ ΑΓ παρὰ δοθεῖσαν τὴν ΑΔ παραβεβλήσθω ἐλλεῖπον εἴδει δεδομένῳ τῷ ΓΔ· λέγω, ὅτι δοθεῖσά ἐστιν ἑκατέρα τῶν ΒΓ, ΒΔ.

τετμήσθω γὰρ ἡ ΑΔ δίχα κατὰ τὸ Ε σημεῖον· δοθεῖσα ἄρα ἐστὶν ἡ ΕΔ. καὶ ἀναγεγράφθω ἀπὸ τῆς ΕΔ τῷ ΓΔ ὅμοιον καὶ ὁμοίως κείμενον εὐθύγραμμον τὸ ΕΖ, καὶ καταγεγράφθω τὸ σχῆμα· δέδοται ἄρα καὶ τὸ ΕΖ τῷ εἴδει. καὶ ἐπεὶ ἀπὸ δεδομένης εὐθείας τῆς ΕΔ δεδομένον τῷ εἴδει εἶδος ἀναγέγραπται τὸ ΕΖ, δέδοται ἄρα τὸ ΕΖ τῷ μεγέθει. καί ἐστιν ἴσον τοῖς ΑΓ, ΚΘ· δέδοται ἄρα καὶ τὰ ΑΓ, ΚΘ τῷ μεγέθει. καί ἐστι τὸ ΑΓ δοθὲν τῷ μεγέθει· ὑπόκειται γάρ· λοιπὸν ἄρα τὸ ΚΘ δοθέν ἐστι τῷ μεγέθει. ἔστι δὲ καὶ τῷ εἴδει δοθέν· ὅμοιον γάρ ἐστι τῷ ΓΔ· τοῦ ΘΚ ἄρα δεδομέναι εἰσὶν αἱ πλευραί· δοθεῖσα ἄρα ἐστὶν ἡ ΚΓ· καί ἐστιν ἴση τῇ ΕΒ· δοθεῖσα ἄρα ἐστὶν καὶ ἡ ΕΒ. ἔστι δὲ καὶ ἡ ΕΔ δοθεῖσα· καὶ λοιπὴ ἄρα ἡ ΒΔ δοθεῖσά ἐστιν. καὶ λόγος τῆς ΒΔ πρὸς τὴν ΒΓ δοθείς· δοθεῖσα ἄρα ἐστὶ καὶ ἡ ΒΓ.

Dt 58. *If a given [area] be applied to a given [straight line] deficient by a form given in form,* [116] *the length and width* [117] *of the defect are given.*

For, let the given [area] (AC) have been applied to the given [straight line] AD deficient by the given form (CD). I say that each of BC, BD is given.

For, let AD have been bisected at the point E; then ED is given [Dt 2]. And on ED let the rectilineal figure (EZ) have been described similar and similarly situated to

---

[116] A curious idiom, which would be more easily understood if we translate 'a form (or figure) given in shape'. But the Greek uses the same word, even in the same case (*dative*) εἴδει.

[117] τὰ πλάτη here (and in Dt 59, 84, and 85) means length as well as width.

(CD) [VI.18], and let the *schema* [118] have been described. Then (EZ) is given in form.

And since on the given straight line ED the form given in form (EZ) has been described, therefore (EZ) is given in magnitude [Dt 52]. And (EZ) = AC+KQ [VI.27]; therefore AC+KQ is given in magnitude.

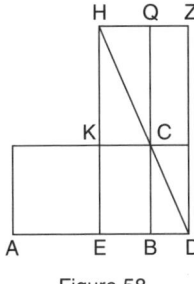

Figure 58

And (AC) is given in magnitude by hypothesis; therefore the remainder (KQ) is given in magnitude [Dt 4].

But it is also given in form, for it is similar to (CD) [VI.24, Def. 3*B]; therefore the sides of (QK) are given [Dt 55], and therefore KC is given; and it is equal to EB; therefore EB is given; and ED is given; therefore the remainder BD is also given. And the ratio BD:BC is given; therefore BC is also given. ∎

Remarks: Dt 58 proves that the deficient parallelogram of the elliptical application of areas (cf. VI.28) has given sides. That it exists seems to be presupposed in the protasis *If .. be applied deficient by a form given in form*. (Cf. footnote 114). Therefore the *Data* is silent about the condition of possibility [119] proved in VI.27. Should Dt 58 be understood as a theorem of uniqueness, then? But in fact there are two 'solutions' to the problem, though the *Data* considers only the smaller *deficiency*.

The protasis is somewhat surprising, not only because the word *platos*, plural *platê*, is used of both dimensions of a parallelogram (its normal meaning being 'breadth' as opposed to *mekos*, 'length'). But some of us would think that the 'unknowns' to be proved given were the sides of the applied parallelogram, and not those of that phantom, *the deficient parallelogram*. However, the givenness of the sides is postponed till Dt 85 after its analogue Dt 84 (corresponding to Dt 59), appended – as if forgotten – at the end of the rectilineal section (Dt 39–80). [120]

It may be useful to insert a *nomenclatura*, a couple of definitions about what I call *Gnomonic Situations*, and remind of some important propositions from the *Elements*.

---

[118] τὸ σχῆμα in the theory of 'application of area' always means a gnomonic figure, *i.e.* a gnomon (EDZQCK[E]) containing the significant rectangle (KQ).

[119] That the given area must not be greater than the parallelogram of the given form that can be described on half of the given line.

[120] Dt 81–83 are about proportional lines, and Dt 86 is a queer lone wolf, perhaps dealing with conic sections. The last eight propositions treat circles.

Definition 58*A, *Schema*: If a parallelogram is cut into four parallelograms by two straight lines parallel to its adjacent sides and intersecting on a diagonal of the parallelogram, let the whole configuration be called a *schema*.

In the parallelogram EDZH (figure 58) KC and BQ, parallel to ED, EH respectively, intersect at C, which is on the diagonal HD, and thus the configuration is a *schema*. It appears that the word *schema* – which means "something held together", *i.e.* "figure" – is used idiomatically in this special sense as a kind of password to gnomonic situations.

Definition 58*B, *Parallelograms about the diagonal*: (CD) and (KQ) are called 'The parallelograms about the diagonal'. (*Elements* I.43).

Two important theorems about them are proved in Book VI of the *Elements*:

VI.24   The parallelograms about the diagonal of any parallelogram are similar to the whole and to one another.
VI.26   If from a parallelogram a parallelogram similar to the whole, similarly situated and having a common angle with it, is subtracted, it is about the same diagonal as the whole.

The three similar parallelograms in Dt 58, (EZ), (KQ), and (CD), represent the *eidos* or *form* of the *schema*.

Definition 58*C, *Parapleromata, Complements*: (EC) and (CZ) in figure 58 are called the complements, *parapleromata*, of the parallelograms (KQ) and (CD).

I.43   The complements of the parallelograms about the diagonal of any parallelogram are equal.

Definition 58*D, *Gnomon*: Let any one of the parallelograms about the diagonal of any parallelogram together with the two complements be called a *gnomon* (II.def 2).

In figure 58, EDZQCK[E] is a gnomon. The parallelogram (KQ) 'contained by' the gnomon I will call the *'form* in the gnomon'.

Dt 58 assumes known that any parallelogram P with a given angle is equal to a gnomon obtained by adding a parallelogram with the given angle and of a given form to one end of P, and describing a *schema* on half of the combined base.

Actually, (figure 58), the parallelogram (AC) is equal to the gnomon EDZQCK obtained by adding the Form (CD) to (AC), then bisecting AD in E and describing on ED a parallelogram (EZ) of the given form (using the diagonal DCH), and completing the *schema* by producing the line BC. By then the parallelograms (EZ), (KQ), and (CD) are similar and given in form (since (CD) is given in form), and they are situated about the same diagonal HCD [VI.26].

Thus, the analysis does not start from scratch, but supposes both that the *schema* can be completed, and all its consequences. Those are:

1 (EZ) is given in form (being one of the three similar parallelograms in the *schema*)
2 and in magnitude [Dt 52, since ED is given in magnitude].
3 (AC) equals the gnomon (because (EC) = (CZ) [I.43] and (DK) = (AK)).
4 The remainder (KQ) is given in magnitude [Dt 4] and in form [VI.24].

The important Dt 55 ensures that the sides of (KQ) are given in magnitude, and therefore EB (equal to KC). Since ED is given, the remainder BD is also given [Dt 4] and therefore BC, since the ratio BD:CD is given because (CD) is given in form. Which was to be proved.

◆

Θε 59 ' Ἐὰν δοθὲν παρὰ δοθεῖσαν παραβληθῇ ὑπερβάλλον εἴδει δεδομένῳ, δέδοται τὰ πλάτη τῆς ὑπερβολῆς.

δοθὲν γὰρ τὸ ΑΒ παρὰ δοθεῖσαν τὴν ΑΓ παραβεβλήσθω ὑπερβάλλον εἴδει δεδομένῳ τῷ ΓΒ· λέγω, ὅτι δοθεῖσά ἐστιν ἑκατέρα τῶν ΘΓ, ΓΕ.
τετμήσθω γὰρ δίχα ἡ ΔΕ κατὰ τὸ Ζ σημεῖον, καὶ ἀναγεγράφθω ἀπὸ τῆς ΕΖ τῷ ΓΒ ὅμοιον καὶ ὁμοίως κείμενον τὸ ΖΗ· περὶ τὴν αὐτὴν ἄρα διάμετρόν ἐστι τὸ ΖΗ τῷ ΓΒ. ἤχθω αὐτῶν διάμετρος ἡ ΘΕΜ, καὶ καταγεγράφθω τὸ σχῆμα. καὶ ἐπεὶ ὅμοιόν ἐστι τὸ ΓΒ τῷ ΖΗ, δέδοται δὲ τὸ ΓΒ τῷ εἴδει, δέδοται ἄρα καὶ τὸ ΖΗ τῷ εἴδει· καὶ ἀναγέγραπται ἀπὸ δεδομένης εὐθείας τῆς ΖΕ· δοθὲν ἄρα ἐστὶ τὸ ΖΗ τῷ μεγέθει. ἔστι δὲ καὶ τὸ ΑΒ δοθέν· δοθέντα ἄρα ἐστὶ τὰ ΑΒ, ΖΗ. καί ἐστιν ἴσα τῷ ΚΛ· δοθὲν ἄρα ἐστὶ τὸ ΚΛ τῷ μεγέθει. ἔστι δὲ καὶ τῷ εἴδει· ὅμοιον γάρ ἐστι τῷ ΓΒ· τοῦ ΚΛ ἄρα αἱ πλευραὶ δεδομέναι εἰσίν· δοθεῖσα ἄρα ἐστὶν ἡ ΚΘ, ὧν ἡ ΚΓ δοθεῖσά ἐστιν· ἴση γάρ ἐστι τῇ ΕΖ· λοιπὴ ἄρα ἡ ΓΘ ἐστι δοθεῖσα· καὶ λόγον ἔχει πρὸς τὴν ΘΒ δοθέντα· δοθεῖσα ἄρα καὶ ἡ ΘΒ.

Dt 59. *If a given* [area] *be applied to a given* [straight line] *exceeding by a figure given in form, the length and width of the excess are given.*

For, let the given area (AB) have been applied to the given straight line AC exceeding by the given form (CB). I say that each of QC, CE is given.

For, let DE have been bisected at the point Z [I.10], and on EZ let (ZH) be described similar and similarly situated to (CB) [VI.18]; then (ZH) is about the same diagonal as (CB) [VI.26].

Let their diagonal QEM have been drawn and let the *schema* have been described. And since (CB) is similar to (ZH), and (CB) is given in form, (ZH) is given in form [Dt 3*B] ; and it is described on the given line ZE; therefore (ZH) is given in magnitude [Dt 52].

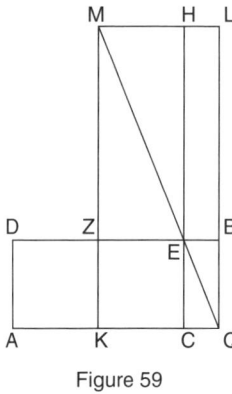

Figure 59

And (AB) is also given; therefore (AB)+(ZH) is given, and it is equal to (KL) [I.36, 43]; therefore (KL) is given in magnitude.

But it is also given in form, for it is similar to (CB) [VI.24, Def. 3*B]; therefore the sides of (KL) are given [Dt 55]; therefore KQ is given, [part] of which [121] KC is given, for it is equal to EZ [I.34].

Therefore the remainder CQ is given; and it has a given ratio to QB [Def. 3]; therefore QB is also given. ∎

Remarks: My comments on the hyperbolical application cannot help being a repetition of those on the elliptical application. In Dt 59 the construction from VI.29 is easily recognized: the applied parallelogram AB exceeds the given line AC by the given form (CB). This time the gnomon is obtained by subtracting the given form from (AB), bisecting the given line DE (= AC) and describing the *schema* on KQ, using the diagonal QE produced. The *schema* implies that

1   (ZH) is given in form
2   and in magnitude (Dt 52, since ZE is given in magnitude).
6   (AB) + (ZH) is given in magnitude.
7   (AB) + (ZH) = (KL) because (AB) equals the gnomon; so
8   (KL) is given in magnitude and in form (being similar to (CB)).

Dt 55 ensures that the sides of (KL) are given, *i.e.* KQ is given. Since KC (= AC/2) is given, CQ is given by subtraction [Dt 4], and therefore also QB, because the ratio CQ:QB is given. Which was to be proved.

Clemens Thaer, of course, makes algebra of Dt 58 and 59 with square root canopy and fractions and the rest of it. The Greek text has nothing of that, and I can only re-

---

[121] The text has the plural, KQ being considered as the sum of KC and CQ.

peat what was said about calculation and trigonometry in the comments on Dt 57.

We shall meet the applications again in Dt 84–85.

◆

Θε 60 'Εὰν παραλληλόγραμμον δεδομένον τῷ εἴδει καὶ τῷ μεγέθει δεδομένῳ γνώμονι αὐξηθῇ ἢ μειωθῇ, δέδοται τὰ πλάτη τοῦ γνώμονος.

παραλληλόγραμμον γὰρ τὸ ΑΒ δεδομένον τῷ εἴδει καὶ τῷ μεγέθει ηὐξήσθω πρότερον δεδομένῳ γνώμονι τῷ ΕΓΒΔΖΗ· λέγω, ὅτι δοθεῖσά ἐστιν ἑκατέρα τῶν ΓΕ, ΔΖ.

ἐπεὶ γὰρ δοθέν ἐστι τὸ ΑΒ, ἔστι δὲ καὶ ὁ Ε{Γ}ΒΔΗΖ γνώμων δοθείς, καὶ ὅλον ἄρα τὸ ΑΗ δοθέν ἐστιν· ἀλλὰ καὶ τῷ εἴδει· ὅμοιον γάρ ἐστι τῷ ΑΒ· τοῦ ΑΗ ἄρα δεδομέναι εἰσὶν αἱ πλευραί· δοθεῖσα ἄρα ἐστὶν ἑκατέρα τῶν ΑΕ, ΑΖ. ἔστι δὲ καὶ ἑκατέρα τῶν ΓΑ, ΑΔ δοθεῖσα· λοιπὴ ἄρα ἑκατέρα τῶν ΕΓ, ΔΖ ἐστι δοθεῖσα.

πάλιν δὴ παραλληλόγραμμον τὸ ΑΗ δεδομένον τῷ εἴδει καὶ τῷ μεγέθει μεμειώσθω δεδομένῳ γνώμονι τῷ ΕΓΒΔΖΗ· λέγω, ὅτι δοθεῖσά ἐστιν ἑκατέρα τῶν ΓΕ, ΔΖ. ἐπεὶ γὰρ δοθέν ἐστι τὸ ΑΗ, οὗ ὁ ΕΓΒΔΖΗ γνώμων δοθείς ἐστιν, λοιπὸν ἄρα τὸ ΑΒ δοθέν ἐστιν· ἀλλὰ καὶ τῷ εἴδει· τοῦ ΑΒ ἄρα αἱ πλευραὶ δεδομέναι εἰσίν· δοθεῖσα ἄρα ἐστὶν ἑκατέρα τῶν ΓΑ, ΑΔ. ἔστι δὲ καὶ ἑκατέρα τῶν ΕΑ, ΑΖ δοθεῖσα· καὶ λοιπὴ ἄρα ἑκατέρα τῶν ΕΓ, ΔΖ δοθεῖσά ἐστιν.

Dt 60. *If a parallelogram given in form and in magnitude be augmented or diminished by a given gnomon, the length and width of the gnomon are given.*

[1] Let the parallelogram (AB), given in form and in magnitude, have been first augmented by the given gnomon ECBDZH; I say that each of CE, DZ are given.

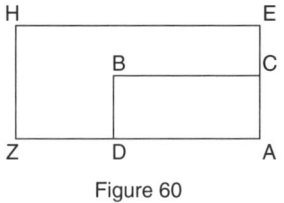
Figure 60

Since (AB) is given, and the gnomon ECBDZH is given [in magnitude], therefore the whole [parallelogram] (AH) is given [in magnitude, Dt 3]; but also in form, for it is similar to (AB) [II.Def. 2 and VI.24]; therefore the sides of (AH) are given [Dt 55]; therefore each of AE, AZ is given. And each of CA, AD is given; therefore each of EC, DZ is given [Dt 4].

[2] Again, let the parallelogram (AH), given in form and in magnitude, have been diminished by the given gnomon ECBDZH; I say that each of CE, DZ are given.

Since (AH) is given, and the gnomon ECBDZH is given, therefore the remaining [parallelogram] (AB) is given [Dt 4]; but also in form; therefore the sides of (AB) are given [Dt 55]; therefore each of CA, AD is given.

And each of EA, AZ is given; therefore each of EC, DZ is given [Dt 4]. ∎

Remarks: Nowhere in the *Data* or the *Elements* do we find any definition of the terms *be augmented or diminished by a gnomon* (γνώμονι αὐξηθῇ ἢ μειωθῇ, aorist passive subjunctives of αὐξάνω and μειόω); it can be inferred from *Elements* II def.2 (Def. 58*D above) together with VI.24, and should be something like

If a parallelogram is augmented or diminished by a gnomon, the resulting parallelogram is similar to the original one.

This is quite in accordance with Heron of Alexandria, who in his definition 58 (probably following an early source) defines a *gnomon in general* as any figure which, when added to any figure whatever, makes the whole figure similar to that to which it is added.

Nor are the *widths* (πλάτη) of a gnomon well defined, but it appears from Dt 60 what is meant. From our definition of *gnomon* 58*D the *widths* of a gnomon can be defined as the *sides of the parallelogram that is part of the gnomon but not a complement (parapleroma)*, in the corner of the gnomon ((HB) in figure 60). Since that parallelogram is around the same diagonal as the one that is 'grasped' by the gnomon, it is similar to that one, and its sides are in the same ratio as the one contained. (The latter property is not applied in Dt 60, although it might have been.) In figure 60 the widths of the gnomon ZDBCEH are (equal to) the (unseen) sides of the parallelogram BH, equal to EC and ZD respectively.

That the gnomon is given must mean that it is given *in magnitude*; by definition the gnomon has a given form, since it produces a figure similar to the given one. So when he says that it is given, he means 'in magnitude'.

Once these supplements are made, Dt 60 is easy to follow: we are in the *Elements* VI.28–29; a rectilineal figure, given in magnitude, is turned into a gnomon and (in VI.28) subtracted from or (in VI.29) added to the figure, given in form and magnitude, which is described on half the given line segment. What Dt 60 calls 'the dimensions' (τὰ πλάτη) of the gnomon, are the 'dimensions' of the defect or excess from Dt 58 and 59. So the three propositions nicely ensure the givenness, that is uniqueness of the segments needed for elliptical and hyperbolical application of area.

As ever, I do not see any textual basis for Clemens Thaer's formulae,

If $C = \lambda(a+x)(\kappa a+\kappa x) - \lambda a \kappa a$, then $x = \text{sqrt}\{(\lambda a \kappa a + C)/\lambda \kappa\} - a$.

If $C = \lambda a \kappa a - \lambda(a-x)(\kappa a - \kappa x)$, then $x = a - \text{sqrt}\{(\lambda a \kappa a - C)/\lambda \kappa\}$.

The text has none of that. All it says is that the output (the length and width of the gnomon) are as given as the input (the sides of the parallelogram and the area of the gnomon), that is: they are unique and can be 'provided'. Calculation had to wait for the invention of trigonometry.

◆

CHAPTER 8. APPLICATION OF AREAS I. (57-62)

Θε 61 ' Ἐὰν δεδομένου τῷ εἴδει εἴδους παρὰ μίαν τῶν πλευρῶν παραλληλόγραμμον χωρίον παραβληθῇ ἐν δεδομένῃ γωνίᾳ, ἔχῃ δὲ τὸ εἶδος πρὸς τὸ παραλληλόγραμμον λόγον δεδομένον, δέδοται τὸ παραλληλόγραμμον τῷ εἴδει.

δεδομένου γὰρ τῷ εἴδει εἴδους τοῦ ΑΖΓΒ παρὰ μίαν τῶν πλευρῶν τὴν ΓΒ παραλληλόγραμμον χωρίον παραβεβλήσθω τὸ ΓΔ ἐν δεδομένῃ γωνίᾳ τῇ ὑπὸ τῶν ΛΓΒ, λόγος δὲ ἔστω τοῦ ΑΓΒ εἴδους πρὸς τὸ ΓΔ παραλληλόγραμμον δοθείς· λέγω, ὅτι δέδοται τὸ ΓΔ τῷ εἴδει.

ἤχθω γὰρ διὰ μὲν τοῦ Β τῇ ΖΓ παράλληλος ἡ ΒΗ, διὰ δὲ τοῦ Ζ τῇ ΓΒ παράλληλος ἡ ΖΗ, καὶ διήχθωσαν αἱ ΖΓ, ΗΒ ἐπὶ τὰ Θ, Κ σημεῖα. ἐπεὶ δοθεῖσά ἐστιν ἡ ὑπὸ τῶν ΖΓΒ γωνία καὶ λόγος ἐστὶ τῆς ΖΓ πρὸς τὴν ΓΒ δοθείς, δοθὲν ἄρα τὸ ΖΒ παραλληλόγραμμον τῷ εἴδει. δέδοται δὲ τῷ εἴδει τὸ ΑΖΒ εἶδος. καὶ ἀναγέγραπται ἀπὸ τῆς αὐτῆς εὐθείας τῆς ΓΒ· λόγος ἄρα ἐστὶ τοῦ ΑΒ εἴδους πρὸς τὸ ΖΒ παραλληλόγραμμον δοθείς. τοῦ δὲ ΖΒ πρὸς τὸ ΓΔ λόγος ἐστὶ δοθείς, ἐπειδὴ καὶ τοῦ ΑΒ πρὸς τὸ ΓΔ ὑπόκειται· ἴσον δὲ τὸ ΓΔ τῷ ΚΒ· λόγος ἄρα καὶ τοῦ ΚΒ πρὸς τὸ ΓΗ ἐστι δοθείς· ὥστε καὶ τῆς ΖΓ πρὸς τὴν ΓΚ λόγος ἐστὶ δοθείς. τῆς δὲ ΖΓ πρὸς τὴν ΓΒ λόγος ἐστὶ δοθείς· καὶ τῆς ΒΓ ἄρα πρὸς τὴν ΓΚ λόγος ἐστὶ δοθείς. καὶ ἐπεὶ δοθεῖσά ἐστιν ἡ ὑπὸ τῶν ΖΓΒ γωνία, καὶ ἡ ἐφεξῆς ἄρα ἡ ὑπὸ τῶν ΒΓΚ ἐστι δοθεῖσα. ἔστι δὲ καὶ ἡ ὑπὸ τῶν ΒΓΛ δοθεῖσα· καὶ λοιπὴ ἄρα ἡ ὑπὸ τῶν ΛΓΚ δοθεῖσά ἐστιν. ἔστι δὲ καὶ ἡ ὑπὸ ΛΚΓ γωνία δοθεῖσα· ἴση γὰρ τῇ ὑπὸ ΚΓΒ· λοιπὴ ἄρα ἡ ὑπὸ ΓΛΚ ἐστι δοθεῖσα· δέδοται ἄρα τὸ ΛΓΚ τρίγωνον τῷ εἴδει· λόγος ἄρα ἐστὶ τῆς ΛΓ πρὸς τὴν ΓΚ δοθείς. τῆς δὲ ΚΓ πρὸς τὴν ΒΓ λόγος ἐστὶ δοθείς· καὶ τῆς ΛΓ ἄρα πρὸς τὴν ΓΒ λόγος ἐστὶ δοθείς. καί ἐστι δοθεῖσα ἡ ὑπὸ τῶν ΛΓΒ γωνία· δέδοται ἄρα τὸ ΓΔ παραλληλόγραμμον τῷ εἴδει.

Dt 61.   *If to one side of a form given in form a parallelogrammic area be applied in a given angle, and [if] the form have a given ratio to the parallelogram, the parallelogram is given in form.*

For, to one side CB of the form AZCB, given in form, let the parallelogrammic area (CD) have been applied in the given angle LCB, and let the ratio of the form ACB to the parallelogram (CD) be given; I say that (CD) is given in form.

For, through B let BH have been drawn parallel to ZC, and through Z ZH parallel to ZB, and let ZC, HB have been drawn through to the points Q, K.

Since △ZCB is given and the ratio ZC:CB is given, therefore the parallelogram (ZB) is given in form.

And the form AZB is given in form, and they have been drawn on the same straight line CB; therefore the ratio of the form (A{Z}B) to the parallelogram (ZB) is given [Dt 49].

The ratio (ZB):(CD) is given, because the ratio (A{Z}B):(CD) is supposed to [be given]; and (CD) = (KB) [I.35]; therefore the ratio (KB):(CH) is given; so the ratio ZC:CK is also given [VI.1].

And the ratio ZC:CB is given; therefore the ratio BC:CK is given [Dt 8].

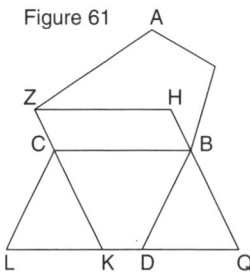

Figure 61

And since ∠ZCB is given, the adjacent ∠BCK is given [I.13, Dt 4].

And ∠BCL is given; therefore the remainder ∠LCK is given.

∠LKC is also given, for it is equal to KCB [I.29]; therefore the third ∠CLK is given [I.32, Dt 3, 4]; therefore ∆LCK is given in form [Dt 40]; therefore the ratio LC:CK is given.

And the ratio KC:BC is given; therefore the ratio LC:CB is given. And ∠LCB is given; therefore the parallelogram (CD) is given in form [Def 3]. ∎

Remarks: Dt 61 is a partial converse of Dt 49 if one of the rectilineal figures in Dt 49 is a parallelogram. Let there be two rectilineal figures P (parallelogram) and Q (any polygon) on one and the same line b. We remember that *frm.P* means that P has given angles and its sides have given ratios:

Dt 49 If *frm.P* & *frm.Q* then *gvn.P:Q*.
Dt 61 If *frm.Q* & (*P has given angles*) & *gvn.P:Q* then *frm.P* (*i.e.* the sides of P have given ratios).

What this proposition has to do with the theory of application of areas is hard to see; in the *ekthesis* the parallelogram (CD) has been applied parabolically to the common line CB in a given angle [I.44], and that is all. The proof depends heavily on the diagram, although only the segments ZC and CB are relevant for the given form ABCZ. If we give short names to the figures involved, ABCZ = M, CBDL = P, CBHZ = R, CBQK = S, the proof can be abbreviated as follows:

frm.R [because it has given angles, and the ratio ZC:BC is given by hypothesis, since M is given in form]
gvn.M:R [Dt 49]
gvn.P:M & gvn.M:R => gvn.P:R [Dt 8]
P = S [I.35], ∴ gvn.S:R
gvn.ZC:CK [VI.1], ∴ [from hypothesis] gvn.BC:CK [Dt 8]
frm. ∆LCK because its angles are given [Dt 40]
∴ gvn.LC:CK, ∴ gvn.LC:CB
Since gvn.∠LCB, therefore frm.P.

CHAPTER 8. APPLICATION OF AREAS I. (57–62)

The proof of the proposition is reduced to proving that *if there be two parallelograms on one and the same line, and one is given in form while the other has given angles, and if the two have a given ratio to each other, then the second parallelogram is given in form.*

◆

Θε 62 ' Ἐὰν δύο εὐθεῖαι πρὸς ἀλλήλας λόγον ἔχωσι δεδομένον καὶ ἀναγραφῇ ἀπὸ μὲν τῆς μιᾶς δεδομένον τῷ εἴδει εἶδος, ἀπὸ δὲ τῆς ἑτέρας χωρίον παραλληλόγραμμον ἐν δεδομένῃ γωνίᾳ, ἔχῃ δὲ τὸ εἶδος πρὸς τὸ παραλληλόγραμμον λόγον δεδομένον, δέδοται τὸ παραλληλόγραμμον τῷ εἴδει.

δύο γὰρ εὐθεῖαι αἱ ΑΒ, ΓΔ πρὸς ἀλλήλας λόγον ἐχέτωσαν δεδομένον, καὶ ἀναγεγράφθω ἀπὸ μὲν τῆς ΑΒ δεδομένον τῷ εἴδει εἶδος τὸ ΑΕΒ, ἀπὸ δὲ τῆς ΓΔ παραλληλόγραμμον τὸ ΔΖ ἐν δεδομένῃ γωνίᾳ τῇ ὑπὸ τῶν ΖΓΔ, λόγος δὲ ἔστω τοῦ ΕΒ εἴδους πρὸς τὸ ΖΔ παραλληλόγραμμον δοθείς· λέγω, ὅτι δέδοται τὸ ΔΖ παραλληλόγραμμον τῷ εἴδει.

ἀναγεγράφθω γὰρ ἀπὸ τῆς ΑΒ τῷ ΔΖ ὅμοιον καὶ ὁμοίως κείμενον παραλληλόγραμμον τὸ ΑΗ. ἐπεὶ λόγος ἐστὶ τῆς ΑΒ πρὸς τὴν ΓΔ δοθείς, καὶ ἀναγέγραπται ἀπὸ τῶν ΑΒ, ΓΔ ὅμοια καὶ ὁμοίως κείμενα εὐθύγραμμα τὰ ΑΗ, ΖΔ, λόγος ἄρα ἐστὶ τοῦ ΑΗ πρὸς τὸ ΖΔ δοθείς. τοῦ δὲ ΖΔ πρὸς τὸ ΕΒ λόγος ἐστὶ δοθείς. καὶ τοῦ ΕΒ ἄρα πρὸς τὸ ΑΗ λόγος ἐστὶ δοθείς. καὶ ἔστι δοθεῖσα ἡ ὑπὸ τῶν ΒΑΘ γωνία· ἴση γάρ ἐστι τῇ ὑπὸ ΖΓΔ. ἐπεὶ οὖν δεδομένου τῷ εἴδει εἴδους τοῦ ΕΒ παρὰ μίαν τῶν πλευρῶν τὴν ΑΒ παραβέβληται τὸ ΑΗ ἐν δεδομένῃ γωνίᾳ τῇ ὑπὸ τῶν ΘΑΒ καὶ λόγος ἐστὶ τοῦ ΕΒ εἴδους πρὸς τὸ ΑΗ παραλληλόγραμμον δοθείς, δέδοται ἄρα τὸ ΑΗ τῷ εἴδει. καί ἐστιν ὅμοιον τῷ ΖΔ· δέδοται ἄρα καὶ τὸ ΖΔ τῷ εἴδει.

Dt 62. *If two straight lines have a given ratio to one another, and on one of them a form, given in form, be described, on the other a parallelogrammic area in a given angle, and [if] the form have a given ratio to the parallelogram, the parallelogram is given in form.*

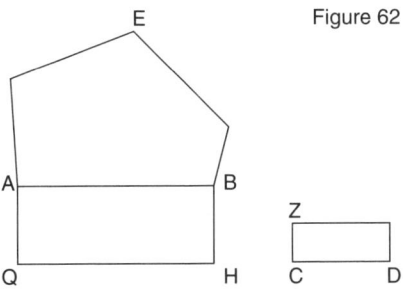

Figure 62

For, let two straight lines AB, CD have a given ratio to one another, and on AB let the form (AEB), given in form, have been described, on CD the parallelogram (DZ) in the given angle ZCD, and let the ratio of the form (EB) to the parallelogram (ZD) be given; I say that the parallelogram (DZ) is given in form.

For, on AB let the parallelogram (AH) have been described similar and similarly situated to the parallelogram (DZ) [VI.18].

Since the ratio AB:CD is given, and similar and similarly situated rectilineal figures (AH), (ZD) have been described on AB, CD, therefore the ratio (AH):(ZD) is given [Dt 50].

And the ratio (ZD):(EB) is given; therefore the ratio (EB):(AH) is given [Dt 8]. And △BAQ is given, for it is equal to △ZCD.

Since, then, to one side AB of the form (EB) that is given in form a parallelogram has been applied in the given angle QAB, and the ratio of the form (EB) to the parallelogram (AH) is given, therefore (AH) is given in form [Dt 61]. And it is similar to (ZD); therefore (ZD) is given in form [Def 3*B]. ∎

Remarks: Dt 62 is an extension of Dt 61: the figures M and P are described on different lines. By an intermediary parallelogram R (= ABHQ) the proof is reduced, via Dt 50, to Dt 61. Its rectangular equivalent is proved in Dt 78.

◆

Chapter 9. Ratio and Angles
Dt 63–67

Θε 63. Ἐὰν τρίγωνον τῷ εἴδει δεδομένον ᾖ, τὸ ἀπὸ ἑκάστης τῶν πλευρῶν αὐτοῦ πρὸς τὸ τρίγωνον λόγον ἕξει δεδομένον.

ἔστω τρίγωνον δεδομένον τῷ εἴδει τὸ ΑΒΓ, καὶ ἀναγεγράφθω ἀπὸ ἑκάστης τῶν πλευρῶν αὐτοῦ τετράγωνα τὰ ΕΒ, ΓΔ, ΓΖ· λέγω, ὅτι ἕκαστον τῶν ΕΒ, ΓΔ, ΓΖ πρὸς τὸ ΑΒΓ τρίγωνον λόγον ἕξει δεδομένον.

ἐπεὶ γὰρ ἀπὸ τῆς αὐτῆς εὐθείας τῆς ΒΓ εὐθύγραμμα δεδομένα τῷ εἴδει ἀναγέγραπται, ἃ ἔτυχεν, τὰ ΑΒΓ, ΓΔ, λόγος ἄρα τοῦ ΑΒΓ πρὸς τὸ ΓΔ δοθείς. διὰ τὰ αὐτὰ δὴ καὶ ἑκατέρου τῶν ΕΒ, ΖΓ πρὸς τὸ ΑΒΓ τρίγωνον λόγος ἐστὶ δοθείς.

Dt 63. *If a triangle be given in form, the squares on each of its sides will have a given ratio to the triangle.*

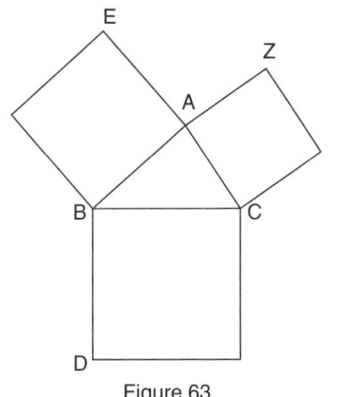

Figure 63

Let ABC be the triangle given in form, and on each of its side let the squares (EB), (CD), (CZ) have been described. I say the each of (EB), (CD), (CZ) will have a given ratio to the triangle ABC.

For, since on the same straight line BC arbitrary rectilineal figures, given in form, ABC, (CD) have been described, the ratio ABC:(CD) is given [Dt 49].

For the same reason the ratios of each of (EB), (ZC) to ABC is given. ∎

Remarks: Dt 63, rather a corollary to Dt 49, is proved merely by reference to that proposition. It could be proved like Dt 48 or Dt 76; in fact it depends on the proof that the ratio of a height to its base is given if the triangle is given in form, which is embedded in Dt 49 and proved in Dt 76. Clemens Thaer thinks that Dt 63 is likely to be spurious, because it is not in the Arabic translation, and Pappus, who counts a total of 90 propositions in contrast to the 94–95 in our manuscripts, seems not to have found it. A nice corollary it is, anyway.

Below I give an alternative proof which puts Dt 63 in line with the other theorems in this chapter. They use three elementary statements, one from the *Elements* I.41, the others derivable from Dt 40. In figure 66, let AC be named 'the base' b, BD 'the height' h, AD 'the projection' p, and let AB be c (as is usual in trigonometry):

1) The triangle is half of the rectangle h,b [I.41]. $T = \frac{1}{2} \sqsubset \sqsupset h,b$.
2) If the angle A is given then the ratio $h:c$ is given.   gvn.$\triangle$A => gvn.$h:c$.
3) If the angle A is given then the ratio $p:c$ is given.   gvn.$\triangle$A => gvn.$p:c$.

All of them invite anachronistic interpretations:

1) That the area of a triangle is half of the product of its base and its height.[122]
2) That sinA is given if the angle A is given.
3) That cosA is given if the angle A is given.

They are anachronistic because 'area', 'sine' and 'cosine' are mathematical 'functions' that presuppose concepts of number and, first and foremost, of *measuring* that are absent from Euclid's geometry.[123] This should be kept in mind when in the following I describe certain statements as having 'equivalent' trigonometrical expressions.

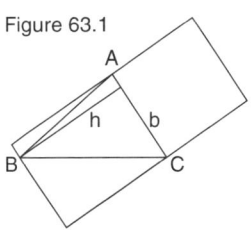

Figure 63.1

An alternative proof of Dt 63, that the ratio of (say) the square on AC to (twice) the triangle is given if the triangle is given in form (that is, if its sides have given ratios to one another and the angles are given), could go as follows, using Dt 76 (figure 63.1):

[I.41] $\square b:2T$ :: $\square b:\sqsubset \sqsupset b,h$
[VI.1] $\square b:2T$ :: $b:h$,

which [Dt 76] is a given ratio.

Since $b:h$ is compounded (see note 63) from $b:c$ and $c:h$, of which the former is given because the triangle is given in form, and the latter depends solely on $\triangle$A (statement 2 above), Dt 8 ensures that $b:h$ is a given ratio, as proved in Dt 76. Trigonometry developed this embryo into a formula for the area of a triangle, $T = \frac{1}{2} bc \sin A$.

Dt 63 is a bit heavier than the following propositions, which need only one given

---

[122] Ian Mueller discusses in detail why that formula does not appear in the *Elements* (Mueller 1981, 152 f.), simply because the concept 'product' of line segments has no place in Euclid's geometry.

[123] Hilbert, in *Grundlagen der Geometrie*, dealt thoroughly with polygons that are equal without being congruent (see e.g. Heath 1956, I.327-28). That the rectangle h,b is twice the triangle means that it is possible to divide the triangle in n parts and the rectangle in $2n$ parts such that any one part of the triangle is congruent with two parts of the rectangle. (This is the easy case of *divisibly-equal, zerlegungsgleiche* figures with $n = 2$).

angle, say △A, to prove that the triangle has a given ratio to a certain rectangle. In Dt 66 (which uses the equivalent of the above mentioned formula 2 T = bc sinA) that rectangle is ⊏⊐b,c. In Dt 64 and 65 it is the excess of the sum of two squares over the third, ▫a + ▫c − ▫b (= 2ac cosB), − known from II.12 and 13.

◆

Θε 64. ' Ἐὰν τρίγωνον ἀμβλεῖαν ἔχῃ γωνίαν δεδομένην, ᾧ μεῖζον δύναται ἡ τὴν ἀμβλεῖαν γωνίαν ὑποτείνουσα πλευρὰ τῶν τὴν ἀμβλεῖαν γωνίαν περιεχουσῶν πλευρῶν, ἐκεῖνο τὸ χωρίον πρὸς τὸ τρίγωνον λόγον ἕξει δεδομένον.

ἔστω τρίγωνον ἀμβλυγώνιον τὸ ΑΒΓ ἀμβλεῖαν γωνίαν ἔχον τὴν ὑπὸ τῶν ΑΒΓ δεδομένην, καὶ διήχθω ἐπ' εὐθείας τῆς ΒΓ εὐθεῖα ἡ ΒΔ, καὶ ἤχθω ἀπὸ τοῦ Α ἐπὶ τὴν ΓΔ κάθετος ἡ ΑΔ· λέγω, ὅτι, ᾧ μεῖζόν ἐστι τὸ ἀπὸ τῆς ΑΓ τῶν ἀπὸ τῶν ΑΒ, ΒΓ, τουτέστι τὸ δὶς ὑπὸ τῶν ΔΒ, ΒΓ, ἐκεῖνο τὸ χωρίον πρὸς τὸ ΑΒΓ τρίγωνον λόγον ἕξει δεδομένον.
ἐπεὶ γὰρ δοθεῖσά ἐστιν ἡ ὑπὸ ΑΒΓ, καὶ ἡ ὑπὸ τῶν ΑΒΔ δοθεῖσά ἐστιν. ἔστι δὲ καὶ ἡ ὑπὸ τῶν ΑΔΒ δοθεῖσα. καὶ λοιπὴ ἄρα ἡ ὑπὸ τῶν ΔΑΒ δοθεῖσά ἐστιν. δέδοται ἄρα τὸ ΔΑΒ τρίγωνον τῷ εἴδει· λόγος ἄρα τῆς ΑΔ πρὸς τὴν ΔΒ δοθείς. καί ἐστιν ὡς ἡ ΑΔ πρὸς τὴν ΔΒ, οὕτως τὸ ὑπὸ τῶν ΑΔ, ΒΓ πρὸς τὸ ὑπὸ τῶν ΔΒ, ΒΓ· ὥστε καὶ τοῦ ὑπὸ τῶν ΔΑ, ΒΓ πρὸς τὸ ὑπὸ τῶν ΔΒ, ΒΓ λόγος ἐστὶ δοθείς· καὶ τοῦ δὶς ὑπὸ τῶν ΔΒ, ΒΓ ἄρα πρὸς τὸ ὑπὸ τῶν ΑΔ, ΒΓ λόγος ἐστὶ δοθείς. ἀλλὰ τοῦ ὑπὸ τῶν ΔΑ, ΒΓ πρὸς τὸ ΑΒΓ τρίγωνον λόγος ἐστὶ δοθείς· καὶ τοῦ δὶς ὑπὸ τῶν ΔΒΓ ἄρα πρὸς τὸ ΑΒΓ τρίγωνον λόγος ἐστὶ δοθείς. καί ἐστι τὸ δὶς ὑπὸ τῶν ΔΒ, ΒΓ, ᾧ μεῖζόν ἐστι τὸ ἀπὸ τῆς ΑΓ τῶν ἀπὸ τῶν ΑΒ, ΒΓ· ἐκεῖνο ἄρα τὸ χωρίον πρὸς τὸ ΑΒΓ τρίγωνον λόγον ἔχει δεδομένον.

Dt 64. *If a triangle have a given obtuse angle, the area by which the square on the side subtending the obtuse angle is greater* [124] *than* [the sum of] *the squares on the sides containing the given angle,* [that area] *will have a given ratio to the triangle.*

Let ABC be the obtuse-angled triangle having the given obtuse angle ABC, and let BD have been produced in a straight line with BC, and from the point A let the perpendicular AD have been drawn to CD; I say that the area by which the square on AC is greater than [the sum of] the squares on AB, BC, that is, twice ⊏⊐DB,BC [II.12], that area will have a given ratio to △ABC.

For, since △ABC is given, △ABD is also given [I.13, Dt 4]. And △ ADB is given, too. Therefore the third △DAB is given [I.32].

---

[124] μεῖζον δύναται literally means *can more than,* the idiomatic expression for one square area being greater than another; the difference is expressed by the dative of the relative pronoun.

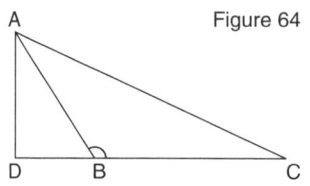

Figure 64

Therefore ∆DAB is given in form [Dt 40].
Therefore the ratio AD:DB is given. And AD:DB ::
⊏⊐AD,BC:⊏⊐DB,BC [VI.1]; and so the ratio
⊏⊐DA,BC:⊏⊐DB,BC is given [Def 2*]; and
therefore the ratio 2⊏⊐DB,BC:⊏⊐AD,BC is given
[Dt 8].
But the ratio ⊏⊐DA,BC:∆ABC is given [I.41];
therefore the ratio 2⊏⊐DBC:∆ABC is given [Dt 8].
And 2⊏⊐DB,BC is the area by which ▫AC is greater than ▫AB + ▫BC;[125] therefore that area has a given ratio to ∆ABC. ∎

Remarks: It is well-known that *Elements* II.12 and 13 prove the equivalent of the cosine law: for any triangle $a^2 + c^2 - 2ac \cos B = b^2$. Dt 64 and 65 prove that the ratio of the 'pythagorean difference' 2⊏⊐DB,BC ($2ac \cos B$) to the triangle depends only on ∠B. (No wonder, if – which we believed in Dt 63 and shall meet again in Dt 66 – the area of the triangle is ½ $ac \sin B$!) The relevant magnitudes are in the text, waiting for trigonometry to be invented. We may list the given ratios as they occur when ∆ABD has been shown to be given in form:

[Dt 40]   gvn. AD:BD $c \sin B : c \cos B$
[VI.1]   gvn. ⊏⊐AD,BC:⊏⊐DB,BC $ac \sin B : ac \cos B$
            gvn. 2⊏⊐DB,BD:⊏⊐AD,BC $2ac \cos B : ac \sin B$
[I.41]   gvn. ⊏⊐AD,BC:∆ABC $ac \sin B : \triangle ABC$ [= 2:1]
[Dt 8]   gvn. 2⊏⊐DB,BD:∆ABC $2ac \cos B : \triangle ABC$

Clemens Thaer calculates further and gives the 'output' as 4 cot B. But the only conclusion drawn in the text is that the relevant (pythagorean) difference has a *given* ratio to the triangle.

◆

Θε 65. ‛Ἐὰν τρίγωνον ὀξεῖαν ἔχῃ γωνίαν δεδομένην, ᾧ ἔλασσον δύναται ἡ τὴν ὀξεῖαν γωνίαν ὑποτείνουσα πλευρὰ τῶν τὴν ὀξεῖαν γωνίαν περιεχουσῶν πλευρῶν, ἐκεῖνο τὸ χωρίον πρὸς τὸ τρίγωνον λόγον ἕξει δεδομένον.

ἔστω τρίγωνον ὀξυγώνιον τὸ ΑΒΓ, ὀξεῖαν ἔχον γωνίαν δεδομένην τὴν ὑπὸ τῶν ΑΒΓ, καὶ ἤχθω ἀπὸ τοῦ Α ἐπὶ τὴν ΒΓ κάθετος ἡ ΑΔ· λέγω, ὅτι, ᾧ ἔλασσόν ἐστι τὸ ἀπὸ τῆς ΑΓ τῶν ἀπὸ τῶν ΑΒ, ΒΓ, τουτέστι τὸ δὶς ὑπὸ τῶν ΓΒ, ΒΔ πρὸς τὸ ΑΒΓ τρίγωνον λόγον ἔχει δεδομένον.

---

[125] McDowell-Sokolik translate wrongly 'greater than the rectangle AB, BC'. The mistake is repeated in Dt 65.

CHAPTER 9. RATIO AND ANGLES. (63–67)

ἐπεὶ γὰρ δοθεῖσά ἐστιν ἡ ὑπὸ τῶν ΑΒΔ γωνία, ἔστι δὲ καὶ ἡ ὑπὸ τῶν ΑΔΒ δοθεῖσα, καὶ λοιπὴ ἄρα ἡ ὑπὸ τῶν ΒΑΔ ἐστι δοθεῖσα· δέδοται ἄρα τὸ ΑΒΔ τρίγωνον τῷ εἴδει· λόγος ἄρα τῆς ΒΔ πρὸς τὴν ΔΑ δοθείς· ὥστε καὶ τοῦ ὑπὸ τῶν ΓΒΔ πρὸς τὸ ὑπὸ τῶν ΓΒ, ΑΔ λόγος ἐστὶ δοθείς· καὶ τοῦ δὶς ὑπὸ τῶν ΓΒ, ΒΔ ἄρα. ἀλλὰ τοῦ ὑπὸ τῶν ΒΓ, ΑΔ πρὸς τὸ ΑΒΓ λόγος ἐστὶ δοθείς· καὶ τοῦ δὶς ὑπὸ τῶν ΓΒ, ΒΔ ἄρα πρὸς τὸ ΑΒΓ τρίγωνον λόγος ἐστὶ δοθείς. καί ἐστι τὸ δὶς ὑπὸ τῶν ΓΒ, ΒΔ, ᾧ ἔλασσόν ἐστι τὸ ἀπὸ τῆς ΑΓ τῶν ἀπὸ τῶν ΑΒ, ΒΓ· ᾧ ἄρα ἔλασσόν ἐστι τὸ ἀπὸ τῆς ΑΓ τῶν ἀπὸ τῶν ΑΒ, ΒΓ, ἐκεῖνο τὸ χωρίον πρὸς τὸ ΑΒΓ τρίγωνον λόγον ἔχει δεδομένον.

Dt 65. *If a triangle have a given acute angle, the area by which the square on the side subtending the acute angle is less than* [the sum of] *the squares on the sides containing the given angle,* [that area] *will have a given ratio to the triangle.*

For, let ABC be the acute-angled triangle having the given acute angle ABC, and from A let the perpendicular AD have been drawn to CD; I say that the area by which ▫AC is less than [the sum of] the squares on AB, BC, that is, twice ⊏⊐CB, BD [II.13], [that area] has a given ratio to △ABC.

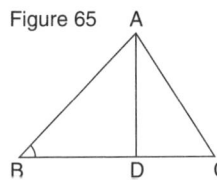

Figure 65

For, since △ABD is given, and △ADB is given, therefore the third angle BAD is given [I.32, Dt 3, 4]; therefore △ABD is given in form [Dt 40].

Therefore the ratio BD:DA is given [Def 3], so that the ratio ⊏⊐CBD : ⊏⊐CB,AD is given [VI.1, Def 2*]; therefore the ratio of 2⊏⊐CB,BD [:⊏⊐AD,BC] is given [Dt 8].

But the ratio ⊏⊐BC,AD : △ABC is given [I.41]; therefore the ratio 2⊏⊐CB,BD : △ABC is given. And 2⊏⊐CB,BD is the area by which ▫AC is less than ▫AB + ▫BC [II.13]; therefore the area by which ▫AC is less than ▫AB + ▫BC, that area has a given ratio to △ABC. ∎

Remarks: While the protasis speaks about a triangle having an acute angle, the ekthesis and the diagram treats an acute-angled triangle, which (according to *Elements* I.def.21) has three acute angles. But only one angle is relevant, so the wording of the protasis is more logical.

◆

Θε 66. Ἐὰν τρίγωνον δεδομένην ἔχῃ γωνίαν, τὸ ὑπὸ τῶν τὴν δεδομένην γωνίαν περιεχουσῶν εὐθειῶν ὀρθογώνιον πρὸς τὸ τρίγωνον λόγον ἔχει δεδομένον.

ἔστω τρίγωνον τὸ ΑΒΓ δεδομένην ἔχον γωνίαν τὴν πρὸς τῷ Α· λέγω, ὅτι τὸ ὑπὸ τῶν ΒΑΓ πρὸς τὸ ΑΒΓ τρίγωνον λόγον ἔχει δεδομένον.

ἤχθω γὰρ ἀπὸ τοῦ Β ἐπὶ τὴν ΑΓ κάθετος ἡ ΒΔ. ἐπεὶ οὖν δοθεῖσά ἐστιν ἡ ὑπὸ τῶν ΒΑΓ γωνία, ἔστι δὲ καὶ ἡ ὑπὸ τῶν ΑΔΒ γωνία δοθεῖσα, καὶ λοιπὴ ἄρα ἡ ὑπὸ ΑΒΔ γωνία δέδοται· δέδοται ἄρα τὸ ΑΒΔ τρίγωνον τῷ εἴδει. λόγος ἄρα ἐστὶ τῆς ΑΒ πρὸς τὴν ΒΔ δοθείς. ὡς δὲ ἡ ΑΒ πρὸς ΒΔ, οὕτως τὸ ὑπὸ τῶν ΒΑΓ πρὸς τὸ ὑπὸ τῶν ΒΔ, ΑΓ· ὥστε καὶ τοῦ ὑπὸ τῶν ΒΑΓ πρὸς τὸ ὑπὸ τῶν ΒΔ, ΑΓ λόγος ἐστὶ δοθείς. τοῦ δὲ ὑπὸ τῶν ΑΓ, ΒΔ πρὸς τὸ ΑΒΓ τρίγωνον λόγος ἐστὶ δοθείς· καὶ τοῦ ὑπὸ τῶν ΒΑΓ ἄρα πρὸς τὸ ΑΒΓ τρίγωνον λόγος ἐστὶ δοθείς.

Dt 66. *If a triangle have a given angle, the rectangle contained by the lines that contain the given angle has a given ratio to the triangle.*

Let ABC be a triangle having a given angle at A; [126] I say that ⊏⊐BAC has a given ratio to △ABC.

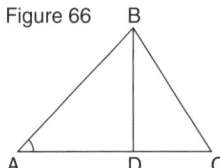
Figure 66

For, let the perpendicular BD from B have been drawn to AC. Now, since ∠BAC is given, and ∠ADB is given, therefore the third angle ABD is also given [I.32, Dt 3, 4]; therefore △ABD is given in form [Dt 40]. Therefore the ratio AB:BD is given [Def. 3].

But AB:BD :: ⊏⊐BAC:⊏⊐BD,AC [VI.1]; so the ratio ⊏⊐BAC:⊏⊐BD,AC is also given [Def 2*]. But the ratio ⊏⊐AC,BD:△ABC is given [I.41]; therefore the ratio ⊏⊐BAC:△ABC is given [Dt 8].   ∎

Remarks: If we put AC = b (the base), BD = h (the height), and AB = c, the theorem first proves that the ratio $c:h$ is given since the right triangle ABD is given in form when ∠A is given. Therefore the ratio ⊏⊐b,c:⊏⊐b,h is given, too, that is the ratio of the said rectangle to twice the triangle. In due course trigonometry would interpret the theorem as showing that the triangle is to the rectangle ⊏⊐bc as sinA to 2, that is, the area of the triangle is ½bc sinA. (cf. remarks on Dt 63) In the context, Dt 66 serves to prove Dt 67

---

[126] Ms. *b* has the usual τὴν ὑπὸ τῶν ΒΑΓ, obviously corrected by a scribe who did not like the "ancient" (Aristolean or pre-Euclidean) notation. Dt 66 may be older than its surroundings. The same notation πρὸς τῷ Α is used in the series Dt 68–75.

CHAPTER 9. RATIO AND ANGLES. (63–67)

Θε 67. Ἐὰν τρίγωνον δεδομένην ἔχῃ γωνίαν, ᾧ μεῖζον δύνανται αἱ τὴν δεδομένην γωνίαν περιέχουσαι πλευραὶ ὡς μία τοῦ ἀπὸ τῆς λοιπῆς, ἐκεῖνο τὸ χωρίον πρὸς τὸ τρίγωνον λόγον ἕξει δεδομένον.

ἔστω τρίγωνον τὸ ΑΒΓ δεδομένην ἔχον γωνίαν τὴν ὑπὸ τῶν ΒΑΓ· λέγω, ὅτι, ᾧ μεῖζόν ἐστι τὸ ἀπὸ συναμφοτέρου τῆς ΒΑΓ τοῦ ἀπὸ τῆς ΒΓ, ἐκεῖνο τὸ χωρίον πρὸς τὸ ΑΒΓ τρίγωνον λόγον ἔχει δεδομένον.

διήχθω γὰρ ἐπ' εὐθείας τῆς ΑΒ εὐθεῖα ἡ ΑΔ, καὶ κείσθω τῇ ΑΓ ἴση ἡ ΑΔ, καὶ ἐπιζευχθεῖσα ἡ ΔΓ διήχθω ἐπὶ τὸ Ε, καὶ ἤχθω διὰ τοῦ Β τῇ ΑΓ παράλληλος ἡ ΒΕ. καὶ ἐπεὶ ἴση ἐστὶν ἡ ΑΔ τῇ ΑΓ, ἴση ἄρα ἐστὶ καὶ ἡ ΔΒ τῇ ΒΕ. καὶ διῆκταί τις ἡ ΒΓ· τὸ ἄρα ὑπὸ τῶν ΔΓΕ μετὰ τοῦ ἀπὸ τῆς ΒΓ ἴσον ἐστὶ τῷ ἀπὸ τῆς ΒΔ. ἴση δὲ ἡ ΔΑ τῇ ΑΓ· τὸ ἄρα ἀπὸ συναμφοτέρου τῆς ΒΑΓ ἴσον ἐστὶ τῷ ὑπὸ τῶν ΔΓΕ μετὰ τοῦ ἀπὸ τῆς ΒΓ· ὥστε τὸ ἀπὸ συναμφοτέρου τῆς ΒΑΓ τοῦ ἀπὸ τῆς ΒΓ μεῖζόν ἐστι τῷ ὑπὸ τῶν ΔΓΕ.

λέγω δή, ὅτι τοῦ ὑπὸ τῶν ΔΓΕ πρὸς τὸ ΑΒΓ τρίγωνον λόγος ἐστὶ δοθείς.

ἐπεὶ γὰρ δοθεῖσά ἐστιν ἡ ὑπὸ τῶν ΒΑΓ γωνία, καὶ ἡ ἐφεξῆς ἄρα ἡ ὑπὸ τῶν ΔΑΓ ἐστι δοθεῖσα. ἔστι δὲ καὶ ἑκατέρα τῶν ὑπὸ τῶν ΑΔΓ, ΔΓΑ δοθεῖσα· ἡμίσειαι γάρ εἰσι τῆς ὑπὸ τῶν ΒΑΓ· [δέδοται γὰρ ἡ ὑπὸ ΒΑΓ·] δέδοται ἄρα τὸ ΔΑΓ τρίγωνον τῷ εἴδει· λόγος ἄρα ἐστὶ τῆς ΔΑ πρὸς τὴν ΔΓ δοθείς· ὥστε καὶ τοῦ ἀπὸ τῆς ΑΔ πρὸς τὸ ἀπὸ τῆς ΔΓ λόγος ἐστὶ δοθείς. καὶ ἐπεὶ ὡς ἡ ΒΑ πρὸς τὴν ΑΔ, οὕτως ἡ ΕΓ πρὸς τὴν ΓΔ, ἀλλ' ὡς μὲν ἡ ΒΑ πρὸς ΑΔ, οὕτως τὸ ὑπὸ ΒΑ, ΑΔ πρὸς τὸ ἀπὸ ΑΔ, ὡς δὲ ἡ ΕΓ πρὸς ΓΔ, οὕτως τὸ ὑπὸ τῶν ΕΓ, ΓΔ πρὸς τὸ ἀπὸ ΓΔ, καὶ ὡς ἄρα τὸ ὑπὸ τῶν ΒΑ, ΑΔ πρὸς τὸ ἀπὸ τῆς ΔΑ, οὕτως τὸ ὑπὸ τῶν ΕΓΔ πρὸς τὸ ἀπὸ τῆς ΓΔ· καὶ ἐναλλάξ, ὡς ἄρα τὸ ὑπὸ τῶν ΒΑΔ πρὸς τὸ ὑπὸ τῶν ΕΓΔ, οὕτως τὸ ἀπὸ τῆς ΑΔ πρὸς τὸ ἀπὸ τῆς ΔΓ. λόγος δὲ τοῦ ἀπὸ τῆς ΑΔ πρὸς τὸ ἀπὸ τῆς ΔΓ δοθείς· λόγος ἄρα καὶ τοῦ ὑπὸ τῶν ΒΑΔ πρὸς τὸ ὑπὸ τῶν ΕΓΔ δοθείς. ἴση δὲ ἡ ΔΑ τῇ ΑΓ· λόγος ἄρα τοῦ ὑπὸ τῶν ΒΑΓ πρὸς τὸ ὑπὸ τῶν ΕΓΔ δοθείς. τοῦ δὲ ὑπὸ τῶν ΒΑΓ πρὸς τὸ ΑΒΓ τρίγωνον λόγος ἐστὶ δοθείς, διὰ τὸ δοθεῖσαν εἶναι τὴν ὑπὸ τῶν ΒΑΓ· καὶ τοῦ ὑπὸ τῶν ΔΓΕ ἄρα πρὸς τὸ ΑΒΓ λόγος ἐστὶ δοθείς. καί ἐστι τὸ ὑπὸ ΔΙ Ε, ᾧ μεῖζόν ἐστι τὸ ἀπὸ συναμφοτέρου τῆς ΒΑΓ τοῦ ἀπὸ τῆς ΒΓ· ᾧ ἄρα μεῖζόν ἐστι τὸ ἀπὸ συναμφοτέρου τῆς ΒΑΓ τοῦ ἀπὸ τῆς ΒΓ, ἐκεῖνο τὸ χωρίον πρὸς τὸ τρίγωνον λόγον ἕξει δεδομένον.

Dt 67. *If a triangle have a given angle, a certain area will have a given ratio to the triangle, namely that area by which the square on the sum of the sides* [127] *containing the given angle is greater than the square on the third side.*

Let ABC be a triangle having the given angle BAC; I say that the area by which □BAC taken as one straight line is greater than □BC, that area has a given ratio to △ABC.

For, let AD be drawn through in a straight line with AB, and let AD be laid out equal to AC, and let DC be joined and drawn through [128] to E, and through B let BE be drawn parallel to AC.

---

[127] *literally* the sides ... as one.

[128] I follow Heath's translation of διήχθω, cf. his note on I.16.

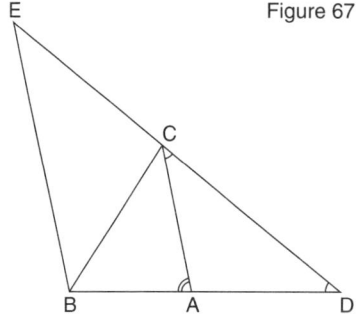

Figure 67

Since AD is equal to AC, therefore DB is equal to BE.[129] And a certain line BC has been drawn through [the isosceles triangle EDB]; therefore ⊏⊐DCE + ▫BC = ▫BD.[130]

And AD is equal to AC; therefore ▫BAC [131] = ⊏⊐(DCE) + ▫BC; so that ▫BAC is greater than ▫BC by ⊏⊐DCE.

Now I say that the ratio ⊏⊐DCE:△ABC is given.

Since ∡BAC is given, the adjacent ∡DAC is given [I.13, Dt 4]. And each of the angles ADC, DCA is given, for they are halves of ∡BAC [I.32, I.5].[132] Therefore △DAC is given in form [Dt 40]; therefore the ratio of DA:DC is given [Def 3], so the ratio ▫AD:▫DC is given [Dt 50 or Dt 24*D converse].

And since BA:AD :: EC:CD [VI.2], but BA:AD :: ⊏⊐BA,AD:▫AD, and EC:CD = ⊏⊐EC,CD:▫CD [VI.1], therefore ⊏⊐BA,AD:▫AD :: ⊏⊐ECD:▫CD; and *enallax* ⊏⊐BAD:⊏⊐ECD :: ▫AD:▫CD [V.16].

But the ratio ▫AD:▫DC is given; therefore the ratio ⊏⊐BAD: ⊏⊐ECD is also given.

And DA is equal to AC; therefore the ratio ⊏⊐BAC:⊏⊐ECD is given.

But the ratio ⊏⊐BAC:△ABC is given, because ∡BAC is given [Dt 66]; therefore the ratio ⊏⊐DCE : △ABC is given [Dt 8].

And ⊏⊐DCE is the area by which the square on BAC, taken as one line, is greater than ▫BC; therefore the area by which ▫BAC is greater than ▫BC, that area will have a given ratio to the triangle. ∎

---

[129] Since the triangles CAD and EBD are similar and both isosceles.

[130] *lemma* proved in scholium no. 133, cf. (Heath 1956) II, 224 on VI.16. See our second *Remark* below.

[131] Read 'the square on the sum of BA+AC taken as one straight line'.

[132] Menge deleted five words, 'the angle BAC is given'. We had that information at the beginning of the paragraph.

Remarks: Whatever was the context and the analysis which inspired Dt 67, it certainly illustrates the tricks that can be performed with the isosceles adjuncts (see p. 131). After constructing the lesser ($\triangle$ACD) and the greater ($\triangle$BED) isosceles adjunct, he asserts (from the blue, because we never learned the proposition of scholium 133) that since BC is an arbitrary transversal line through one vertex of the isosceles triangle EBD, therefore

$$\square BD = \sqsubset\!\!\sqsupset DC,CE + \square BC \text{ [scholium 133]}$$
And $\quad BD = BA + AD = BA + AC$
$\therefore \quad \square BD = \square(BA+AC) = \sqsubset\!\!\sqsupset DC,CE + \square BC.$

Scholium 133, which (as far as I know) is met in no other source,[133] obviously belongs in Book II of the *Elements*.

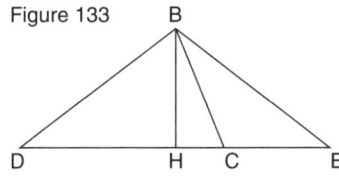

Figure 133

If in an isosceles triangle a straight line is drawn arbitrarily from the apex to the base or the base produced, the square on the segment drawn plus the rectangle contained by the parts of the base is equal to the square on one of the equal sides.

The proof runs easily. If we drop the perpendicular from the apex and apply II.5 on the parts of the base, and finally add the square on the perpendicular, we have

[II.5] $\quad \sqsubset\!\!\sqsupset EC,DC + \square CH \quad = \square HD$
$\quad\quad\quad \sqsubset\!\!\sqsupset EC,DC + \square CH + \square BH = \square HD + \square BH$
[I.47] $\quad \sqsubset\!\!\sqsupset EC,DC + \square BC \quad = \square BD$

Since (in figure 67) $\triangle$CAD is the lesser isosceles adjunct, BD = BA+AC, so we have

$$\sqsubset\!\!\sqsupset EC,DC + \square BC \quad = \square(BA+AC)$$

Dt 67 then proves, using the adjuncts and propositions from Book VI, that the rectangle $\sqsubset\!\!\sqsupset DC,CE$ (which is the excess area mentioned in the protasis) has a given ratio to the triangle.

Dt 67 is enigmatic in several ways: What is its information worth? What kind of analysis preceded it? In which mathematical context? We observe that the vocabulary, *meizon dunatai*, 'is greater in square', is of the sort that is primarily found in book X

---

[133] (Heath 1956) II,73, in his note on III.35 quotes it 'as remarked by Todhunter'.

of the Elements. I feel sure, though I cannot prove it, that this proposition was invented in a context of geometric analysis where isosceles adjuncts were used frequently and scholium 133 would be a useful auxiliary. But at the moment I fail to see the worth of the information that □(AC+AB) − □BC has a given ratio to the triangle ABC. I can see its vague connexion with II.12 and 13, but the main difficulty in our appreciating the statement is that we do not get what we are interested in: namely the value of the ratio, which is not expressed or expressible in any Greek measure, but amounts (in trigonometric idiom) to 4 cot (A/2). [134]

This is the sort of frustration that affects us everywhere in the Data: we get very little information, hardly any 'knowledge' of the Givens. And why not? Probably because 'knowing' geometrical objects was problematic in those days when the concept of 'given' came into being, and the consequences of incommensurability were just being understood. Next to nothing is known of these items, and very little that is worth knowing: length, size, distance, - any of the *rheta*, the attributes that can be spoken of by means of numbers.

◆

---

[134] Put AB = c, BC = a, AC = b, CD = d, CE = e.
Then $ed = (b + c)^2 − a^2 = b^2 + c^2 − a^2 + 2bc = 2bc \cos A + 2bc = 2bc (1+ \cos A)$.
ABC = $\tfrac{1}{2}bc \sin A$.
Therefore ed/ABC = $4(1 + \cos A)/\sin A = 8 \cos^2(A/2) / 2 \sin(A/2) \cos(A/2) = 4 \cot (A/2)$.

Deductive Structure of Dt 56–67
From the *Data*

		Dt 56	57	58	59	60	61	62	63	64	65	66	67
Data													
Def.	1*	.	+	+	+	.	+	.	.	.	.	.	.
	2*	.	+	.	.	.	+	.	.	+	+	+	.
	3	.	+	+	+	+	+	+	.	+	+	+	+
Dt	1	.	+	.	.	.	.	.	.	.	.	.	.
	2	.	+	+	+	.	.	.	.	.	.	.	.
	3	.	+	.	+	+	+	.	.	+	+	+	.
	4	.	+	+	+	+	+	.	.	+	+	+	+
	7	.	.	+	.	.	.	.	.	.	.	.	.
	8	+	+	.	.	.	+	+	.	+	+	+	+
	40	.	+	.	.	.	+	.	.	+	+	+	+
	49	.	.	.	.	.	+	.	+	.	.	.	.
	50	.	.	.	.	.	.	+	.	.	.	.	+
	52	.	+	+	+	.	.	.	.	.	.	.	.
	55	.	.	+	+	+	.	.	.	.	.	.	.
	61	.	.	.	.	.	.	+	.	.	.	.	.
	66	.	.	.	.	.	.	.	.	.	.	.	+
		Dt 56	57	58	59	60	61	62	63	64	65	66	67

Within this group only Dt 61 and 66 are used to establish any of the others. Dt 63 is a corollary to Dt 49 and consequently applies no propositions from the *Elements*. They have done their work in Dt 48 already.

Deductive Structure of Dt 56–67
From the *Elements*

		Dt 56	57	58	59	60	61	62	63	64	65	66	67
post I.	4	.	+	.	.	.	.	.	.	.	.	.	.
	5	.	.	.	.	.	.	.	.	.	.	.	+
	10	.	.	+	+	.	.	.	.	.	.	.	.
	12	.	.	.	.	.	.	.	.	.	.	+	.
	13	.	.	.	.	.	+	.	.	+	.	.	+
	29	.	.	.	.	.	+	.	.	.	.	.	+
	32	.	+	.	.	.	+	.	.	+	+	+	+
	34	+	.	+	+	.	+	.	.	.	.	.	.
	35	.	+	.	.	.	+	.	.	.	.	.	.
	36	.	.	+	+	.	.	.	.	.	.	.	.
	41	.	.	.	.	.	.	.	+	+	+	+	.
	43	.	.	+	+	.	.	.	.	.	.	.	.
	46	.	+	.	.	.	.	.	.	.	.	.	.
II.	def.2	.	.	.	.	+	.	.	.	.	.	.	.
	12	.	.	.	.	.	.	.	+	.	.	.	.
	13	.	.	.	.	.	.	.	.	+	.	.	.
V.	7	+	.	.	.	.	.	.	.	.	.	.	.
	14	.	.	.	.	.	.	.	.	.	.	.	+
	16	.	.	.	.	.	.	.	.	.	.	.	+
VI.	1	+	+	.	.	.	+	.	.	+	+	+	+
	2	.	.	.	.	.	.	.	.	.	.	.	+
	4	.	.	.	.	.	.	.	.	.	.	.	+
	12	+	.	.	.	.	.	.	.	.	.	.	.
	14	+	.	.	.	.	.	.	.	.	.	.	.
	18	.	.	+	+	.	.	+	.	.	.	.	.
	24	.	.	+	+	+	.	.	.	.	.	.	.
	26	.	.	+	+	.	.	.	.	.	.	.	.
		Dt 56	57	58	59	60	61	62	63	64	65	66	67

Chapter 10. Equiangular parallelograms II
Reciprocal proportion. Dt (56) & 68–75

Preliminary remarks: Common to this series of theorems are the following premises, configurations, and conclusions, partly already known from Dt 56*A (Figure 68.1 top):

P and Q are equiangular parallelograms with sides a, b and c, d respectively.
R is a third parallelogram, equiangular with P and Q and equal in area to Q.
R is applied to a with the breadth e in a straight line with b.

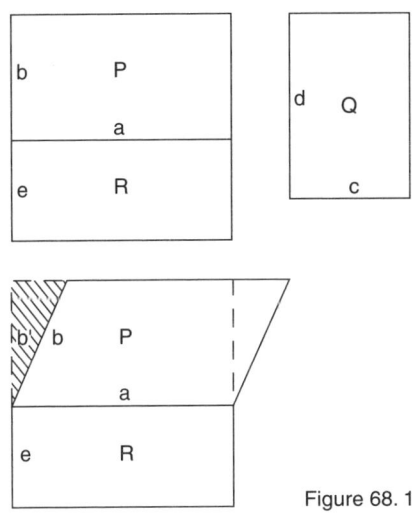

Figure 68. 1

In Dt 56 it was proved, in a very verbose way, that

since R = Q, e is the fourth proportional to a, c, and d (so that $a:c :: d:e$) [VI.14],

and since P:R :: $b:e$ [VI.1], therefore P:Q :: $b:e$ [V.7].
So that, if gvn P:Q then gvn.$b:e$.

If P (a,b) and Q (c,d) are not equiangular, but have given angles, we have (figure 68.1 bottom):
if P′ and R is equiangular with Q and $a:c :: d:e$, then R = Q, and so P:Q :: P:R :: P′:R :: $b′:e$.

Because of the given angles, the shaded triangle is given in form [Dt 40], and therefore gvn.$b′:b$, so that if gvn.$b′:e$ then also gvn.$b:e$ [Dt 8] and *vice versa*. This leads to a couple of cognate conditional statements of interest in the given-idiom:

1) If gvn.P:Q & gvn.$a:c$, then gvn.$b:d$.
2) If gvn.$a:c$ & gvn.$b:d$, then gvn.P:Q.

Dt 68 proves   1).
Dt 69 proves the same for non-equiangular parallelograms with given angles.

Dt 70.1 proves 2), partial converse of Dt 68.

Dt 70.2 proves the same for non-equiangular parallelograms with given angles.

Dt 71 and 72 applies Dt 70 to triangles, considered as half parallelograms.

Dt 73 purports to prove the partial converse of Dt 56: that if $a:c :: d:e$ and gvn.$b:e$ then gvn.$P:Q$. But the proposition is hopelessly corrupt in the manuscripts, as we shall see.

Dt 74.1 proves the same as Dt 56, in almost the same words, Dt 56 being somewhat more verbose and a mixture of two propositions.

Dt 75 applies Dt 74 to triangles.

◆

Θε 68. ' Ἐὰν δύο ἰσογώνια παραλληλόγραμμα πρὸς ἄλληλα λόγον ἔχῃ δεδομένον, καὶ μία πλευρὰ πρὸς μίαν πλευρὰν λόγον ἔχῃ δεδομένον, καὶ ἡ λοιπὴ πλευρὰ πρὸς τὴν λοιπὴν πλευρὰν λόγον ἕξει δεδομένον.

δύο γὰρ Ἰσογώνια παραλληλόγραμμα τὰ ΑΒ, ΓΔ πρὸς ἄλληλα λόγον ἐχέτω δεδομένον, ἐχέτω δὲ καὶ μία πλευρὰ πρὸς μίαν πλευρὰν λόγον δεδομένον, καὶ ἔστω τῆς ΒΕ πρὸς τὴν ΖΔ λόγος δοθείς· λέγω, ὅτι καὶ τῆς ΑΕ πρὸς τὴν ΖΓ λόγος ἐστὶ δοθείς.

παραβεβλήσθω γὰρ παρὰ τὴν ΕΒ τῷ ΓΔ ἴσον παραλληλόγραμμον τὸ ΕΗ, καὶ κείσθω, ὥστε ἐπ' εὐθείας εἶναι τὴν ΑΕ τῇ ΕΘ· ἐπ' εὐθείας ἄρα ἐστὶ καὶ ἡ ΚΒ τῇ ΒΗ.

ἐπεὶ οὖν λόγος ἐστὶ τοῦ ΑΒ πρὸς τὸ ΓΔ δοθείς, ἴσον δὲ τὸ ΓΔ τῷ ΕΗ, λόγος ἄρα τοῦ ΑΒ πρὸς τὸ ΕΗ δοθείς· ὥστε καὶ τῆς ΑΕ πρὸς τὴν ΕΘ λόγος ἐστὶ δοθείς. καὶ ἐπεὶ ἴσον ἐστὶ τὸ ΕΗ τῷ ΓΔ, ἔστι δὲ καὶ ἰσογώνιον, τῶν ΕΗ, ΓΔ ἄρα ἀντιπεπόνθασιν αἱ πλευραὶ αἱ περὶ τὰς ἴσας γωνίας· ἔστιν ἄρα ὡς ἡ ΕΒ πρὸς τὴν ΖΔ, οὕτως ἡ ΓΖ πρὸς τὴν ΕΘ. λόγος δὲ τῆς ΕΒ πρὸς τὴν ΖΔ δοθείς· καὶ τῆς ΓΖ ἄρα πρὸς τὴν ΕΘ λόγος ἐστὶ δοθείς. τῆς δὲ ΕΘ πρὸς τὴν ΑΕ λόγος ἐστὶ δοθείς· καὶ τῆς ΑΕ ἄρα πρὸς τὴν ΓΖ λόγος ἐστὶ δοθείς.

Dt 68. *If two equiangular parallelograms have a given ratio to one another, and one side have a given ratio to one side, the other side will have a given ratio to the other side.*

For, let two equiangular parallelograms (AB) and (CD) have a given ratio to one another, and one side have a given ratio to one side, and let it be BE:ZD. I say that the ratio of AE to ZC is also given.

For, let the parallelogram (EH), equal to the parallelogram (CD), have been applied to the straight line EB, and let it lie such that AE is in a straight line with EQ [I.45]. Then KB is also in a straight line with BH [I.29, 14].

Then, since the ratio (AB):(CD) is given, and (CD) is equal to (EH), therefore the ratio (AB):(EH) is given [V.7, Def 2*]; so the ratio AE:EQ is given [VI.1].

CHAPTER 10. EQUIANGULAR PARALLELOGRAMS II. (68–75)

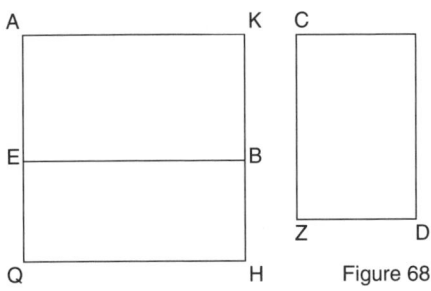

Figure 68

And since (EH) and (CD) are equal and equiangular, therefore the sides about the equal angles of (EH) and (CD) are reciprocally proportional [VI.14]. Therefore EB:ZD :: CZ:EQ.
And the ratio EB:ZD is given; therefore the ratio CZ:EQ is also given [Def 2*].
And the ratio EQ:AE is given; therefore the ratio AE:CZ is also given [Dt 8]. ■

Remarks: In terms of figure 68.1 (with Q = R), Dt 68 proves the same thing as Dt 56 plus a little more: Since P:Q :: b:e, gvn.P:Q & gvn.a:c => gvn.b:d, because b:d is compounded from b:e and e:d, that is from b:e and c:a (see footnote 63); and given ratios compound to a given ratio [Dt 8].

◆

Θε 69. ’ Ἐὰν δύο παραλληλόγραμμα δεδομένας ἔχῃ γωνίας καὶ λόγον πρὸς ἄλληλα ἔχῃ δεδομένον, καὶ μία πλευρὰ πρὸς μίαν πλευρὰν λόγον ἔχῃ δεδομένον, καὶ ἡ λοιπὴ πλευρὰ πρὸς τὴν λοιπὴν πλευρὰν λόγον ἕξει δεδομένον.

δύο γὰρ παραλληλόγραμμα τὰ ΑΒ, ΗΕ δεδομένας ἔχοντα γωνίας τὰς πρὸς τοῖς Δ, Ζ πρὸς ἄλληλα λόγον ἐχέτω δεδομένον, λόγος δὲ ἔστω τῆς ΔΒ πρὸς τὴν ΖΗ δοθείς· λέγω, ὅτι καὶ τῆς ΑΔ πρὸς τὴν ΕΖ λόγος δέδοται.

εἰ μὲν οὖν Ἰσογώνιόν ἐστι τὸ ΑΒ παραλληλόγραμμον τῷ ΕΗ παραλληλογράμμῳ, φανερόν.

εἰ δὲ οὔ, συνεστάτω πρὸς τῇ ΔΒ καὶ τῷ πρὸς αὐτῇ σημείῳ τῷ Δ τῇ ὑπὸ τῶν ΕΖΗ γωνίᾳ ἴση ἡ ὑπὸ τῶν ΒΔΚ, καὶ συμπεπληρώσθω τὸ ΔΛ παραλληλόγραμμον.

ἐπεὶ δοθεῖσά ἐστιν ἑκατέρα τῶν ὑπὸ ΔΑΓ, ΑΚΔ, καὶ λοιπὴ ἄρα ἡ ὑπὸ τῶν ΑΔΚ ἐστὶ δοθεῖσα· δέδοται ἄρα τὸ ΑΔΚ τρίγωνον τῷ εἴδει· λόγος ἄρα ἐστὶ τῆς ΑΔ πρὸς τὴν ΔΚ δοθείς. καὶ ἐπεὶ λόγος ἐστὶ τοῦ ΔΓ πρὸς τὸ ΖΘ δοθείς· ὑπόκειται γάρ· καί ἐστιν ἴσον τὸ ΔΓ τῷ ΔΛ, λόγος ἄρα καὶ τοῦ ΔΛ πρὸς τὸ ΖΘ δοθείς. καί ἐστιν Ἰσογώνιον τὸ ΔΛ τῷ ΖΘ, καὶ λόγος ἐστὶ τοῦ ΔΛ πρὸς τὸ ΕΗ δοθείς, { καί ἐστι τῆς ΔΒ πρὸς τὴν ΖΗ· ὑπόκειται γάρ·} λόγος ἄρα ἐστὶ καὶ τῆς ΔΚ πρὸς τὴν ΕΖ δοθείς. τῆς δὲ ΔΚ πρὸς τὴν ΔΑ λόγος ἐστὶ δοθείς· καὶ τῆς ΑΔ ἄρα πρὸς τὴν ΕΖ λόγος ἐστὶ δοθείς.

Dt 69. *If two parallelograms have given angles and a given ratio to one another, and one side have a given ratio to one side, the other side will have a given ratio to the other side.*

Let two parallelograms (AB), (HE), having given angles at the points D and Z, have a given ratio to one another, and let the ratio DB:ZH be given; I say that the ratio AD:EZ is also given.

Figure 69

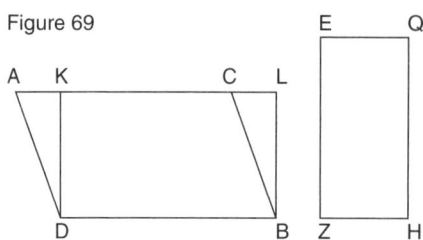

If the parallelogram (AB) is equiangular with (EH), it is evident [Dt 68].[135]

If not, on the straight line DB and at the point D on it let △BDK have been constructed equal to △EZH [I.23], and let the parallelogram (DL) have been completed.

Since, then, each of the angles DAK, AKD is given [I.29, Dt 4], the remaining △ADK is given [I.32, Dt 3, 4]; therefore △ADK is given in form [Dt 40]; therefore the ratio AD:DK is given [Def 3].

And since the ratio (DC):(ZQ) is given (for that is hypothezised), and (DC) = (DL) [I.35], therefore the ratio (DL):(ZQ) is given [V.7, Def 2*].

And (DL) is equiangular with (ZQ), and the ratio (DL):(EH) is given, {and [also] the ratio DB:ZH, for that is hypothezised,[136]} therefore the ratio DK:EZ is given [Dt 68]. And the ratio DK:DA is given; therefore the ratio AD:EZ is given [Dt 8]. ∎

Remarks: Dt 69 proves the same thing as Dt 68, but about parallelograms that are not equiangular but have given angles; this extension of VI.14 is also seen in the following propositions. Has it has any significant mathematical consequences? I do not see them.

Θε 70. 'Εὰν δύο παραλληλογράμμων περὶ ἴσας γωνίας ἢ περὶ ἀνίσους μέν, δεδομένας δέ, αἱ πλευραὶ πρὸς ἀλλήλας λόγον ἔχωσι δεδομένον, καὶ αὐτὰ τὰ παραλληλόγραμμα πρὸς ἄλληλα λόγον ἕξει δεδομένον.

δύο γὰρ παραλληλογράμμων τῶν ΑΒ, ΕΗ περὶ ἴσας γωνίας τὰς πρὸς τοῖς Γ, Ζ ἢ περὶ ἀνίσους μέν, δεδομένας δέ, αἱ πλευραὶ πρὸς ἀλλήλας λόγον ἐχέτωσαν δεδομένον, τουτέστι λόγος ἔστω τῆς μὲν ΑΓ πρὸς τὴν ΕΖ δοθείς, τῆς δὲ ΒΓ πρὸς τὴν ΖΗ· λέγω, ὅτι καὶ τοῦ ΓΔ πρὸς τὸ ΖΘ λόγος ἐστὶ δοθείς.

ἔστω γὰρ ἰσογώνιον τὸ ΓΔ τῷ ΖΘ, καὶ παραβεβλήσθω παρὰ τὴν ΓΒ εὐθεῖαν τῷ ΖΘ παραλληλογράμμῳ ἴσον παραλληλόγραμμον τὸ ΓΜ, καὶ κείσθω ὥστε ἐπ' εὐθείας εἶναι τὴν ΑΓ τῇ ΓΝ· καὶ ἡ ΔΒ ἄρα τῇ ΒΜ ἐστιν ἐπ' εὐθείας. καὶ ἴσον ἐστὶ τὸ ΒΝ τῷ ΖΘ· ἔστι δὲ καὶ ἰσογώνιον· τῶν ΒΝ, ΘΖ ἄρα ἀντιπεπόνθασιν αἱ πλευραὶ αἱ περὶ τὰς ἴσας γωνίας· ἔστιν ἄρα ὡς ἡ ΓΒ πρὸς τὴν ΖΗ, οὕτως ἡ ΖΕ πρὸς τὴν ΓΝ. λόγος δὲ τῆς ΓΒ πρὸς τὴν ΖΗ δοθείς· λόγος ἄρα καὶ τῆς ΕΖ πρὸς τὴν ΓΝ δοθείς. τῆς δὲ ΕΖ πρὸς τὴν

---

[135] 'Evident' (*phaneron*) means that we have just seen it in a diagram.

[136] McDowell-Sokolik translates 'is [the same as the ratio]' – which is obviously wrong. I put the passage in braces, since it is superfluous information and is missing in ms *b*.

CHAPTER 10. EQUIANGULAR PARALLELOGRAMS II. (68–75)

ΑΓ λόγος ἐστὶ δοθείς· καὶ τῆς ΑΓ ἄρα πρὸς τὴν ΓΝ λόγος ἐστὶ δοθείς· ὥστε καὶ τοῦ ΓΔ πρὸς τὸ ΓΜ λόγος ἐστὶ δοθείς. ἔστι δὲ τὸ ΓΜ τῷ ΖΘ ἴσον· λόγος ἄρα καὶ τοῦ ΓΔ πρὸς τὸ ΕΗ δοθείς.

μὴ ἔστω δὴ ἰσογώνιον τὸ ΑΒ τῷ ΖΘ, καὶ συνεστάτω πρὸς τῇ ΒΓ εὐθείᾳ καὶ τῷ πρὸς αὐτῇ σημείῳ τῷ Γ τῇ ὑπὸ τῶν ΕΖΗ γωνίᾳ ἴση γωνία ἡ ὑπὸ ΒΓΚ, καὶ συμπεπληρώσθω τὸ ΓΛ παραλληλόγραμμον. καὶ ἐπεὶ δοθεῖσά ἐστιν ἡ ὑπὸ τῶν ΑΓΒ, ἔστι δὲ καὶ ἡ ὑπὸ ΚΓΒ δοθεῖσα, καὶ λοιπὴ ἄρα ἡ ὑπὸ ΑΓΚ ἐστὶ δοθεῖσα. ἔστι δὲ καὶ ἡ ὑπὸ τῶν ΓΑΚ δοθεῖσα· καὶ λοιπὴ ἄρα ἡ ὑπὸ τῶν ΑΚΓ ἐστὶ δοθεῖσα· δέδοται ἄρα τὸ ΑΓΚ τρίγωνον τῷ εἴδει· λόγος ἄρα ἐστὶ τῆς ΑΓ πρὸς τὴν ΓΚ δοθείς· τῆς δὲ ΑΓ πρὸς τὴν ΕΖ λόγος ἐστὶ δοθείς· καὶ τῆς ΓΚ ἄρα πρὸς τὴν ΕΖ λόγος ἐστὶ δοθείς. ἔστι δὲ καὶ τῆς ΓΒ πρὸς τὴν ΖΗ λόγος δοθείς, καί ἐστιν ἴση ἡ ὑπὸ τῶν ΚΓΒ γωνίᾳ τῇ ὑπὸ τῶν ΕΖΗ· λόγος ἄρα ἐστὶ τοῦ ΓΛ πρὸς τὸ ΖΘ δοθείς. ἴσον δὲ τὸ ΓΛ τῷ ΓΔ· λόγος ἄρα ἐστὶ τοῦ ΓΔ πρὸς τὸ ΖΘ δοθείς.

Dt 70. *If in two parallelograms the sides about equal angles, or about unequal but given angles have a given ratio to one another, the parallelograms themselves will have a given ratio to one another.*

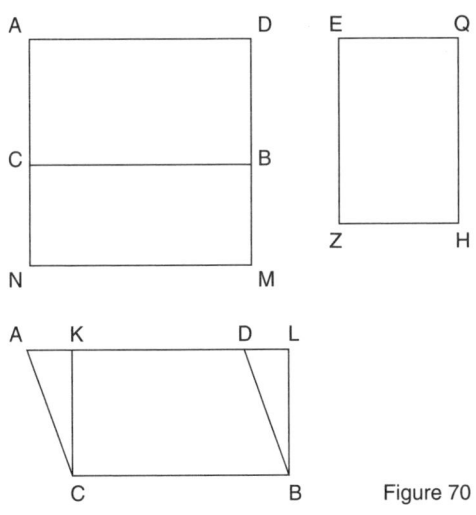

Figure 70

In two parallelograms (AB), (EH) let the sides about the equal angles at the points C, Z, or about unequal but given angles, have a given ratio to one another, that is let the ratios AC:EZ and BC:ZH be given; I say that the ratio (CD):(ZQ) is also given.

[1] (Figure 70 top) Let (CD) be equiangular with (ZQ), and let the parallelogram (CM), equal to the parallelogram (ZQ), have been applied to the straight line CB, and let it lie such that AC is in a straight line with CN; then DB is also in a straight line with BM.

And (BN) and (ZQ) are equal and also equiangular; therefore the sides about the equal angles of (BN), (QZ) are reciprocally proportional [VI.14]; therefore CB:ZH :: ZE:CN; but the ratio CB:ZH is given; therefore the ratio EZ:CN is also given [Def 2*].

But the ratio EZ:AC is given; therefore the ratio AC:CN is given [Dt 8]. So that the ratio (CD):(CM) is given [VI.1, Def 2*]. And (CM) = (ZQ); therefore the ratio (CD):(EH) is given.

[2] (Figure 70 bottom) Then let (AB) be not equiangular with (ZQ), and on the straight line BC and at the point C on it let △BCK have been constructed equal to △EZH [I.23], and let the parallelogram (CL) have been completed.

Since △ACB is given, and △KCB is given, the remaining △ACK is given [Dt 4]. And △CAK is given [I.29, Dt 4]; therefore the remaining △AKC is also given [I.32]; therefore △ACK is given in form [Dt 40]; therefore the ratio AC:CK is given [Def 3]; and the ratio AC:EZ is given; therefore the ratio CK:EZ is given [Dt 8].

And the ratio CB:ZH is given, and △KCB = △EZH.; therefore the ratio (CL):(ZQ) is given [Dt 70, part 1]. And (CL) = (CD); therefore the ratio (CD):(ZQ) is given [V.7, Def 2*]. ∎

Remarks: Dt 70 is the partial converse to Dt 68 and 69: If gvn.a:c & gvn.b:d, then gvn.P:Q. Proof of equiangular case (figure 56.1):
Since Q = R [by construction], $a:c :: d:e$ [VI.14], and therefore gvn.$d:e$.
Since gvn.$b:d$, therefore [Dt 8] gvn.$b:e$.
But P:Q :: P:R :: $b:e$, and therefore gvn.P:Q.
The non-equiangular case is reduced to the equiangular case as in Dt 69.

◆

Θε 71. ' Ἐὰν δύο τριγώνων περὶ ἴσας γωνίας ἢ περὶ ἀνίσους μέν, δεδομένας δέ, αἱ πλευραὶ πρὸς ἀλλήλας λόγον ἔχωσι δεδομένον, καὶ αὐτὰ τὰ τρίγωνα πρὸς ἄλληλα λόγον ἔχει δεδομένον.

δύο γὰρ τριγώνων τῶν ΑΒΓ, ΔΕΘ περὶ ἴσας γωνίας τὰς πρὸς τοῖς Α, Δ ἢ περὶ ἀνίσους μέν, δεδομένας δέ, αἱ πλευραὶ πρὸς ἀλλήλας λόγον ἐχέτωσαν δεδομένον, καὶ ἔστω λόγος τῆς μὲν ΒΑ πρὸς τὴν ΕΔ δοθείς, τῆς δὲ ΑΓ πρὸς τὴν ΔΘ· λέγω, ὅτι καὶ τοῦ ΑΒΓ τριγώνου λόγος ἐστὶ δοθεὶς πρὸς τὸ ΕΔΘ τρίγωνον.
συμπεπληρώσθω γὰρ τὰ ΑΗ, ΔΖ παραλληλόγραμμα. ἐπεὶ οὖν δύο παραλληλογράμμων τῶν ΑΗ, ΔΖ περὶ τὰς ἴσας γωνίας ἢ περὶ ἀνίσους μέν, δεδομένας δὲ τὰς πρὸς τοῖς Α, Δ αἱ πλευραὶ πρὸς ἀλλήλας λόγον ἔχουσι δεδομένον, καὶ τὰ παραλληλόγραμμα λόγον ἕξει δεδομένον πρὸς ἄλληλα· λόγος ἄρα τοῦ ΑΗ πρὸς τὸ ΔΖ δοθείς. καί ἐστι τοῦ μὲν ΑΗ ἥμισυ τὸ ΑΒΓ τρίγωνον, τοῦ δὲ ΔΖ τὸ ΔΕΘ· λόγος ἄρα τοῦ ΑΒΓ πρὸς τὸ ΔΕΘ τρίγωνον δοθείς.

Dt 71. *If in two triangles the sides about equal angles, or about unequal but given angles, have a given ratio to one another, the triangles themselves have a given ratio to one another.*

Let there be two triangles ABC, DEQ, and let the sides about the equal angles at A, D, or unequal but given angles, have a given ratio to one another, and let the ratios BA:ED and AC:DQ be given. I say that the ratio △ABC:△EDQ is given.

CHAPTER 10. EQUIANGULAR PARALLELOGRAMS II. (68–75)

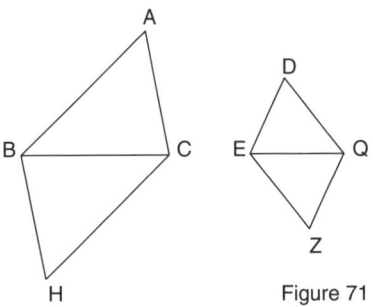

Figure 71

Let the parallelograms (AH), (DZ) have been completed.[137] And since in two parallelograms (AH), (DZ) the sides about the equal angles, or unequal but given angles at A, D, have a given ratio to one another, the parallelograms will have a given ratio to one another [Dt 70]; therefore the ratio (AH):(DZ) is given.

But △ABC is half of the parallelogram (AH), and △DEQ is half of (DZ) [I.34]; therefore the ratio △ABC:△DEQ is given [V.15, Def 2*].

Remarks: Dt 71 proves Dt 70 for triangles, which are doubled into parallelograms (reminiscent of I.37–38) for which Dt 70 holds.

◆

Θε 72. ʼΕὰν δύο τριγώνων αἵ τε βάσεις ἐν δεδομένῳ λόγῳ ὦσι καὶ αἱ ἐπ' αὐτὰς ἠγμέναι ἀπὸ τῶν γωνιῶν ἤτοι ἴσας γωνίας ποιοῦσαι ἢ ἀνίσους μέν, δεδομένας δέ, τὰς πρὸς ταῖς βάσεσιν, καὶ αὐτὰ τὰ τρίγωνα πρὸς ἄλληλα λόγον ἕξει δεδομένον.

ἔστω δύο τρίγωνα τὰ ΑΒΓ, ΔΕΖ, καὶ ἤχθωσαν αἱ ΑΗ, ΔΘ ἤτοι ἴσας γωνίας ποιοῦσαι τὰς ὑπὸ τῶν ΑΗΓ, ΔΘΖ ἢ ἀνίσους μέν, δεδομένας δέ, καὶ ἔστω λόγος τῆς μὲν ΒΓ πρὸς ΕΖ δοθείς, τῆς δὲ ΑΗ πρὸς τὴν ΔΘ δοθείς· λέγω, ὅτι καὶ τοῦ ΑΒΓ τριγώνου πρὸς τὸ ΔΕΖ τρίγωνον λόγος ἐστὶ δοθείς.

συμπεπληρώσθω γὰρ τὰ ΚΓ, ΛΖ παραλληλόγραμμα. καὶ ἐπεὶ αἱ ὑπὸ τῶν ΑΗΓ, ΔΘΖ γωνίαι ἤτοι ἴσαι εἰσίν, ἢ ἄνισοι μέν, δεδομέναι δέ, ἴση δὲ ἡ μὲν ὑπὸ τῶν ΑΗΓ τῇ ὑπὸ ΚΒΓ, ἡ δὲ ὑπὸ τῶν ΔΘΖ τῇ ὑπὸ τῶν ΛΕΖ, καὶ αἱ πρὸς τοῖς Β, Ε ἄρα γωνίαι ἤτοι ἴσαι εἰσὶν ἢ ἄνισοι μέν, δεδομέναι δέ. καὶ ἐπεὶ λόγος ἐστὶ τῆς ΑΗ πρὸς τὴν ΔΘ δοθείς, ἴση δὲ ἡ μὲν ΑΗ τῇ ΚΒ, ἡ δὲ ΔΘ τῇ ΛΕ, λόγος ἄρα ἐστὶ καὶ τῆς ΚΒ πρὸς τὴν ΛΕ δοθείς. ἔστι δὲ καὶ τῆς ΒΓ πρὸς τὴν ΕΖ λόγος δοθείς, καὶ αἱ πρὸς τοῖς Β, Ε σημείοις γωνίαι ἤτοι ἴσαι εἰσίν, ἢ ἄνισοι μέν, δεδομέναι δέ· καὶ τοῦ ΓΚ ἄρα παραλληλογράμμου πρὸς τὸ ΛΖ παραλληλόγραμμον λόγος ἐστὶ δοθείς· ὥστε καὶ τοῦ ΑΒΓ τριγώνου πρὸς τὸ ΔΕΖ τρίγωνον λόγος ἐστὶ δοθείς.

Dt 72. *If in two triangles the bases be in a given ratio, and the straight lines drawn to them from the [subtending] angles making, at the bases, either equal angles, or unequal but given angles,* [if those lines] *be in a given ratio, the triangles themselves will have a given ratio to one another.*

---

[137] We are supposed to know how to 'complete' a triangle into a parallelogram. The diagram saves us, as so often in Greek geometry. See next footnote.

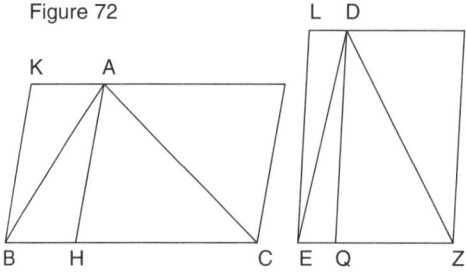

Figure 72

Let there be two triangles ABC, DEZ, and let the straight lines AH, DQ have been drawn making either the equal angles AHC, DQZ, or unequal but given angles, and let the ratios BC:EZ and AH:DQ be given. I say that the ratio △ABC:△DEZ is given.

Let the parallelograms (KC), (LZ) have been completed.[138]

And since the angles AHC, DQZ are either equal or unequal but given, and ∠AHC = ∠KBC, ∠DQZ = ∠LEZ [I.29], therefore the angles at B, E are either equal or unequal but given.

And since the ratio AH:DQ is given, and AH = KB, DQ = LE, therefore the ratio KB:LE is given [V.7, Def 2].

And the ratio BC:EZ is given [by hypothesis], and the angles at the points B, E are either equal or unequal but given; therefore the ratio (CK):(LZ) is given [Dt 70]; so the ratio △ABC:△DEZ is given [I.41, V.15]. ∎

Remarks: Dt 72 is also about triangles, the bases of which are in a given ratio while transversals from the top angles making given angles with the bases are also in a given ratio. The proof is reduced to Dt 70.

◆

Θε 73. ’Εὰν δύο παραλληλογράμμων περὶ ἴσας γωνίας ἢ περὶ ἀνίσους μέν, δεδομένας δέ, αἱ πλευραὶ οὕτως ἔχωσιν, ὥστε εἶναι ὡς τὴν τοῦ πρώτου πλευρὰν πρὸς τὴν τοῦ δευτέρου πλευράν, οὕτως τὴν λοιπὴν τοῦ δευτέρου πλευρὰν πρὸς ἄλλην τινά, ἔχῃ δὲ ἡ λοιπὴ τοῦ πρώτου πλευρὰ πρὸς αὐτὴν λόγον δεδομένον, καὶ αὐτὰ τὰ παραλληλόγραμμα πρὸς ἄλληλα λόγον ἕξει δεδομένον.

δύο γὰρ παραλληλογράμμων τῶν ΑΒ, ΕΗ περὶ ἴσας γωνίας ἢ περὶ ἀνίσους μέν, δεδομένας δέ, τὰς πρὸς τοῖς Γ, Ζ αἱ πλευραὶ οὕτως ἐχέτωσαν πρὸς ἀλλήλας, ὥστε εἶναι ὡς τὴν ΓΒ πρὸς τὴν ΖΗ, οὕτως τὴν ΕΖ πρὸς τὴν ΓΚ, τῆς δὲ ΑΓ πρὸς τὴν ΓΚ λόγος ἔστω δοθείς· λέγω, ὅτι καὶ τοῦ ΓΔ παραλληλογράμμου πρὸς τὸ ΕΗ παραλληλόγραμμον λόγος ἐστὶ δοθείς.

ἔστω γὰρ πρότερον τὸ ΑΒ τῷ ΕΗ ἰσογώνιον, καὶ παραβεβλήσθω παρὰ τὴν ΒΓ εὐθεῖαν τῷ ΕΗ παραλληλογράμμῳ ἴσον παραλληλόγραμμον τὸ ΓΘ, καὶ κείσθω ὥστε

---

[138] By drawing KB, KA parallel to AH, BC respectively. One would hardly understand the text without a diagram, since the text does not make it clear which lines are parallel. In most cases a triangle is doubled by drawing lines parallel to two sides of the triangle.

CHAPTER 10. EQUIANGULAR PARALLELOGRAMS II. (68–75) 185

ἐπ' εὐθείας εἶναι τὴν ΑΓ τῇ ΚΓ· ἐπ' εὐθείας ἄρα ἐστὶ καὶ ἡ ΔΒ τῇ ΘΒ. καὶ ἐπεὶ ἴσον ἐστὶ τὸ ΓΘ τῷ ΕΗ, ἔστι δὲ καὶ ἰσογώνιον, τῶν ΓΘ, ΕΗ ἄρα ἀντιπεπόνθασιν αἱ πλευραὶ αἱ περὶ τὰς ἴσας γωνίας· ἔστιν ἄρα ὡς ἡ ΓΒ πρὸς τὴν ΖΗ, οὕτως ἡ ΕΖ πρὸς τὴν ΓΚ· ὡς δὲ ἡ ΓΒ πρὸς τὴν ΖΗ, οὕτως ἡ ΕΖ καὶ πρὸς ἣν ἡ ΑΓ λόγον ἔχει δεδομένον· λόγος ἄρα τῆς ΑΓ πρὸς τὴν ΓΚ δοθείς· ὥστε καὶ τοῦ ΑΒ πρὸς τὸ ΓΘ, τουτέστι πρὸς τὸ ΕΗ λόγος ἐστὶ δοθείς.

μὴ ἔστω δὴ ἰσογώνιον, καὶ συνεστάτω πρὸς τῇ ΓΒ εὐθείᾳ καὶ τῷ πρὸς αὐτῇ σημείῳ τῷ Γ τῇ ὑπὸ τῶν ΕΖΗ γωνίᾳ ἴση ἡ ὑπὸ τῶν ΒΓΛ, καὶ συμπεπληρώσθω τὸ ΓΜ παραλληλόγραμμον. ἐπεὶ δοθεῖσά ἐστιν ἑκατέρα τῶν ὑπὸ τῶν ΑΓΒ, ΛΓΒ, καὶ λοιπὴ ἄρα ἡ ὑπὸ τῶν ΑΓΛ ἐστι δοθεῖσα. δέδοται δὲ καὶ ἡ ὑπὸ τῶν ΓΑΛ· καὶ λοιπὴ ἄρα ἡ ὑπὸ ΓΛΑ δέδοται· ὥστε δέδοται τὸ ΑΓΛ τρίγωνον τῷ εἴδει· λόγος ἄρα ἐστὶ τῆς ΑΓ πρὸς τὴν ΓΛ δοθείς. καὶ ἐπεί ἐστιν, ὡς ἡ ΓΒ πρὸς τὴν ΖΗ, οὕτως ἡ ΕΖ πρὸς ἣν ἡ ΑΓ λόγον ἔχει δεδομένον, τῆς δὲ ΑΓ πρὸς τὴν ΓΛ λόγος ἐστὶ δοθείς, ἔστιν ἄρα ὡς ἡ ΓΒ πρὸς τὴν ΖΗ, οὕτως ἡ ΖΕ πρὸς τὴν ΓΛ. καί ἐστιν ἴση ἡ ὑπὸ ΒΓΛ γωνία τῇ ὑπὸ τῶν ΕΖΗ· λόγος ἄρα τοῦ ΓΜ παραλληλογράμμου πρὸς τὸ ΕΗ παραλληλόγραμμον δοθείς. ἴσον δέ ἐστι τὸ ΓΜ τῷ ΓΔ· λόγος ἄρα τοῦ ΓΔ πρὸς τὸ ΕΗ δοθείς.

Dt 73. *If in two parallelograms the sides about equal angles, or about unequal but given angles, be such that, as one side of the first parallelogram is to one side of the second, so is the other side of the second parallelogram to some straight line, and* [if] *the other side of the first parallelogram have a given ratio to that line, then the parallelograms themselves will have a given ratio to one another.*

In two parallelograms (AB), (EH) let the sides about equal angles, or about unequal but given angles, at C, Z, be such that, as CB is to ZH, so is EZ to CX,[139] and let the ratio AC:CX be given. I say that the ratio (CD):(EH) is given.[140]

[1] (Figure 73 top) First, let (AB) be equiangular with (EH), and let the parallelogram (CQ), equal to the parallelogram (EH), have been applied to the straight line BC, and let it lie such that AC is in a straight line with KC; then DB is also in a straight line with QB.

And since (CQ) and (EH) are equal and also equiangular, the sides about the equal angles of (CQ), (EH) are reciprocally proportional; therefore CB:ZH :: EZ:CK. But as CB is to ZH, so is EZ to that line [CX] to which AC has the given ratio; therefore [CK = CX, and] the ratio AC:CK is given; so [VI.1] the ratio (AB):(CQ),

---

[139] The *mss* have CK. Euclid's anticipation of the construction (below) of the parallelogram CK,KQ leads to catastrophe. CX is ἄλλην τινά, 'some other straight line', and should not be named CK. We have renamed it to mend the error. CX does not appear in the figure.

[140] Euclid's off-hand manner of naming his figures is nowhere more conspicuous than here; the parallelogram (CD) is identical with (AB).

that is (AB):(EH) is given.[141]

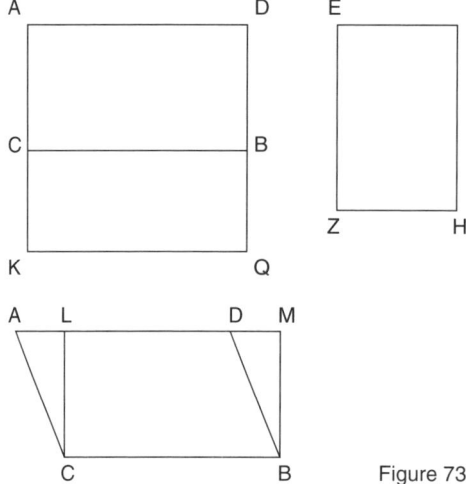

Figure 73

[2] Then let it be not equiangular (figure 73 bottom), and on the straight line CB and at the point C on it let △BCL have been constructed equal to △EZH [I.23], and let the parallelogram (CM) have been completed.

Since each of the angles ACB, LCB is given, the remaining △ACL is given [Dt 4]. And △CAL is given [I.29, Dt 4]; therefore the remaining △CLA is given [I.32, Dt 3, 4]; so △ACL is given in form [Dt 40]; therefore the ratio AC:CL is given [Def 3].

And since [by hypothesis], as CB is to ZH, so is EZ to that line [CX] to which AC has a given ratio, and the ratio AC:CL is given, therefore CB:ZH :: ZE:CL [*nonsense, because (CM) is not equal to (EH) but only equiangular; see my remarks*].

And △BCL = △EZH; therefore the ratio (CM):(EH) is given. And (CM) = (CD) [I.35]; therefore the ratio (CD):(EH) is given [V.7, Def 2*]. ∎

Remarks: The corruption of the first argument (which Menge observed), is quite easily repaired, as I have suggested by introducing the line CX. – The second argument is more complicated; I suggest to replace the first paragraph by imitating the construction from the first argument, again introduce CX, and rewrite the conclusion:

[2] Then let it be not equiangular, and at the straight line CB let the parallelogram (CQ) [*i.e.* (BCK), not visible in figure 73, bottom] have been applied, equal to and equiangular with (EH), and let KC be produced to L and QB to M, so that (CM) is equiangular with (EH). Now CB:ZH :: EZ:CK [VI.14]. But [by hypothesis] CB:ZH :: EZ:*CX* ('some straight line'); therefore [V.9] CK = *CX*. And [also by hypothesis] the ratio AC:*CX* is given, whence the ratio AC:CK is also given.

---

[141] Menge wrote in a footnote (p. 139 f., I translate from Latin) 'This part of the proof seems to be false: First, the side of the other parallelogram (CQ) is supposed to be equal to CK, which was to be proved. Secondly, Euclid proves that the ratio AC:CK is given, which was supposed.' See footnote 139.

CHAPTER 10. EQUIANGULAR PARALLELOGRAMS II. (68–75)

Since each of the angles ACB, LCB is given, the remaining ∠ACL is given [Dt 4]. And ∠CAL is given [I.29, Dt 4]; therefore the remaining ∠CLA is given [I.32, Dt 3, 4]; so △ACL is given in form [Dt 40]; therefore the ratio AC:CL is given [Def 3]. And AC:CK was proved to be given; therefore CL:CK is a given ratio [Dt 8].

And CL:CK :: (CM):CQ) [VI.1]; therefore gvn.(CM):(CQ). And (CM) = (AB) [I.35], (EH) = (CQ) [by construction]. Therefore gvn.(AB):EH). ∎

Dt 73, corrupt as it is, is the partial converse of Dt 68 (the equiangular case) and Dt 69 (the non-equiangular case):   If a:c :: d:e & gvn.b:e, then gvn.P:Q.

◆

Θε 74. ' Ἐὰν δύο παραλληλόγραμμα λόγον ἔχῃ δεδομένον, ἤτοι ἐν ἴσαις γωνίαις ἢ ἀνίσοις μέν, δεδομέναις δέ, ἔσται ὡς ἡ τοῦ πρώτου πλευρὰ πρὸς τὴν τοῦ δευτέρου πλευράν, οὕτως ἡ ἑτέρα τοῦ δευτέρου πλευρὰ πρὸς ἣν ἡ λοιπὴ τοῦ πρώτου λόγον ἔχει δεδομένον.

δύο γὰρ παραλληλόγραμμα τὰ ΑΒ, ΕΗ πρὸς ἄλληλα λόγον ἐχέτω δεδομένον ἤτοι ἐν ἴσαις γωνίαις ἢ ἐν ἀνίσοις μέν, δεδομέναις δέ, ταῖς πρὸς τοῖς Γ, Ζ· λέγω, ὅτι ἐστὶν ὡς ἡ ΓΒ πρὸς τὴν ΖΗ, οὕτως ἡ ΕΖ πρὸς ἣν ἡ ΑΓ λόγον ἔχει δεδομένον.

τὸ γὰρ ΑΒ τῷ ΕΗ ἤτοι ἰσογώνιόν ἐστιν ἢ οὔ.

ἔστω πρότερον ἰσογώνιον, καὶ παραβεβλήσθω παρὰ τὴν ΓΒ εὐθεῖαν τῷ ΕΗ παραλληλογράμμῳ ἴσον παραλληλόγραμμον τὸ ΓΘ, καὶ κείσθω ὥστε ἐπ' εὐθείας εἶναι τὴν ΑΓ τῇ ΓΚ· ἐπ' εὐθείας ἄρα ἐστὶ καὶ ἡ ΔΒ τῇ ΒΘ. καὶ ἐπεὶ λόγος ἐστὶ τοῦ ΑΒ πρὸς τὸ ΕΗ δοθείς, ἴσον δὲ τὸ ΕΗ τῷ ΓΘ, λόγος ἄρα ἐστὶ τοῦ ΑΒ πρὸς τὸ ΓΘ δοθείς· ὥστε καὶ τῆς ΑΓ πρὸς τὴν ΓΚ λόγος ἐστὶ δοθείς. καὶ ἐπεὶ ἴσον ἐστὶ τὸ ΓΘ τῷ ΕΗ, ἔστι δὲ καὶ ἰσογώνιον, τῶν ΓΘ, ΕΗ ἄρα ἀντιπεπόνθασιν αἱ πλευραὶ αἱ περὶ τὰς ἴσας γωνίας· ἔστιν ἄρα ὡς ἡ ΓΒ πρὸς τὴν ΖΗ, οὕτως ἡ ΕΖ πρὸς τὴν ΓΚ. τῆς δὲ ΓΚ πρὸς τὴν ΑΓ λόγος ἐστὶ δοθείς· ἔστιν ἄρα ὡς ἡ ΓΒ πρὸς τὴν ΖΗ, οὕτως ἡ ΕΖ πρὸς ἣν ἡ ΑΓ λόγον ἔχει δεδομένον.

μὴ ἔστω δὴ ἰσογώνιον, καὶ συνεστάτω πρὸς τῇ ΓΒ εὐθείᾳ καὶ τῷ πρὸς αὐτῇ σημείῳ τῷ Γ τῇ ὑπὸ ΕΖΗ γωνίᾳ ἴση ἡ ὑπὸ τῶν ΛΓΒ, καὶ συμπεπληρώσθω τὸ ΓΜ παραλληλόγραμμον. ἐπεὶ οὖν λόγος ἐστὶ τοῦ ΓΔ πρὸς τὸ ΕΗ δοθείς, ἴσον δὲ τὸ ΓΔ τῷ ΓΜ, λόγος ἄρα ἐστὶ τοῦ ΓΜ πρὸς τὸ ΕΗ δοθείς. καί ἐστιν ἴση ἡ ὑπὸ τῶν ΛΓΒ γωνία τῇ ὑπὸ τῶν ΕΖΗ· ἔστιν ἄρα ὡς ἡ ΓΒ πρὸς τὴν ΖΗ, οὕτως ἡ ΕΖ πρὸς ἣν ἡ ΓΛ λόγον ἔχει δεδομένον. τῆς δὲ ΓΑ πρὸς τὴν ΓΛ λόγος ἐστὶ δοθείς· ἔστιν ἄρα ὡς ἡ ΓΒ πρὸς τὴν ΖΗ, οὕτως ἡ ΕΖ πρὸς ἣν ἡ ΑΓ λόγον ἔχει δεδομένον.

Dt 74. *If two parallelograms, either in equal angles, or unequal but given angles, have a given ratio, then as the side of the first parallelogram is to the side of the second, so the other side of the second will be to that straight line to which the other side of the first parallelogram has the given ratio.*

Let two parallelograms (AB), (EH) have a given ratio to one another, either in equal angles, or unequal but given angles, namely C, Z. I say that as CB is to ZH, so is EZ to that straight line to which AC has the given ratio.

(AB) is either equiangular with (EH) or not.

[1] (Figure 74 top) Let it first be equiangular, and let the parallelogram (CQ), equal to the parallelogram (EH), have been applied to the straight line CB, and let it lie such that AC is in a straight line with CK; then DB is also in a straight line with BQ.

And since the ratio (AB):(EH) is given, and (EH) = (CQ), the ratio (AB):(CQ) is given; so the ratio AC:CK is given [VI.1]. And since CQ and EH are equal and equiangular, the sides about the equal angles of (CQ), (EH) are reciprocally proportional [VI.14]; therefore CB:ZH :: EZ:CK. And the ratio CK:AC is given; therefore as CB is to ZH, so is EZ to that line to which AC has the given ratio.

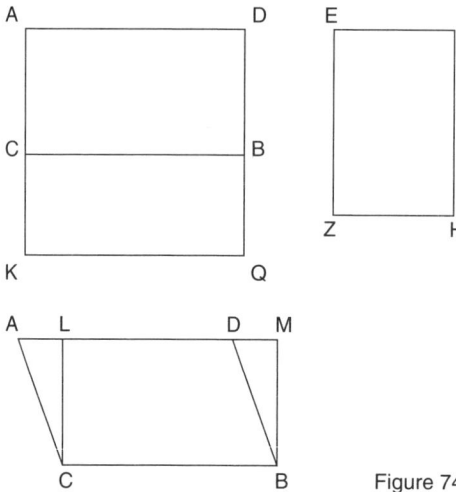

[2] (Figure 74 bottom) Then let it be not equiangular, and at the point C on the straight line CB let ∠LCB have been constructed equal to ∠EZH [I.23], and let the parallelogram (CM) have been completed.

Then, since the ratio (CD):(EH) is given, and (CD) = (CM) [I.35], the ratio (CM):(EH) is given; and ∠LCB = ∠EZH; therefore as CB is to ZH, so is EZ to that line to which CL has the given ratio [as above].

Figure 74

And the ratio CA:CL is given [frm. △ACL]; therefore as CB is to ZH, so is EZ to that line to which AC has *a* given ratio [Dt 8]. ∎

Remarks: The meaning of λόγον δεδομένον in the last part is not *the* given ratio, but *a* given ratio. Whereas in Dt 56, which is proved once again here (now for both the equiangular and the non-equiangular case), the ratio of the second to the fifth line is explicitly said to be the same as the ratio of the parallelograms, this cannot hold for the non-equiangular case. But (as we saw in the chapter about 'by a given greater than in ratio') Euclid seldom cares about 'same' ratio when he is treating 'given' ones. – Dt 74 seems to be a more sophisticated treatment than Dt 56.

◆

The series is rounded off by Dt 75, which applies Dt 74 on triangles.

CHAPTER 10. EQUIANGULAR PARALLELOGRAMS II. (68–75)

Θε 75. ᾽Εὰν δύο τρίγωνα πρὸς ἄλληλα λόγον ἔχῃ δεδομένον, ἤτοι ἐν ἴσαις γωνίαις ἢ ἐν ἀνίσοις μέν, δεδομέναις δέ, ἔσται ὡς ἡ τοῦ πρώτου πλευρὰ πρὸς τὴν τοῦ δευτέρου πλευράν, οὕτως ἡ ἑτέρα τοῦ δευτέρου πλευρὰ πρὸς ἣν ἡ λοιπὴ τοῦ πρώτου λόγον ἔχει δεδομένον.

ἔστω δύο τρίγωνα τὰ ΑΒΓ, ΔΕΖ πρὸς ἄλληλα λόγον ἔχοντα δεδομένον, καὶ ἔστωσαν αἱ πρὸς τοῖς Α, Δ γωνίαι ἤτοι ἴσαι ἢ ἄνισοι μέν, δεδομέναι δέ· λέγω, ὅτι ἐστὶν ὡς ἡ ΑΒ πρὸς τὴν ΔΕ, οὕτως ἡ ΔΖ πρὸς ἣν ἡ ΑΓ λόγον ἔχει δεδομένον.

συμπεπληρώσθω γὰρ τὰ ΑΗ, ΔΘ παραλληλόγραμμα. καὶ ἐπεὶ λόγος ἐστὶ τοῦ ΑΒΓ τριγώνου πρὸς τὸ ΔΕΖ τρίγωνον δοθείς, λόγος ἄρα καὶ τοῦ ΑΗ παραλληλογράμμου πρὸς τὸ ΔΘ παραλληλόγραμμον δοθείς. ἐπεὶ οὖν δύο παραλληλόγραμμα τὰ ΑΗ, ΔΘ πρὸς ἄλληλα λόγον ἔχει δεδομένον ἤτοι ἐν ἴσαις γωνίαις ἢ ἀνίσοις μέν, δεδομέναις δέ, ἔστιν ἄρα ὡς ἡ ΑΒ πρὸς τὴν ΔΕ, οὕτως ἡ ΔΖ πρὸς ἣν ἡ ΑΓ λόγον ἔχει δοθέντα.

Dt 75. *If two triangles, either in equal angles, or unequal but given angles, have a given ratio to one another then as the side of the first triangle is to the side of the second, so the other side of the second will be to that straight line to which the other side of the first triangle has the given ratio.*

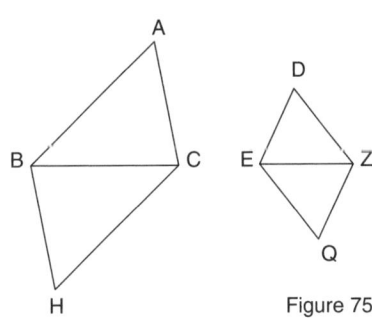

Figure 75

Let two triangles ABC, DEZ have a given ratio to one another, and let the angles at A and D be either equal or unequal but given. I say the as AB is to DE, so is DZ to that line to which AC has the given ratio.

For, let the parallelograms (AH), (DQ) have been completed. And since the ratio △ABC:△DEZ. is given, therefore the ratio (AH):(DQ) is given.

Since, then, two parallelograms (AH), (DQ) have a given ratio to one another, either in equal angles, or unequal but given angles, therefore as AB is to DE, so is DZ to that line to which AC has the/a [142] given ratio. ∎

---

[142] See remarks on Dt 74.

Chapter 11. Duplicates and Outsiders
Dt 76–83

Θε 76. ' Ἐὰν τριγώνου δεδομένου τῷ εἴδει ἀπὸ τῆς κορυφῆς ἐπὶ τὴν βάσιν κάθετος ἀχθῇ, ἡ ἀχθεῖσα πρὸς τὴν βάσιν λόγον ἔχει δεδομένον.

ἔστω τρίγωνον δεδομένον τῷ εἴδει τὸ ΑΒΓ, καὶ ἤχθω ἀπὸ τοῦ Α ἐπὶ τὴν ΒΓ κάθετος ἡ ΑΔ· λέγω, ὅτι λόγος ἐστὶ τῆς ΑΔ πρὸς τὴν ΒΓ δοθείς.

ἐπεὶ γὰρ δέδοται τὸ ΑΒΓ τρίγωνον τῷ εἴδει, δοθεῖσα ἄρα ἐστὶ καὶ ἡ ὑπὸ ΑΒΔ γωνία. ἔστι δὲ καὶ ἡ ὑπὸ τῶν ΒΔΑ δοθεῖσα· καὶ λοιπὴ ἄρα ἡ ὑπὸ τῶν ΒΑΔ ἐστι δοθεῖσα· δέδοται ἄρα τὸ ΑΒΔ τρίγωνον τῷ εἴδει· λόγος ἄρα ἐστὶ τῆς ΒΑ πρὸς τὴν ΑΔ δοθείς. τῆς δὲ ΑΒ πρὸς τὴν ΒΓ λόγος ἐστὶ δοθείς· καὶ τῆς ΑΔ ἄρα πρὸς τὴν ΒΓ λόγος ἐστὶ δοθείς.

Dt 76. *If in a triangle given in form a perpendicular be drawn from the vertex to the base, the line drawn has a given ratio to the base.*

Let ABC be a triangle given in form, and let the perpendicular AD have been drawn from A to BC. I say that the ratio AD:BC is given.

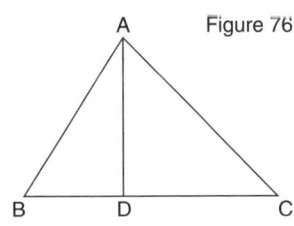

Figure 76

For, since △ABC is given in form, therefore △ABD is also given [Def. 3]. And △BDA is also given [by construction]; therefore the remaining △BAD is also given [I.32]; therefore △ABD is given in form [Dt 40].

Therefore the ratio BA:AD is given [Def. 3]. And the ratio AB:BC is given; therefore the ratio AD:BC is also given [Dt 8]. ∎

Remarks: The equivalent of Dt 76 was proved already in Dt 48 ('the ratio of EA to AB is given'). Several of the intermediate theorems would have profited from Dt 76.

Θε 77. ' Ἐὰν δύο εἴδη δεδομένα τῷ εἴδει πρὸς ἄλληλα λόγον ἔχῃ δεδομένον, καὶ μία πλευρὰ ὁποιαοῦν ἑνὸς τῶν εἰδῶν πρὸς ὁποιανοῦν τῶν τοῦ ἑτέρου λόγον ἕξει δεδομένον.

δύο γὰρ εἴδη τὰ ΑΒΓ, ΔΕΖ δεδομένα τῷ εἴδει πρὸς ἄλληλα λόγον ἐχέτω δεδομένον· λέγω, ὅτι καὶ μία πλευρὰ ὁποιαοῦν τοῦ ΑΒΓ πρὸς μίαν πλευρὰν ὁποιανοῦν τοῦ ΔΕΖ λόγον ἔχει δεδομένον.

ἀναγεγράφθω γὰρ ἀπὸ τῶν ΒΓ, ΕΖ τετράγωνα τὰ ΒΗ, ΕΘ. ἐπεὶ ἀπὸ τῆς αὐτῆς εὐθείας τῆς ΒΓ δύο εἴδη ἀναγέγραπται, ἃ ἔτυχεν, δεδομένα τῷ εἴδει τὰ ΑΒΓ, ΒΗ, λόγος ἄρα τοῦ ΑΒΓ πρὸς τὸ ΒΗ δοθείς. διὰ τὰ αὐτὰ δὴ πάλιν καὶ τοῦ ΔΕΖ πρὸς τὸ ΕΘ λόγος ἐστὶ δοθείς. ἐπεὶ οὖν λόγος ἐστὶ τοῦ ΑΒΓ πρὸς τὸ ΔΕΖ δοθείς, ἀλλὰ τοῦ μὲν ΑΒΓ πρὸς τὸ ΒΗ λόγος ἐστὶ δοθείς, τοῦ δὲ ΔΕΖ πρὸς τὸ ΕΘ λόγος ἐστὶ δοθείς, καὶ τοῦ ΒΗ ἄρα πρὸς τὸ ΕΘ λόγος ἐστὶ δοθείς· ὥστε καὶ τῆς ΒΓ πρὸς τὴν ΕΖ λόγος ἐστὶ δοθείς.

Dt 77. *If two forms, given in form, have a given ratio to one another, any one side of one of the forms will also have a given ratio to any one side of the other form.*

For, let the two forms ABC, DEZ, given in form, have a given ratio to one another. I say that any one side of the form ABC also has a given ratio to any one side of the form DEZ.

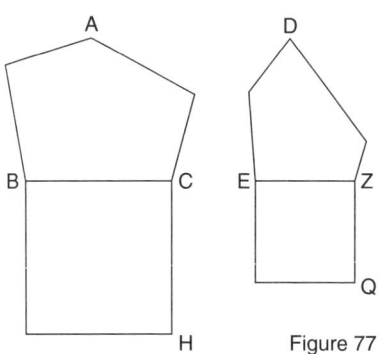

Figure 77

For, on the straight lines BC, EZ let the squares (BH), (EQ) have been described.

Since, on the same straight line BC two arbitrary rectilineal figures (ABC), (BH) given in form have been described, therefore the ratio (ABC):(BH) is given [Dt 49]. For the same reasons the ratio (DEZ):(EQ) is also given.

Then, since the ratio (ABC):(DEZ) is given, and the ratio (ABC):(BH) is given and the ratio (DEZ):(EQ) is given, therefore the ratio (BH):(EQ) is also given [Dt 8]; so that the ratio BC:EZ is given [Dt 54, i.e. 24*D].[143] ∎

Remarks: Dt 77 proves the same statement as Dt 54 (even using the idiom 'two forms' for 'two rectilineal figures') , but in a slightly different way. The last argument, that BC:EZ is given, uses the problematic 24*D, *if the ratio of two squares is given, the ratio of their sides is also given*. With Clemens Thaer I suspect this and the two following propositions to be spurious, because they are not in Pappus' collection, and Dt 79 is missing in the Arabic tradition. Dt 77 and 78 are repetitions, and Dt 79 does not belong in the *Data* at all.

◆

---

[143] The usual generalization is missing: 'For the same reasons the ratios of the other sides to the other sides are given'.

CHAPTER 11. DUPLICATES AND OUTSIDERS. (76–83)

Θε 78. ' Ἐὰν δοθὲν εἶδος πρός τι ὀρθογώνιον λόγον ἔχῃ δεδομένον, καὶ μία πλευρὰ πρὸς μίαν πλευρὰν λόγον ἔχῃ δοθέντα, δέδοται τὸ ὀρθογώνιον τῷ εἴδει.

δοθὲν γὰρ εἶδος τὸ ΑΖΒ πρός τι ὀρθογώνιον τὸ ΓΔ λόγον ἐχέτω δεδομένον, καὶ ἔστω λόγος τῆς ΖΒ πρὸς τὴν ΕΔ δοθείς· λέγω, ὅτι δέδοται τὸ ΓΔ τῷ εἴδει.

ἀναγεγράφθω γὰρ ἀπὸ τῆς ΖΒ τετράγωνον τὸ ΖΗ, καὶ παραβεβλήσθω παρὰ τὴν ΕΔ τῷ ΖΗ ἴσον παραλληλόγραμμον τὸ ΕΚ, καὶ κείσθω ὥστε ἐπ' εὐθείας εἶναι τὴν ΓΕ τῇ ΕΘ· ἐπ' εὐθείας ἄρα ἐστὶ καὶ ἡ ΜΔ τῇ ΔΚ. καὶ ἐπεὶ ἀπὸ τῆς αὐτῆς εὐθείας τῆς ΖΒ δύο εὐθύγραμμα, ἃ ἔτυχεν, δεδομένα τῷ εἴδει ἀναγέγραπται τὰ ΑΖΒ, ΖΗ, λόγος ἄρα ἐστὶ τοῦ ΑΖΒ πρὸς τὸ ΖΗ δοθείς. τοῦ δὲ ΑΖΒ πρὸς τὸ ΓΔ λόγος ἐστὶ δοθείς· καὶ τοῦ ΖΗ ἄρα πρὸς τὸ ΓΔ λόγος ἐστὶ δοθείς. ἀλλὰ τὸ ΖΗ τῷ ΕΚ ἐστὶ ἴσον· καὶ τοῦ ΓΔ ἄρα πρὸς τὸ ΕΚ λόγος ἐστὶ δοθείς· ὥστε καὶ τῆς ΓΕ πρὸς τὴν ΕΘ λόγος ἐστὶ δοθείς. καὶ ἐπεὶ ἴσον ἐστὶ καὶ ἰσογώνιον τὸ ΖΗ τῷ ΕΚ, [ἔστι δὲ καὶ ὀρθογώνιον·] ἀντιπεπόνθασιν ἄρα αὐτῶν αἱ πλευραί, καί ἐστιν ὡς ἡ ΖΒ πρὸς ΕΔ, οὕτως ἡ ΕΘ πρὸς ΖΛ. λόγος δὲ ὑπόκειται τῆς ΖΒ πρὸς τὴν ΕΔ δοθείς· λόγος ἄρα καὶ τῆς ΕΘ πρὸς τὴν ΖΛ δοθείς. τῆς δὲ ΕΘ πρὸς τὴν ΓΕ λόγος ἐστὶ δοθείς· καὶ τῆς ΓΕ ἄρα πρὸς τὴν ΖΛ λόγος ἐστὶ δοθείς. ἴση δὲ ἡ ΛΖ τῇ ΖΒ· [τετράγωνον γάρ· τῆς ΛΖ ἄρα πρὸς ΕΔ λόγος δοθείς· σύγκειται γάρ·] καὶ τῆς ΓΕ ἄρα πρὸς τὴν ΕΔ λόγος ἐστὶ δοθείς. καί ἐστιν ὀρθὴ ἡ πρὸς τῷ Ε γωνία· δέδοται ἄρα τὸ ΓΔ τῷ εἴδει.

Dt 78. *If a given form have a given ratio to some rectangle, and one side* [of the form] *have a given ratio to one side* [of the rectangle], *the rectangle is given in form.*

For, let the given form (AZB) have a given ratio to the rectangle (CD), and let the ratio ZB:ED be given. I say that (CD) is given in form.

For, let the square (ZH) have been described on ZB [I.46], and to ED let the parallelogram (EK) have been applied equal to (ZH) [I.44]. And let it lie such that CE is in a straight line with EQ;[144] then MD is also in a straight line with DK [I.29, 14].

And since on the same straight line ZB two arbitrary rectilineal figures given in form (AZB), (ZH) have been described, therefore the ratio (AZB):(ZH) is given [Dt 49].

But the ratio (AZB):(CD) is given; therefore the ratio (ZH):(CD) is also given [Dt 8]. But (ZH) = (EK); therefore the ratio (CD):(EK) is given; so that the ratio CE:EQ is also given [VI.1].

---

[144] This is nothing more or less than saying that (EK) is a rectangle. And yet Euclid gives a spurious sense of generality by requiring(EK) to be a parallelogram. Much of the parallelogram talk is purely ornamental, since Dt 40 will ensure givenness in the cognate rectangle, as we see it time and again.

And since (ZH) = (EK), and they are equiangular, [145] therefore their sides are reciprocally proportional, and ZB:ED :: EQ:ZL.

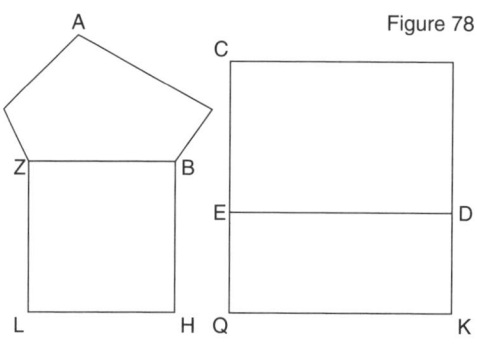

Figure 78

But by hypothesis the ratio ZB:ED is given; therefore the ratio EQ:ZL is also given.

But the ratio EQ:CE is given; therefore the ratio CE:ZL is also given. And LZ = ZB; [146] therefore the ratio CE:ED is given.

And the angle at [the point] E is right. Therefore (CD) is given in form. ∎

Remarks: The Greek δοθέν εἶδος means a (rectilineal figure) given (in) form. In Dt 53 it was explicitly (and pleonastically) called 'a form given in form'. Dt 78 repeats that case of Dt 62 where the parallelogram is explicitly a rectangle.

Θε 79. ' Ἐὰν δύο τρίγωνα μίαν γωνίαν μιᾷ γωνίᾳ ἴσην ἔχῃ, καὶ ἀπὸ τῶν ἴσων γωνιῶν ἐπὶ τὰς βάσεις κάθετοι εὐθεῖαι γραμμαὶ ἀχθῶσιν, ᾖ δέ, ὡς ἡ τοῦ πρώτου τριγώνου βάσις πρὸς τὴν κάθετον, οὕτως ἡ τοῦ ἑτέρου τριγώνου βάσις πρὸς τὴν κάθετον, ἰσογώνια ἔσται τὰ τρίγωνα.

ἔστω δύο τρίγωνα τὰ ΑΒΓ, ΘΖΗ ἴσας ἔχοντα γωνίας τὰς πρὸς τοῖς Ζ, Β, καὶ ἤχθωσαν ἀπὸ τῶν Ζ, Β κάθετοι αἱ ΒΔ, ΖΚ· ἔστω δέ, ὡς ἡ ΑΓ πρὸς τὴν ΒΔ, οὕτως ἡ ΘΗ πρὸς τὴν ΚΖ· λέγω, ὅτι ἰσογώνιόν ἐστι τὸ ΑΒΓ τρίγωνον τῷ ΘΖΗ τριγώνῳ.

περιγεγράφθω γὰρ περὶ τὸ ΘΖΗ τρίγωνον κύκλος, οὗ τμῆμα ἔστω τὸ ΘΖΗ, καὶ συνεστάτω πρὸς τῇ ΘΗ εὐθείᾳ καὶ τῷ πρὸς αὐτῇ σημείῳ τῷ Θ τῇ ὑπὸ τῶν ΒΑΓ γωνίᾳ ἴση ἡ ὑπὸ τῶν ΗΘΛ, καὶ ἐπεζεύχθωσαν αἱ ΖΛ, ΛΗ, καὶ ἤχθω κάθετος ἡ ΛΜ. ἐπεὶ ἴση

---

[145] Menge bracketed the passage 'and (ZH) is also a rectangle', perhaps suspecting that some scrupulous scribe felt called on to share his knowledge that squares are special rectangles.

[146] The passage bracketed by Menge says 'for it is a square; therefore the ratio LZ:ED is given, for it is compounded'. Menge gives a footnote on p. 153, which I translate from Latin: *Here Euclid is a bit briefer than he usually is; he would normally conclude as follows: Since the ratio CE:LZ is given, and LZ = ZB, the ratio CE:ZB is given. But the ratio ZB:ED is given (by hypothesis); therefore the ratio CE:ED is given.*

ἐστὶν ἡ ὑπὸ τῶν ΒΑΔ τῇ ὑπὸ τῶν ΛΘΗ, ἔστι δὲ καὶ ἡ ὑπὸ τῶν ΘΛΗ τῇ ὑπὸ ΑΒΓ ἴση, καὶ λοιπὴ ἄρα ἡ ὑπὸ τῶν ΒΓΑ λοιπῇ τῇ ὑπὸ τῶν ΘΗΛ ἐστιν ἴση· ὅμοιον ἄρα ἐστὶ τὸ ΒΑΓ τρίγωνον τῷ ΘΗΛ τριγώνῳ. καὶ κάθετοι ἠγμέναι εἰσὶν αἱ ΒΔ, ΛΜ· ἔστιν ἄρα ὡς ἡ ΑΓ πρὸς τὴν ΒΔ, οὕτως ἡ ΘΗ πρὸς τὴν ΛΜ· ἦν δέ, ὡς ἡ ΑΓ πρὸς τὴν ΒΔ, οὕτως ἡ ΘΗ πρὸς τὴν ΖΚ· ὑπόκειται γάρ· καὶ ὡς ἄρα ἡ ΘΗ πρὸς τὴν ΛΜ, οὕτως ἡ ΘΗ πρὸς τὴν ΖΚ· ἴση ἄρα ἐστὶν ἡ ΖΚ τῇ ΛΜ· ἔστι δὲ καὶ παράλληλος· καὶ ἡ ΖΛ ἄρα τῇ ΘΗ παράλληλός ἐστιν· ἴση ἄρα ἐστὶν ἡ ὑπὸ τῶν ΖΛΘ γωνία τῇ ὑπὸ τῶν ΛΘΗ. ἀλλ' ἡ μὲν ὑπὸ τῶν ΛΘΗ τῇ ὑπὸ τῶν ΒΑΓ ἐστιν ἴση· ἡ δὲ ὑπὸ ΖΛΘ τῇ ὑπὸ τῶν ΖΗΘ ἐστιν ἴση· καὶ ἡ ὑπὸ τῶν ΒΑΓ ἄρα τῇ ὑπὸ τῶν ΖΗΘ ἐστιν ἴση. ἔστι δὲ καὶ ἡ ὑπὸ τῶν ΑΒΓ τῇ ὑπὸ τῶν ΘΖΗ ἴση· λοιπὴ ἄρα ἡ ὑπὸ τῶν ΒΓΑ λοιπῇ τῇ ὑπὸ τῶν ΖΘΗ ἐστιν ἴση· ἰσογώνιον ἄρα ἐστὶ τὸ ΑΒΓ τρίγωνον τῷ ΖΘΗ τριγώνῳ.

Dt 79. *If two triangles have one angle equal to one angle, and from the equal angles straight lines be drawn perpendicularly to the bases, and if the base of the first triangle be to the perpendicular as the base of the second triangle is to the perpendicular, then the two triangles will be equiangular.*

Let there be two triangles ABC, QZH having the equal angles Z and B, and from Z, B, let the perpendiculars BD, ZK have been drawn. And let AC:BD :: QH:KZ. I say that △ABC is equiangular with △QZH.

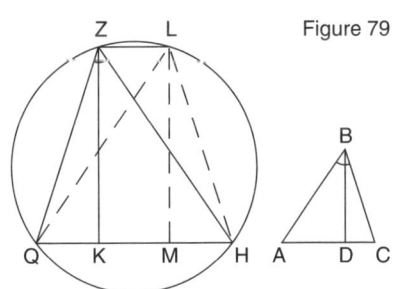

Figure 79

For, let a circle have been circumscribed about △QZH [IV.5], of which QZH shall be a segment, and at the point Q on the straight line QH let ∡HQL have been constructed equal to ∡BAC, and let ZL, LH be joined, and let the perpendicular LM have been drawn.

Since ∡BAD = ∡LQH, and also ∡QLH = ∡ABC [III.21], therefore the remaining ∡BCA is equal to the remaining ∡QHL [I.32].

Therefore △BAC is similar to △QHL [VI.4; VI.Def. 1]. And the lines BD, LM are drawn perpendicularly; therefore AC:BD :: QH:LM [VI.4, V.22]; but AC:BD :: QH:ZK, for that was hypothesized.

Therefore also QH:LM :: QH:ZK [V.9]; therefore ZK is equal to LM [V.11], and it is also parallel [to it] [I.28]. Therefore ZL is also parallel to QH [I.33]. Therefore the angle ZLQ = the angle LQH [I.29].

But ∡ LQH = ∡BAC; and ∡ZLQ = ∡ZHQ [III.21]; therefore ∡BAC = ∡ZHQ. And ∡ABC = ∡QZH; therefore the remaining ∡BCA = the remaining ∡ZQH [I.32]. Therefore △ABC is equiangular with △ZQH. ∎

Remarks: The absence of the word 'given' and the fact that *two* triangles are involved shows that Dt 79 does not belong in the *Data*, but would fit perfectly into Book VI of the *Elements*. Clemens Thaer suspects that it was not in Pappus' source (since it is missing in the Arabic version) and that he may have found it as a lemma without number. If it is spurious (says C.T.), the alternate proof of Dt 80 (which is all that the Arabic text has) must be genuine. I give a shorthand version of that one in the remarks on Dt 80; it is most interesting in employing the otherwise useless (?) Dt 45 and 67.

Dt 79 can, however, be stated in terms of 'Givens' as a partial converse of Dt 76:

Dt 79* *If a triangle have a given angle, and the ratio of the perpendicular from the given angle to the base is given, the triangle will be given in form.*

Using figure 80 we have

Dt 76   frm.△ABC (∴ gvn.∡A & gvn.∡C) => gvn.AE:BC
Dt 79*   gvn.∡A & gvn.AE:BC => gvn.∡C ∴ frm.△ABC

Dt 79* is embedded in Dt 80 as we shall see. Before analysing that theorem, let us compare the sequence Dt 79–80 with, say, Dt 41: *If a triangle have one given angle, and the sides about the given angle have a given ratio to one another, the triangle is given in form.* This is shown to be true because *two triangles that have one angle equal to one angle and the sides about the equal angles proportional are similar*, VI.6. That is, the statement about the individual triangle in the *Data* is proved by a cognate statement from the *Elements*, which involves two triangles. In the *Data* one triangle is present with the given properties, while another, with the same properties, is constructed *in position* and therefore *given in form*; having the same properties it is similar to the first one, which is therefore also given in form.

Now, to prove Dt 79* in the same way (which is *the* way such things are proved in the *Data*), we need a proposition from the *Elements*. And since no such is found there, it must be proved in the *Data*, – which explains the alien character of Dt 79, in which the word 'given' does not occur. As a preliminary to a proof of Dt 80 we may note (if a, b, and c are the sides, h_A, h_B, h_C the heights):

(1)   To any triangle are associated three equal rectangles, ⊏⊐a,h_A, ⊏⊐b,h_B, ⊏⊐c,h_C, each of them being equal to 2△ABC.

(2)   If the triangle is given in form, the ratios h_A:b, h_A:c [Dt 40], h_A:a [Dt 76] are given, as well as their analogues (involving h_B and h_C).

The Givens of 79*, the partial converse to Dt 76, can now be expanded:

CHAPTER 11. DUPLICATES AND OUTSIDERS. (76–83)

gvn.$\triangle A$ =Dt 40=> gvn.h_B:c =VI.1=> gvn.$\sqsubset\sqsupset b,h_B$:$\sqsubset\sqsupset b,c$ =I.41=> gvn.$\sqsubset\sqsupset a,h_A$:$\sqsubset\sqsupset b,c$
gvn.h_A:a =VI.1=> gvn.$\sqsubset\sqsupset a,h_A$:$\square a$ =Dt 8=> gvn.$\sqsubset\sqsupset b,c$:$\square a$

These are the hypotheses of Dt 80; here follows a short-hand version of that proposition (with my notation in the right column):

Hypoth.	gvn.$\triangle A$	
	gvn.$\sqsubset\sqsupset BA,AC$:$\square BC$	$\sqsubset\sqsupset c,b$:$\square a$
Assert.	[gvn.$\triangle B$ & gvn.$\triangle C$ ∴]	frm.$\triangle ABC$
Constr.	BD, AE [heights]	h_B, h_A
Proof		
[Dt 40]	gvn.AB:BD	gvn.c:h_B
[VI.1]	gvn.$\sqsubset\sqsupset BA,AC$:$\sqsubset\sqsupset AC,BD$	gvn.$\sqsubset\sqsupset c,b$:$\sqsubset\sqsupset b,h_B$
[I.41]	$\sqsubset\sqsupset AC,BD$ = $\sqsubset\sqsupset BC,AE$ = 2$\triangle ABC$	$\sqsubset\sqsupset b,h_B$ = $\sqsubset\sqsupset a,h_A$ = 2$\triangle$
	gvn.$\sqsubset\sqsupset BA,AC$:$\sqsubset\sqsupset BC,AE$	gvn.$\sqsubset\sqsupset c,b$:$\sqsubset\sqsupset a,h_A$
[Hyp]	gvn.$\sqsubset\sqsupset BA,AC$:$\square BC$	gvn.$\sqsubset\sqsupset c,b$:$\square a$
[Dt 8]	gvn.$\sqsubset\sqsupset BC,AE$:$\square BC$	gvn.$\sqsubset\sqsupset a,h_A$:$\square a$
[VI.1]	gvn.AE:BC	gvn.h_A:a

By means of the 'circle that admits the given angle' [III.33] he constructs a triangle QZH on the line ZH given in position and magnitude with $\triangle Q$ = $\triangle A$, so that the height QI, has to the base ZH the given ratio AE:BC. Since the vertices of △QZH are given in position, the triangle is given in form, and from Dt 79 follows that it is equiangular with △ABC, which, therefore, is also given in form (cp. the argument in Dt 39 ff., discussed on p. 122). – A scholium (no. 175) disproves by an *reductio ad absurdum* the objection that the parallel KQ might not cut or touch the circle.

As I said, Dt 79 serves Dt 80 as VI.6 serves Dt 41. To prove that the two triangles in figure 79 are equiangular, Euclid describes the circumcircle about △QZH, and inscribes (in a subcontrary way) in the same circle a third triangle equiangular with △ABC. From propositions in *Elements* III and from the theory of parallels in *Elements* I it follows that the two triangles in the circle are equiangular. Whence △QZH is equiangular with △ABC.

◆

Θε 80. 'Εὰν τρίγωνον μίαν ἔχῃ γωνίαν δεδομένην, καὶ τὸ ὑπὸ τῶν τὴν δεδομένην γωνίαν περιεχουσῶν εὐθειῶν πρὸς τὸ ἀπὸ τῆς λοιπῆς πλευρᾶς τετράγωνον λόγον ἔχῃ δεδομένον, δέδοται τὸ τρίγωνον τῷ εἴδει.

ἔστω τρίγωνον τὸ ΑΒΓ δεδομένην ἔχον γωνίαν τὴν πρὸς τῷ Α, καὶ τὸ ὑπὸ τῶν ΒΑΓ πρὸς τὸ ἀπὸ τῆς ΒΓ λόγον ἐχέτω δεδομένον· λέγω, ὅτι δέδοται τὸ ΑΒΓ τρίγωνον τῷ εἴδει.

ἤχθωσαν γὰρ ἀπὸ τῶν Α, Β ἐπὶ τὰς ΒΓ, ΓΑ κάθετοι αἱ ΒΔ, ΑΕ. ἐπεὶ οὖν δοθεῖσά ἐστιν ἡ ὑπὸ ΒΑΔ γωνία, ἔστι δὲ καὶ ἡ ὑπὸ τῶν ΑΔΒ δοθεῖσα, δέδοται ἄρα τὸ ΑΔΒ τρίγωνον τῷ εἴδει· λόγος ἄρα ἐστὶ τῆς ΑΒ πρὸς τὴν ΒΔ δοθείς· ὥστε καὶ τοῦ ὑπὸ τῶν ΒΑΓ πρὸς τὸ ὑπὸ τῶν ΑΓ, ΒΔ λόγος ἐστὶ δοθείς. τῷ δὲ ὑπὸ τῶν ΑΓ, ΒΔ ἴσον ἐστὶ τὸ ὑπὸ τῶν ΒΓ, ΑΕ· ἑκάτερον γὰρ αὐτῶν διπλάσιόν ἐστι τοῦ ΑΒΓ τριγώνου· λόγος ἄρα καὶ τοῦ ὑπὸ τῶν ΒΑΓ πρὸς τὸ ὑπὸ τῶν ΒΓ, ΑΕ δοθείς· τοῦ δὲ ὑπὸ τῶν ΒΑΓ πρὸς τὸ ἀπὸ τῆς ΒΓ λόγος ἐστὶ δοθείς· καὶ τοῦ ὑπὸ τῶν ΒΓ, ΑΕ ἄρα πρὸς τὸ ἀπὸ τῆς ΒΓ λόγος ἐστὶ δοθείς, καὶ τῆς ΒΓ πρὸς ΑΕ λόγος ἐστὶ δοθείς.

ἐκκείσθω τῇ θέσει καὶ τῷ μεγέθει δεδομένη εὐθεῖα ἡ ΖΗ, καὶ γεγράφθω ἐπὶ τῆς ΖΗ τμῆμα τὸ ΖΘΗ δεχόμενον γωνίαν ἴσην τῇ ὑπὸ τῶν ΒΑΓ· δοθεῖσα δὲ ἡ ὑπὸ τῶν ΒΑΓ γωνία· δοθεῖσα ἄρα καὶ ἡ ἐν τῷ ΖΘΗ τμήματι γωνία· θέσει ἄρα ἐστὶ τὸ ΖΘΗ τμῆμα. ἤχθω ἀπὸ τοῦ Η τῇ ΖΗ πρὸς ὀρθὰς ἡ ΗΚ· θέσει ἄρα ἐστὶν ἡ ΗΚ. καὶ πεποιήσθω, ὡς ἡ ΒΓ πρὸς τὴν ΑΕ, οὕτως ἡ ΖΗ πρὸς τὴν ΗΚ. λόγος δὲ τῆς ΒΓ πρὸς τὴν ΑΕ δοθείς· λόγος ἄρα καὶ τῆς ΖΗ πρὸς τὴν ΗΚ δοθείς· δοθεῖσα δὲ ἡ ΖΗ· δοθεῖσα ἄρα καὶ ἡ ΗΚ. ἀλλὰ καὶ τῇ θέσει· καί ἐστι δοθὲν τὸ Η· δοθὲν ἄρα καὶ τὸ Κ.

ἤχθω διὰ τοῦ Κ τῇ ΖΗ παράλληλος ἡ ΚΘ· θέσει ἄρα ἐστὶν ἡ ΘΚ· θέσει δὲ καὶ τὸ ΖΘΗ τμῆμα· δοθὲν ἄρα ἐστὶ τὸ Θ σημεῖον. ἐπεζεύχθωσαν αἱ ΖΘ, ΘΗ, καὶ ἤχθω κάθετος ἡ ΘΛ· δοθεῖσα ἄρα ἐστὶν ἡ ΘΛ. ἔστι δὲ καὶ τὸ Θ σημεῖον δοθέν, καὶ ἑκάτερον τῶν Ζ, Η· δέδοται ἄρα ἑκάστη τῶν ΘΖ, ΖΗ, ΘΗ τῇ θέσει καὶ τῷ μεγέθει· δέδοται ἄρα τὸ ΖΘΗ τρίγωνον τῷ εἴδει. καὶ ἐπεί ἐστιν, ὡς ἡ ΒΓ πρὸς τὴν ΑΕ, οὕτως ἡ ΖΗ πρὸς τὴν ΗΚ, ἴση δὲ ἡ ΗΚ τῇ ΘΛ, ἔστιν ἄρα ὡς ἡ ΒΓ πρὸς τὴν ΑΕ, οὕτως ἡ ΖΗ πρὸς τὴν ΘΛ. καί ἐστιν ἴση ἡ ὑπὸ τῶν ΒΑΓ γωνία τῇ ὑπὸ τῶν ΖΘΗ· ἰσογώνιον ἄρα ἐστὶ τὸ ΑΒΓ τρίγωνον τῷ ΘΖΗ τριγώνῳ. δέδοται δὲ τὸ ΘΖΗ τρίγωνον τῷ εἴδει· δέδοται ἄρα καὶ τὸ ΑΒΓ τρίγωνον τῷ εἴδει.

Dt 80. *If a triangle have one angle given, and the rectangle contained by the lines about the given angle have a given ratio to the square on the third side, the triangle is given in form.*

Let ABC be a triangle with a given angle at A, let the rectangle (BAC) have a given ratio to the square on BC. I say that the triangle ABC is given in form.

For, from A, B let the perpendiculars BD, AE have been drawn to BC, CA. Then, since △BAD is given, and the angle ADB is also given, therefore △ADB is given in form [Dt 40]; therefore the ratio AB:BD is given. So the ratio ⊏⊐BAC:⊏⊐AC,BD is given [VI.1].

But ⊏⊐BC,AE is equal to ⊏⊐AC,BD, for each of them is twice △ABC; therefore the ratio ⊏⊐BAC:⊏⊐BC,AE is also given.

And the ratio ⊏⊐BAC:▫BC is given; therefore the ratio ⊏⊐BC,AE:▫BC is also given, and the ratio BC:AE is given.

Let the straight line ZH lie given in position and magnitude, and on ZH let the segment ZQH have been described admitting an angle equal to BAC [III.33]; and △BAC is given; therefore the angle in the segment ZQH is also given; therefore the segment ZQH is [given] in position [Def. 8].

Let HK have been drawn from H at right angles to ZH; then HK is [given] in position [Dt 29]. And let be made ZH:HK :: BC:AE; but the ratio BC:AE is given; therefore the ratio ZH:HK is also given. And ZH is given; therefore HK is also given. But in position, too. And H is given; therefore K is given, too.

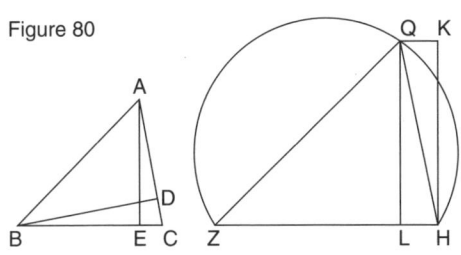

Figure 80

Let KQ have been drawn through K parallel to ZH; then QK is [given] in position [Dt 28]; and the segment ZQH is also [given] in position; therefore the point Q is given [Dt 25].

Let ZQ,QH have been joined, and let the perpendicular QL have been drawn; then QL is given.

And the point Q is given, and each of Z,H; therefore each of QZ, ZH, QH is given in position and in magnitude; therefore △ZQH is given in form.

And since ZH:HK :: BC:AE, and HK = QL, therefore ZH:QL :: BC:AE. And ∠BAC = ∠ZQH; therefore △ABC is equiangular with △QZH. But △QZH is given in form; therefore △ABC is given in form. ∎

See *remarks* on Dt 79. Here is a shorthand version of the alternate proof of Dt 80:

Hypotheses: In △ABC gvn.∠A, gvn.⊏⊐BA,AC:▫BC.
Assertion: frm.△ABC.
Proof:
Make ▫(BA+AC) − ▫BC = ▫D. (The text says *chorion* D, 'area' D).   (*)
Dt 67 => gvn.▫D:△ABC.
Dt 66 => gvn.△ABC:⊏⊐BA,AC.
Dt 8   => gvn.▫D:⊏⊐BA,AC.
Hypot.   gvn.⊏⊐BA,AC:▫BC.
Dt 8   => gvn.▫D:▫BC.
Dt 6   => gvn.(▫D+▫BC):▫BC.
(*)   => gvn.▫(BA+AC):▫BC.
Dt 54 => gvn.(BA+AC):BC.
Hyp & Dt 45 => frm.△ABC ∎

◆

Θε 81. ' Ἐὰν τρεῖς εὐθεῖαι ἀνάλογον οὖσαι τρισὶν εὐθείαις ἀνάλογον οὔσαις τὰς ἄκρας ἐν δεδομένῳ λόγῳ ἔχωσιν, καὶ τὰς μέσας ἐν δεδομένῳ λόγῳ ἕξουσιν· καὶ ἐὰν ἡ ἄκρα πρὸς τὴν ἄκραν λόγον ἔχῃ δεδομένον καὶ ἡ μέση πρὸς τὴν μέσην, καὶ ἡ λοιπὴ ἄκρα πρὸς τὴν λοιπὴν ἄκραν λόγον ἕξει δεδομένον.

τρεῖς γὰρ εὐθεῖαι ἀνάλογον οὖσαι αἱ Α, Β, Γ τρισὶν εὐθείαις ἀνάλογον οὔσαις ταῖς Δ, Ε, Ζ τὰς ἄκρας ἐν δεδομένῳ λόγῳ ἐχέτωσαν, καὶ ἔστω λόγος τῆς μὲν Α πρὸς τὴν Δ δοθείς, τῆς δὲ Γ πρὸς τὴν Ζ λόγος δοθείς· λέγω, ὅτι καὶ τῆς Β πρὸς τὴν Ε λόγος ἐστὶ δοθείς.

ἐπεὶ γὰρ λόγος ἐστὶ τῆς μὲν Α πρὸς τὴν Δ δοθείς, τῆς δὲ Γ πρὸς τὴν Ζ δοθείς, λόγος ἄρα τοῦ ὑπὸ τῶν Α, Γ πρὸς τὸ ὑπὸ τῶν Δ, Ζ δοθείς. ἀλλὰ τῷ μὲν ὑπὸ τῶν Α, Γ ἴσον ἐστὶ τὸ ἀπὸ τῆς Β, τῷ δὲ ὑπὸ τῶν Δ, Ζ ἴσον ἐστὶ τὸ ἀπὸ τῆς Ε. λόγος ἄρα ἐστὶ τοῦ ἀπὸ τῆς Β πρὸς τὸ ἀπὸ τῆς Ε δοθείς· ὥστε καὶ τῆς Β πρὸς τὴν Ε λόγος ἐστὶ δοθείς.

ἔστω δὴ πάλιν τῆς μὲν Α πρὸς τὴν Δ λόγος δοθείς, τῆς δὲ Β πρὸς τὴν Ε λόγος δοθείς· λέγω, ὅτι καὶ τῆς Γ πρὸς τὴν Ζ λόγος ἐστὶ δοθείς. ἐπεὶ λόγος ἐστὶ τῆς μὲν Α πρὸς τὴν Δ, τῆς δὲ Β πρὸς τὴν Ε δοθείς, λόγος ἐστὶ καὶ τοῦ ἀπὸ τῆς Β πρὸς τὸ ἀπὸ τῆς Ε δοθείς. ἀλλὰ τῷ μὲν ἀπὸ τῆς Β ἴσον τὸ ὑπὸ τῶν Α, Γ, τῷ δὲ ἀπὸ τῆς Ε ἴσον ἐστὶ τὸ ὑπὸ τῶν Δ, Ζ· λόγος ἄρα ἐστὶ τοῦ ὑπὸ τῶν Α, Γ πρὸς τὸ ὑπὸ τῶν Δ, Ζ δοθείς. καὶ μιᾶς πλευρᾶς τῆς Α πρὸς μίαν πλευρὰν τὴν Δ λόγος ἐστὶ δοθείς· καὶ λοιπῆς ἄρα τῆς Γ πρὸς λοιπὴν τὴν Ζ λόγος ἐστὶ δοθείς.

Dt 81. *If three proportional straight lines have to three proportional straight lines the extremes in given ratio, the means will also be in a given ratio. And if the extreme have a given ratio to the extreme and also the mean to the mean, then the other extreme will have a given ratio to the other extreme.*

For, let three proportional straight lines A, B, C have to three proportional straight lines D, E, Z the extremes in given ratio, and

[1] let the ratio A:D and the ratio C:Z be given. I say that the ratio B:E is also given.

For, since the ratio A:D is given, and the ratio C:Z is given, therefore the ratio ⊏⊐A,C:⊏⊐D:Z is given [Dt 70]. But ⬜B = ⊏⊐A,C, and ⬜E = ⊏⊐D,Z [VI.17]. Therefore the ratio ⬜B:⬜E is given; so that the ratio B:E is given [Dt 54 or Dt 24*D].

A_____   D_____
B_____   E_____
C_____   Z_____

[2] Then, again, let the ratio A:D and the ratio B:E be given. I say that the ratio C:Z is also given.

Since the ratio A:D is given, and the ratio B:E is given, therefore the ratio ▭B:▭E is given [Dt 50]. But ▭B = ⊏⊐A,C, and ▭E= ⊏⊐D,Z [VI.17]; therefore the ratio ⊏⊐A,C:⊏⊐D,Z is given.

And the ratio of one side A to one side D is given; therefore the ratio of the remaining side C to the remaining side Z is also given [Dt 68]. ■

Remarks: Dt 81 & 83 are applications of the theorems about equiangular parallelograms (Dt 68 & 70). In the idiom of compound ratio (which is absent from the Data, but mentioned with Dt 56) those theorems can be stated as follows: Since given ratios compound into a given ratio (Dt 8), and since the ratio of equiangular parallelograms is compounded from the ratios of their sides (VI, 23),

Dt 70 can be said to prove that the compound ratio (of the sides) of two equiangular parallelograms is given if its components (*i.e.* the ratios) are given, (which is little more than we knew from Dt 8), and

Dt 68 can be said to prove that if one pair of components in a given compound ratio is given, then the other component ratio is also given.

Dt 81 uses the fact that the square on the mean proportional equals the rectangle of the extremes, and then Dt 70 and Dt 68. Dt 81 is cognate to Dt 24 and uses the unproven Dt 24*D, which through Dt 54 plays an important role in the proof. The second part of Dt 81 is a partial converse of the first.

Hypotheses	A:B :: B:C & D:E :: E:Z
1)	gvn.A:D & gvn.C:Z
Assertion	gvn.B:E
Proof	
[Dt 70]	gvn.⊏⊐A,C:⊏⊐D,Z
[VI.17]	▭B = ⊏⊐A,C, ▭E = ⊏⊐D,Z ∴ gvn.▭B:▭E
[Dt 24*D, Dt 54]	gvn.B:E.
2)	gvn.A:D & gvn.B:E
Assertion	gvn.C:Z
Proof	
[Dt 50, VI.17]	gvn.B:E => gvn.▭B:▭E ∴ gvn.⊏⊐A,C:⊏⊐D,Z
[Dt 68]	gvn.A:D => gvn.C:Z

◆

Θε 82. Ἐὰν τέσσαρες εὐθεῖαι ἀνάλογον ὦσιν, ἔσται, ὡς ἡ πρώτη πρὸς ἣν ἡ δευτέρα λόγον ἔχει δεδομένον, οὕτως ἡ τρίτη πρὸς ἣν ἡ τετάρτη λόγον ἔχει δεδομένον.

ἔστωσαν τέσσαρες εὐθεῖαι ἀνάλογον αἱ Α, Β, Γ, Δ, ὡς ἡ Α πρὸς τὴν Β, οὕτως ἡ Γ πρὸς τὴν Δ· λέγω, ὅτι ἐστίν, ὡς ἡ Α πρὸς ἣν ἡ Β λόγον ἔχει δεδομένον, οὕτως ἡ Γ πρὸς ἣν ἡ Δ λόγον ἔχει δεδομένον.

ἔστω γὰρ πρὸς ἣν ἡ Β λόγον ἔχει δεδομένον ἡ Ε, καὶ πεποιήσθω, ὡς ἡ Β πρὸς τὴν Ε, οὕτως ἡ Δ πρὸς τὴν Ζ. λόγος δὲ τῆς Β πρὸς τὴν Ε δοθείς· λόγος ἄρα καὶ τῆς Δ πρὸς τὴν Ζ ἐστι δοθείς. καὶ ἐπεί ἐστιν, ὡς ἡ Α πρὸς τὴν Β, οὕτως ἡ Γ πρὸς τὴν Δ, ἐστι δὲ καί, ὡς ἡ Β πρὸς τὴν Ε, οὕτως ἡ Δ πρὸς τὴν Ζ, δι'ἴσου ἄρα ἐστίν, ὡς ἡ Α πρὸς τὴν Ε, οὕτως ἡ Γ πρὸς τὴν Ζ. καί ἐστιν ἡ μὲν Ε πρὸς ἣν ἡ Β λόγον ἔχει δεδομένον, ἡ δὲ Ζ πρὸς ἣν ἡ Δ· ἔστιν ἄρα ὡς ἡ Α πρὸς ἣν ἡ Β λόγον ἔχει δεδομένον, οὕτως ἡ Γ πρὸς ἣν ἡ Δ λόγον ἔχει δεδομένον.

Dt 82. *If four lines be proportional, as the first is to that to which the second has a given ratio, so the third will be to that to which the fourth has the* [147] *given ratio.*

Let four lines A, B, C, D be proportional, A:B :: C:D. I say that as A is to that to which B has the given ratio, so C will be to that to which D has the given ratio.

A _____
B _____
C _____   E _____
D _____   Z _____

For, let E be that to which B has the given ratio, and let B:E :: D:Z. The ratio B:E is given; therefore the ratio D:Z is also given.

And since A:B :: C:D, and also B:E :: D:Z ::, therefore *di'isou* A:E :: C:Z [V.22]. And E is that to which B has the given ratio, and Z [is that] to which D [has the given ratio]. Therefore as A is to that to which B has the given ratio, so C will be to that to which D has the given ratio. ∎

Remarks: The proposition proves that *if four lines be proportional, and the second have the same* given *ratio to a fifth as the fourth has to a sixth, then as the first is to the fifth, so the third will be to the sixth.* This is a trivial consequence, nay repetition, of V.22 (the *di'isou* theorem), and stated thus it is easier to follow

A:B :: C:D & B:E :: D:Z =[V.22]=> A:E :: C:Z.

---

[147] I translate the indefinite λόγον δεδομένον as *the* given ratio; it is in fact the *same* given ratio, as is often the case where this idiomatic statement occurs.

The wording should be compared to that of Dt 56; the idiom of 'given ratio' is superfluous, and no givenness is proved of the compound ratios A:E and C:Z. The theorem would fit into some Elements.

◆

Θε 83. Ἐὰν τέσσαρες εὐθεῖαι οὕτως ἔχωσι πρὸς ἀλλήλας, ὥστε τριῶν ληφθεισῶν ἐξ αὐτῶν ὁποιωνοῦν καὶ τετάρτης αὐταῖς προσληφθείσης ἀνάλογον, πρὸς ἣν ἡ λοιπὴ τῶν ἐξ ἀρχῆς τεσσάρων εὐθειῶν λόγον ἔχει δεδομένον, ἀνάλογον γίγνεσθαι τὰς τέσσαρας εὐθείας, ἔσται, ὡς ἡ τετάρτη πρὸς τὴν τρίτην, οὕτως ἡ δευτέρα πρὸς ἣν ἡ πρώτη λόγον ἔχει δεδομένον.

ἔστωσαν τέσσαρες εὐθεῖαι αἱ Α, Β, Γ, Δ οὕτως ἔχουσαι πρὸς ἀλλήλας, ὥστε τριῶν ληφθεισῶν ἐξ αὐτῶν ὁποιωνοῦν τῶν Α, Β, Γ καὶ τετάρτης αὐταῖς προσληφθείσης τῆς Ε, πρὸς ἣν ἡ Δ λόγον ἔχει δεδομένον, ἀνάλογον εἶναι τὰς Α, Β, Γ, Ε εὐθείας· λέγω, ὅτι ἐστίν, ὡς ἡ Δ πρὸς τὴν Γ, οὕτως ἡ Β πρὸς ἣν ἡ Α λόγον ἔχει δεδομένον.

ἐπεὶ γάρ ἐστιν, ὡς ἡ Α πρὸς τὴν Β, οὕτως ἡ Γ πρὸς τὴν Ε, τὸ ἄρα ὑπὸ τῶν Α, Ε ἴσον ἐστὶ τῷ ὑπὸ τῶν Β, Γ. καὶ ἐπεὶ λόγος ἐστὶ τῆς Ε πρὸς τὴν Δ δοθείς, λόγος ἄρα ἐστὶ καὶ τοῦ ὑπὸ τῶν Α, Δ πρὸς τὸ ὑπὸ τῶν Α, Ε δοθείς· τῷ δὲ ὑπὸ τῶν Α, Ε ἐστιν ἴσον τὸ ὑπὸ τῶν Β, Γ· λόγος ἄρα καὶ τοῦ ὑπὸ τῶν Δ, Α πρὸς τὸ ὑπὸ τῶν Β, Γ ἐστι δοθείς. ἔστιν ἄρα ὡς ἡ Δ πρὸς τὴν Γ, οὕτως ἡ Β πρὸς ἣν ἡ Α λόγον ἔχει δεδομένον.

Dt 83. *If four straight lines be to each other in such a way that, any three of them taken together with a fourth to which the last of the original four has a given ratio, the four straight lines be proportional, then as the fourth is to the third, so the second will be to that to which the first has the given ratio.*

Let four straight lines A, B, C, D be to each other in such a way that, any three of them A, B, C taken together with a fourth E, to which D has a given ratio, the straight lines A, B, C, E are proportional. I say that as D is to C, so B will be to that to which A has the given ratio.

For, since C:E :: A:B, therefore ⊏⊐A,E = ⊏⊐B,C [VI.16]. And since the ratio E:D is given, therefore the ratio ⊏⊐A,D:⊏⊐A,E is also given [VI.1]; and ⊏⊐A,E = ⊏⊐B,C; therefore the ratio ⊏⊐D,A:⊏⊐B,C is given.

Therefore as D is to C, so is B to that to which A has the given ratio. ∎

Remarks: The meaning of this proposition is difficult to follow for a couple of reasons: First, the line to which A has a given ratio is not named in the ekthesis and proof; below I let it emerge under the name of Z. Secondly, there is only one given ratio, namely D:E, so A:Z must be the same. Dt 83 expands (and the abrupt conclusion in the last line certainly presupposes) Dt 74, as will appear from figure 83.1, in which Q (⊏⊐B,C) = Q' (⊏⊐A,E) = Q" (⊏⊐D,Z). Here is a short-hand version:

Hypotheses   A:B :: C:E
                 gvn.D:E & [A:Z :: D:E ∴] gvn.A:Z
Assertion   D:C :: B:Z.
Proof
[Hyp, VI.16]   ⊏⊐A,E = ⊏⊐B,C [= ⊏⊐D,Z]   Q' = Q = Q"
[Hyp, VI.1]   gvn.⊏⊐A,D:⊏⊐A,E   gvn.P:Q'
Therefore   gvn.⊏⊐A,D:⊏⊐B,C & gvn.⊏⊐A,D:⊏⊐D,Z   gvn.P:Q & gvn.P:Q"
[VI.1]   gvn.A:Z
[or VI:14]   Q' = Q" => A:Z :: D:E ∴ gvn.A:Z

The proposition proves that *if four lines be proportional, and the first have the same given ratio to a fifth as a sixth has to the fourth, then as the second is to the fifth, so the sixth will be to the third*. This is a trivial consequence, nay repetition, of V.23 (the *di'isou* theorem perturbed), and stated thus it is easier to follow:
A:B :: C:E [∴ B:A :: E:C] & A:Z :: D:E =[V.23]=> B:Z :: D:C.

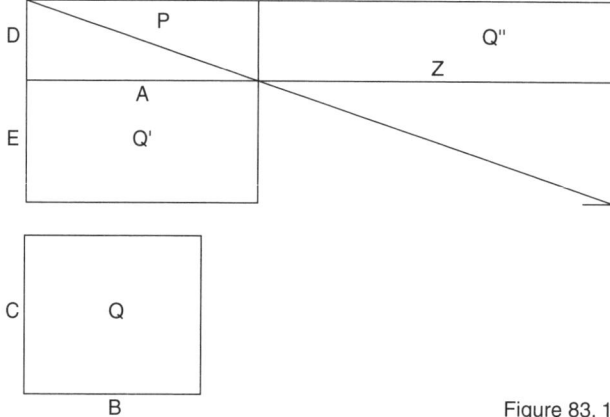

Figure 83. 1

Had Eudoxus Dt 82 and 83 in mind when he proved V.22 and 23? Or was it the other way round? It is really hard to decide with so little evidence. Nobody has previously pointed to this similarity, as far as I know.

As in Dt 82, the idiom of 'given ratio' is superfluous, and the theorem would fit into some Elements.

CHAPTER 11. DUPLICATES AND OUTSIDERS. (76–83)

Deductive Structure of Dt 68–83

From the *Data*

		Dt	68	69	70	71	72	73	74	75	76	77	78	79	80	81	82	83
Data																		
Def.	2*		+	.	+	+	+	+	+	+	.	.	+	.	+	+	+	+
	3		.	+	+	.	.	+	.	.	+	.	+	.	+	.	.	.
	8		.	.	.	.	.	.	.	.	.	.	.	+	.	.	.	.
Dt	2		.	.	.	.	.	.	.	.	.	.	.	+	.	.	.	.
	3		.	+	+	.	.	+	.	.	+	.	.	.	+	.	.	.
	4		.	+	+	.	.	+	.	.	+	.	.	.	+	.	.	.
	8		+	+	+	.	.	.	.	.	+	+	+	.	+	.	.	.
	9		.	.	.	.	.	.	.	.	.	.	.	.	.	+	.	.
	24*D		.	.	.	.	.	.	.	.	+	.	.	.	.	.	.	.
	25		.	.	.	.	.	.	.	.	.	.	.	.	+	.	.	.
	26		.	.	.	.	.	.	.	.	.	.	.	.	+	.	.	.
	27		.	.	.	.	.	.	.	.	.	.	.	.	+	.	.	.
	28		.	.	.	.	.	.	.	.	.	.	.	.	+	.	.	.
	29		.	.	.	.	.	.	.	.	.	.	.	.	+	.	.	.
	30		.	.	.	.	.	.	.	.	.	.	.	.	+	.	.	.
	39		.	.	.	.	.	.	.	.	.	.	.	.	+	.	.	.
	40		.	+	+	.	.	+	.	.	+	.	.	.	+	.	.	.
	49		.	.	.	.	.	.	.	.	.	+	+	.	.	.	.	.
	50		.	.	.	.	.	.	.	.	.	.	.	.	.	+	.	.
	54		.	.	.	.	.	.	.	.	+	.	.	.	+	.	.	.
	56		.	.	.	.	+	.	.	.	.	.	.	.	.	.	.	.
	68		.	+	.	.	.	.	.	.	.	.	.	.	+	.	.	.
	70		.	.	.	+	+	.	.	.	.	.	.	.	.	.	.	.
	74		.	.	.	.	.	.	+	.	.	.	.	.	.	.	.	+
	79		.	.	.	.	.	.	.	.	.	.	.	+	.	.	.	.
		Dt	68	69	70	71	72	73	74	75	76	77	78	79	80	81	82	83

Dt 71 and 72 are corollaries to Dt 70, Dt 75 to 74. The latter, like Dt 56 (which proves the same thing), does not belong in the *Data*, but in some Elements. The same holds for Dt 79, 82, and 83, whereas Dt 80 depends heavily on the *Data*.

Deductive Structure of Dt 68–83

From the *Elements*

		Dt	68	69	70	71	72	73	74	75	76	77	78	79	80	81	82	83
Elements																		
C.N.	1		.	.	.	.	.	.	.	.	.	.	.	+	.	.	.	.
	3		+	.	.	.	.	.	.	.	.	.	.	.	.	.	.	.
I.	11		.	.	.	.	.	.	.	.	.	.	.	.	+	.	.	.
	12		.	.	.	.	.	.	.	.	.	.	.	+	.	.	.	.
	14		+	.	+	.	+	+	.	.	.	+	.	.	.	.	.	.
	23		.	+	+	.	+	+	.	.	.	.	.	+	.	.	.	.
	28		.	.	.	.	.	.	.	.	.	.	.	+	.	.	.	.
	29		+	+	+	.	+	+	+	.	.	.	+	+	.	.	.	.
	31		.	.	.	.	.	.	.	.	.	.	.	.	+	.	.	.
	32		.	+	+	.	.	+	.	+	.	.	.	+	+	.	.	.
	33		.	.	.	.	.	.	.	.	.	.	.	+	.	.	.	.
	34		.	.	.	+	.	.	.	.	.	.	.	+	.	.	.	.
	35		.	+	+	.	.	+	+	.	.	.	.	.	.	.	.	.
	41		.	.	.	.	+	.	.	+	.	.	.	+	.	.	.	.
	45		+	.	+	.	.	+	+	.	.	+	.	.	.	.	.	.
	46		.	.	.	.	.	.	.	.	.	+	+	.	.	.	.	.
III.	21		.	.	.	.	.	.	.	.	.	.	.	+	.	.	.	.
	33		.	.	.	.	.	.	.	.	.	.	.	.	+	.	.	.
IV.	5		.	.	.	.	.	.	.	.	.	.	.	+	.	.	.	.
V.	7		+	.	+	.	+	+	+	.	.	.	+	.	+	.	.	+
	9		.	.	.	.	.	.	.	.	.	.	.	+	.	.	.	.
	11		.	.	.	.	.	.	.	.	.	.	.	+	.	.	.	.
	15		.	.	.	+	+	.	.	.	.	.	.	.	.	.	.	.
	22		.	.	.	.	.	.	.	.	.	.	.	+	.	.	+	.
VI.	Def.1		.	.	.	.	.	.	.	.	.	.	.	+	.	.	.	.
	1		+	.	+	.	.	+	+	.	.	+	.	+	.	.	.	+
	4		.	.	.	.	.	.	.	.	.	.	.	+	.	.	.	.
	12		.	.	.	.	.	.	.	.	.	.	.	.	+	.	.	.
	14		+	.	+	.	.	+	+	.	.	.	+	.	.	.	.	.
	16		.	.	.	.	.	.	.	.	.	.	.	.	.	.	.	+
	17		.	.	.	.	.	.	.	.	.	.	.	.	.	+	.	.
		Dt	68	69	70	71	72	73	74	75	76	77	78	79	80	81	82	83

Chapter 12. Application of areas. II
Dt (57–62) & 84–85

Θε **84.** ΄Ἐὰν δύο εὐθεῖαι δοθὲν χωρίον περιέχωσιν ἐν δεδομένῃ γωνίᾳ, ἡ δὲ ἑτέρα τῆς ἑτέρας δοθείσῃ μείζων ᾖ, καὶ ἑκατέρα αὐτῶν ἔσται δοθεῖσα.

δύο γὰρ εὐθεῖαι αἱ ΑΒ, ΒΓ δοθὲν χωρίον περιεχέτωσαν τὸ ΑΓ ἐν δεδομένῃ γωνίᾳ τῇ ὑπὸ τῶν ΑΒΓ, ἡ δὲ ΓΒ τῆς ΒΑ δοθείσῃ μείζων ἔστω· λέγω, ὅτι δοθεῖσά ἐστιν ἑκατέρα τῶν ΒΑ, ΒΓ.

ἐπεὶ γὰρ ἡ ΒΓ τῆς ΒΑ δοθείσῃ μείζων ἐστίν, ἔστω ἡ δοθεῖσα ἡ ΔΓ· λοιπὴ ἄρα ἡ ΔΒ τῇ ΒΑ ἴση ἐστίν. καὶ συμπεπληρώσθω τὸ ΑΔ. καὶ ἐπεὶ ἴση ἐστὶν ἡ ΑΒ τῇ ΔΒ, λόγος ἄρα ἐστὶ τῆς ΑΒ πρὸς τὴν ΒΔ δοθείς· δοθεῖσα δὲ καὶ ἡ ὑπὸ τῶν ΑΒΔ γωνία· δέδοται ἄρα τὸ ΑΔ τῷ εἴδει. ἐπεὶ οὖν τὸ ΑΓ δοθὲν παρὰ δοθεῖσαν τὴν ΔΓ παραβέβληται ὑπερβάλλον εἴδει δεδομένῳ τῷ ΑΔ, δέδοται ἄρα τὸ πλάτος τῆς ὑπερβολῆς· δοθεῖσα ἄρα ἐστὶν ἡ ΒΔ. ἀλλὰ καὶ ἡ ΔΓ· καὶ ὅλη ἄρα ἡ ΒΓ δοθεῖσά ἐστιν. ἔστι δὲ καὶ ἡ ΑΒ δοθεῖσα· ἑκατέρα ἄρα τῶν ΑΒ, ΒΓ δοθεῖσά ἐστιν.

Dt 84. *If two straight lines contain a given area in a given angle, and one of them be greater than the other by a given* [straight line], *each of them will be given, too.*

For, let the two straight lines AB, BC contain the given area (AC) in the given angle ABC, and let CB be greater than BA by a given [straight line]. I say that each of BA, BC is given.

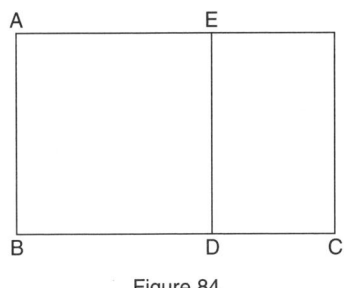

Figure 84

For, since BC is greater than BA by a given [straight line], let the given [straight line] be DC; then the remainder DB is equal to BA [Def. 9].

And let (AD) have been completed.[148] And since AB is equal to DB, the ratio AB:BD is given; and the angle ABD is given; therefore (AD) is given in form [Def. 3].

Then, since, the given [area] (AC) has been applied to the given straight line DC exceeding by the given form (AD), the width of the excess is given [Dt 59]. Therefore BD is given; and DC is also given; therefore the whole BC is given. And AB is also given; thus each of AB, BC is given. ∎

---

[148] *i.e.* let the parallelogram (AD) be constructed. The Greek συμπεπληρώσθω usually means 'let *a parallelogram* be completed'.

Remarks: Dt 84 and 85 are, as was said above (p. 155), extensions of Dt 59 and 58: not only the length and width of the excess and defect are shown to be given, but also (as an evident corollary) the length and width of the given area, which is very easily seen.

Although quadratic problems like '*A rectangle has a known area, and the sum or difference of its length and height is known. To find its length and height*' were well known since Babylonian times, the notion of equations seems to be completely alien to the author of the Data. When Thomas Heath in his comments on II.5-6 writes about Dt 85, 'This proposition then enables us to solve the problem of finding a rectangle the area and perimeter of which are both given', he is right [149] in so far that the proposition proves that the problem has only one solution (the two apparent solutions are simply the result of interchanging the sides). But Dt 85 does not teach us how to find the solutions; that was not even done in Dt 58; for a constructive solution to the problem we must look in the *Elements* VI.28.

To sum up: Dt 58 proves that the length and width of the *defect* (of the elliptical application of a 'given' area) are 'given', and Dt 85 proves the same fact for the length and width of *the given area*. Dt 59 proves that the length and width of the *excess* (of the hyperbolical application of a 'given' area) are 'given', and Dt 84 proves the same fact for the length and width of *the given area*.

Few theorems in the *Data* leave us with so many frustrated expectations as those four propositions. I am inclined to follow (Herz-Fischler 1984); I paraphrase:

> To interpret proposition 85 as a non-constructive 'existence' proof that the set of equations $x-y = b$ and $xy = c^2$ has a unique solution, is not valid in the context of Greek Mathematics, for it requires the *a priori* assumption that we are dealing with equations here, and that the Greeks did consider such things as the existence of solutions to equations. Propositions 84 and 85 did not, at the time of Euclid, have anything to do with the concept of 'equations'. In the context of the *Data* they do not present any unusual aspects.

But then, I cannot help admit my suspicion: that we may all be barking up the wrong tree, looking in the wrong direction.

◆

---

[149] Although Dt 84 and 85 are about any parallelograms and not specificallly rectangles; but the talk of parallelograms may be purely ornamental, since in all such problems parallelograms can be reduced to rectangles.

CHAPTER 12. APPLICATION OF AREAS II. (84–85)

Θε 85. Ἐὰν δύο εὐθεῖαι δοθὲν χωρίον περιέχωσιν ἐν δεδομένῃ γωνίᾳ, ᾗ δὲ συναμφότερος δοθεῖσα, καὶ ἑκατέρα αὐτῶν ἔσται δοθεῖσα.

δύο γὰρ εὐθεῖαι αἱ ΑΒ, ΒΓ δοθὲν χωρίον περιεχέτωσαν τὸ ΑΓ ἐν δεδομένῃ γωνίᾳ τῇ ὑπὸ τῶν ΑΒΓ, καὶ ἔστω συναμφότερος ἡ ΑΒΓ δοθεῖσα· λέγω, ὅτι καὶ ἑκατέρα τῶν ΑΒ, ΒΓ ἐστι δοθεῖσα.

διήχθω γὰρ ἡ ΓΒ ἐπὶ τὸ Δ, καὶ κείσθω τῇ ΑΒ ἴση ἡ ΒΔ, καὶ διὰ τοῦ Δ τῇ ΒΑ παράλληλος ἤχθω ΔΕ, καὶ συμπεπληρώσθω τὸ ΑΔ. καὶ ἐπεὶ ἴση ἐστὶν ἡ ΔΒ τῇ ΒΑ, καί ἐστι δοθεῖσα ἡ ὑπὸ ΑΒΔ γωνία, ἐπεὶ καὶ ἡ ἐφεξῆς αὐτῇ δοθεῖσά ἐστιν, δέδοται ἄρα τὸ ΕΒ τῷ εἴδει. καὶ ἐπεὶ δοθεῖσά ἐστι συναμφότερος ἡ ΑΒΓ, ἴση δὲ ἡ ΑΒ τῇ ΒΔ, δοθεῖσα ἄρα ἐστὶν ἡ ΔΓ. ἐπεὶ οὖν δοθὲν τὸ ΑΓ παρὰ δοθεῖσαν τὴν ΔΓ παραβέβληται ἐλλεῖπον εἴδει δεδομένῳ τῷ ΕΒ, δέδοται τὰ πλάτη τοῦ ἐλλείμματος· δοθεῖσαι ἄρα εἰσὶν αἱ ΑΒ, ΒΔ. ἀλλὰ καὶ συναμφότερος ἡ ΑΒΓ δοθεῖσά ἐστιν· καὶ λοιπὴ ἄρα ἡ ΒΓ δοθεῖσά ἐστιν· δοθεῖσα ἄρα ἐστὶν ἑκατέρα τῶν ΑΒ, ΒΓ.

Dt 85. *If two straight lines contain a given area in a given angle, and their sum be given, each of them will be given.*

For, let the two straight lines AB, BC contain the given area (AC) in the given angle ABC, and let the sum AB+BC [150] be given. I say that each of AB, BC is given.

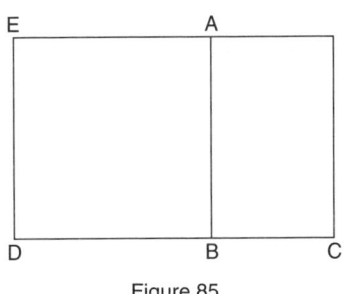

Figure 85

For, let CB have be produced to D, and let BD have been laid out equal to AB, and through D let DE be drawn parallel to BA, and let (AD) have been completed.

And since DB is equal to BA, and the angle ABD is given, because its adjacent angle is given, too, (EB) is given in form [Def. 3].

And since the sum AB+BC is given, and AB is equal to BD, the [straight line] DC is given.

Then, since the given [area] (AC) has been applied to the given straight DC deficient by the given form (EB), the length and width of the defect are given [Dt 58]. Therefore AB and BD are given; but the sum AB+BC is also given; therefore the remainder BC is also given; therefore each of AB, BC is given [Dt 4]. ∎

◆

---

[150] The text has ABC, as if on one straight line.

Chapter 13. Intersecting Hyperbolas
H. G. Zeuthen's conjecture about Dt 86

Θε 86. ' Ἐὰν δύο εὐθεῖαι δοθὲν χωρίον περιέχωσιν ἐν δεδομένῃ γωνίᾳ, δύνηται δὲ ἡ ἑτέρα τῆς ἑτέρας δοθέντι μεῖζον ἢ ἐν λόγῳ, καὶ ἑκατέρα αὐτῶν ἔσται δοθεῖσα.

δύο γὰρ εὐθεῖαι αἱ ΑΒ, ΒΓ δοθὲν χωρίον περιεχέτωσαν τὸ ΑΓ ἐν δεδομένῃ γωνίᾳ τῇ ὑπὸ τῶν ΑΒΓ, τὸ δὲ ἀπὸ τῆς ΓΒ τοῦ ἀπὸ τῆς ΒΑ δοθέντι μεῖζον ἔστω ἢ ἐν λόγῳ· λέγω, ὅτι καὶ ἑκατέρα τῶν ΑΒ, ΒΓ ἐστὶ δοθεῖσα.
ἐπεὶ γὰρ τὸ ἀπὸ τῆς ΓΒ τοῦ ἀπὸ τῆς ΒΑ δοθέντι μεῖζόν ἐστιν ἢ ἐν λόγῳ, ἀφῃρήσθω τὸ δοθὲν τὸ ὑπὸ τῶν ΓΒΔ· λοιποῦ ἄρα τοῦ ὑπὸ τῶν ΔΓΒ πρὸς τὸ ἀπὸ τῆς ΑΒ λόγος ἐστὶ δοθείς. καὶ ἐπεὶ δοθέν ἐστι τὸ ὑπὸ τῶν ΑΒΓ, ἔστι δὲ καὶ τὸ ὑπὸ τῶν ΓΒ, ΒΔ δοθέν, λόγος ἄρα ἐστὶ τοῦ ὑπὸ τῶν ΑΒ, ΒΓ πρὸς τὸ ὑπὸ τῶν ΓΒΔ δοθείς. ὡς δὲ τὸ ὑπὸ τῶν ΑΒΓ πρὸς τὸ ὑπὸ τῶν ΓΒ, ΒΔ, οὕτως ἡ ΑΒ πρὸς τὴν ΒΔ· ὥστε καὶ τῆς ΑΒ πρὸς τὴν ΒΔ λόγος ἐστὶ δοθείς· ὥστε καὶ τοῦ ἀπὸ τῆς ΑΒ πρὸς τὸ ἀπὸ τῆς ΒΔ λόγος ἐστὶ δοθείς. τοῦ δὲ ἀπὸ τῆς ΑΒ πρὸς τὸ ὑπὸ τῶν ΒΓΔ λόγος ἐστὶ δοθείς· καὶ τοῦ ὑπὸ τῶν ΒΓΔ ἄρα πρὸς τὸ ἀπὸ τῆς ΔΒ λόγος ἐστὶ δοθείς· ὥστε καὶ τοῦ τετράκις ὑπὸ τῶν ΒΓΔ πρὸς τὸ ἀπὸ τῆς ΒΔ λόγος ἐστὶ δοθείς· τοῦ τετράκις ὑπὸ τῶν ΒΓΔ ἄρα μετὰ τοῦ ἀπὸ τῆς ΒΔ πρὸς τὸ ἀπὸ τῆς ΒΔ λόγος ἐστὶ δοθείς. ἀλλὰ τὸ τετράκις ὑπὸ τῶν ΒΓΔ μετὰ τοῦ ἀπὸ τῆς ΒΔ τὸ ἀπὸ συναμφοτέρου ἐστὶ τῆς ΒΓ, ΓΔ. λόγος ἄρα ἐστὶ καὶ τοῦ ἀπὸ συναμφοτέρου τῆς ΒΓ, ΓΔ πρὸς τὸ ἀπὸ τῆς ΒΔ δοθείς· ὥστε καὶ συναμφοτέρου τῆς ΒΓΔ πρὸς τὴν ΒΔ λόγος ἐστὶ δοθείς· καὶ συνθέντι ἄρα δύο τῶν ΓΒ πρὸς τὴν ΒΔ λόγος ἐστὶ δοθείς· ὥστε καὶ μιᾶς τῆς ΓΒ πρὸς τὴν ΒΔ λόγος ἐστὶ δοθείς. ὡς δὲ ἡ ΓΒ πρὸς ΒΔ, οὕτως τὸ ὑπὸ τῶν ΓΒΔ πρὸς τὸ ἀπὸ τῆς ΒΔ· καὶ τοῦ ὑπὸ τῶν ΓΒΔ ἄρα πρὸς τὸ ἀπὸ τῆς ΒΔ λόγος ἐστὶ δοθείς. δοθὲν δὲ τὸ ὑπὸ τῶν ΓΒ, ΒΔ· δοθὲν ἄρα καὶ τὸ ἀπὸ τῆς ΒΔ· δοθεῖσα ἄρα ἐστὶν ἡ ΒΔ· ὥστε καὶ ἡ ΒΓ δο θεῖσά ἐστιν· τῆς γὰρ ΓΒ πρὸς τὴν ΒΔ λόγος ἐστὶ δοθείς, καὶ δέδοται ἡ ΒΔ· καί ἐστι δοθὲν τὸ ΑΓ, καὶ δοθεῖσα ἡ Β γωνία· δοθεῖσα ἄρα ἐστὶ καὶ ἡ ΑΒ· ἑκατέρα ἄρα τῶν ΑΒ, ΒΓ δοθεῖσά ἐστιν.

Dt 86. *If two straight lines contain a given area in a given angle and if the square on one of them be by a given [area] greater than in [a given] ratio to the square on the other, then each of them will also be given.*

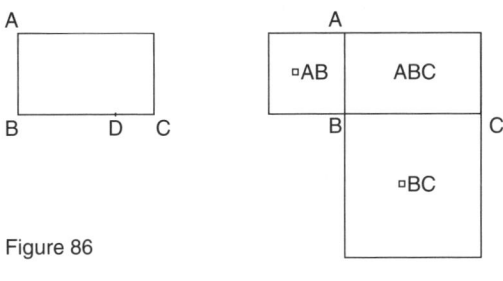

Figure 86

For, let the two straight lines AB, BC contain the given area (AC) in the given angle ABC, and let the square on CB be by a given [area] greater than in [a given] ratio to the square on BA. I say that each of AB, BC is also given.

For, since ⊐CB is by a given [area] greater than in a given ratio to ⊐BA, let the given rectangle (CBD) have been subtracted; then the ratio of the remaining rectangle (DCB) to the square on AB is given [Def. 11].

And since the [rectangle] ⊏⊐ABC is given, and ⊏⊐CB,BD is also given, therefore the ratio ⊏⊐AB,BC:⊏⊐CBD is given [Dt 1].

But ⊏⊐ABC:⊏⊐CB,BD :: AB:BD [VI.1]; [151] therefore the ratio AB:BD is also given [Def. 2*]; therefore the ratio ⊐AB:⊐BD is also given [Dt 50].

The ratio ⊐AB:⊏⊐BCD is given [by hypothesis]; therefore the ratio ⊏⊐BCD:⊐DB is also given [Dt 8]; therefore the ratio 4⊏⊐BCD:⊐BD is also given [Dt 6]; and [*synthenti,*] the ratio (4⊏⊐BCD + ⊐BD) : ⊐BD is given [Dt 6].

But 4⊏⊐BCD + ⊐BD = ⊐(BC+CD) [II.8]. Therefore the ratio ⊐(BC+CD):⊐BD is given. And so the ratio (BC + CD):BD is given [Dt 54 or 24*D]; therefore, *synthenti*, the ratio 2CB:BD is given [Dt 6]; so that the ratio of CB alone to BD is given [Dt 8].

But CB:BD :: ⊏⊐CBD:⊐BD [VI.1]. Therefore the ratio ⊏⊐CBD:⊐BD is given [Def. 2*]. ⊏⊐CB,BD is given; therefore ⊐BD is also given [Dt 2]; and so BD is given; so that BC, too, is given, because the ratio CB:BD is given, <and BD is given>; [and the ratio AB:BD is given; therefore AB is given;] [152] and the parallelogram (AC) is given and the angle B is given ; therefore AB is also given [Dt 57]. Therefore each of AB and BC are given. ∎

Remarks: Dt 86 is the only instance in the *Data* where we meet the relation R'11, 'by a given greater than in ratio'. In fact it serves only as a mnemotechnic idiom to describe that a certain difference between two areas has a given ratio to one of them. The proposition may be said to deal with the givenness of the intersection of two hyperbolas having the same two given straight lines for conjugate diameters and asymptotes respectively. So said *H. G. Zeuthen* in his last publication [153] (1917), in

---

[151] Τὸ ὑπὸ τῶν ΑΒΓ must be understood as 'the rectangle ABC', not the parallelogram, if *Elements* VI.1 is to apply (because CB,BD is a rectangle); a figure like 86.1 will save the text, A̲BC being the parallelogram mentioned in the *protasis*, ABC the rectangle in the argument. A̲B = AB. See my *remarks*.

[152] The argument could end here with the passage in brackets and the conclusion "Therefore ..."; but Euclid wants to revive the parallelogram, which he has ignored all the time (previous note).

[153] *Hvorledes Mathematiken i tiden fra Platon til Euklid blev rationel Videnskab*, p. 115 ff. It was published in Danish only, with a short summary in French: *Sur la réforme qu'a subie la mathematique de Platon à Euclide, et grâce à laquelle elle est devenue science raisonnée.*

CHAPTER 13. INTERSECTING HYPERBOLAS. (86)

some sense his testament and his legacy to the study of Greek mathematics; I suppose he is right in this conjecture (and shall come back to that below, p. 224), though it is hardly evident from the Greek text, and even less from his treatment of the proposition. He says, in my translation from Danish:

> The geometric form conceals a complete algebraic solution of the equations
> $$xy = a \text{ and } y^2 - mx^2 = b.$$
> Below we are going to translate the solution into our present algebraic language, premising only one remark: that the products signify rectangles, x^2 and y^2 signify [geometric] squares, and what we write on the right side of our equations renders the statement that the left side expression is *given*, that is: can be determined step by step, when a, b, and m are given.
> The operations are the following: one puts $b = yz$, and has, successively rendering Euclid's reasonings step by step:

$$xy =_1 a$$

$$y^2 - mx =_2 b$$

$$\frac{y(y-z)}{x^2} =_3 m$$

$$\frac{x}{z} =_4 \frac{a}{b}$$

$$\frac{x^2}{z^2} =_5 \frac{a^2}{b^2}$$

$$\frac{x^2}{y(y-z)} =_6 \frac{1}{m}$$

$$\frac{4y(y-z)}{z^2} =_7 4m\frac{a^2}{b^2}$$

$$\frac{4y(y+z)+z^2}{z^2} =_8 \frac{(2y-z)^2}{z^2} =_9 4m\frac{a^2}{b^2}+1$$

$$\frac{2y-z}{z} =_{10} \sqrt{4m\frac{a^2}{b^2}+1}$$

$$\frac{2y}{z} =_{11} \sqrt{4m\frac{a^2}{b^2}+1}+1$$

$$\frac{y}{z} =_{12} \frac{yz}{z^2} =_{13} \frac{b}{z^2} =_{14} \frac{1}{2}\left(\sqrt{4m\frac{a^2}{b^2}+1}+1\right)$$

$$z =_{15} \sqrt{\frac{2b}{\sqrt{4m\frac{a^2}{b^2}+1}+1}}$$

$$yz =_{16} \frac{b}{\sqrt{4m\frac{a^2}{b^2}+1}+1}$$

$$y =_{17} b\sqrt{\frac{\sqrt{4m\frac{a^2}{b^2}+1}+1}{2b}}$$

What we write is no more than the algebraic symbols for the operations, by which these successive determinations are done according to previous propositions of the book. The use of *geometry plays hardly any role* (Taisbak's emphasis); its one and only function is to mark off the auxiliary magnitude, which we call z, along y so that $y-z$ is also presented as a line segment. (In Euclid's text line segments are marked throughout by letters at their endpoints.)

This is algebra, all of it, and when it is put into words (in Euclid's text), *the geometrical way of exposition plays no role at all* (Taisbak's emphasis), and geometrical intuition nowhere replaces the algebraic reasonings. Each of the equations we have set up rests on a previous proposition or definition in Menge's edition of the *Data*.

◆

Zeuthen quotes Dt 86 as an example of Euclid's ability, nay dexterity, of handling algebraic problems in geometric form. I am going to challenge Zeuthen's algebraic interpretation, which (as far as I know) was fully stated in Danish only, by showing what I see in the Greek text.

I will leave Zeuthen's paper alone for a while, to give a transcription in my own style. First a note about my notation: I, myself, find the arguments easier to follow if line segments are denoted by single letters instead of two, and if congruent segments have the same name when only their lengths and not positions are relevant. (In fact that is what Euclid does.) So letters x, y, z, w denote line segments; $\sqsubset\sqsupset x,y$ is a rectangle; $\square x$ is a *geometric* square. Expressions in braces {...} should be read as 'is given' and are my interpretations, not in the text; letters in those braces denote 'latent coactors', *i.e.* 'known' segments or areas that are neither mentioned nor seen.

An example: when the text says "$\sqsubset\sqsupset$ABC is given" I transcribe it as gvn.$\sqsubset\sqsupset x,y$ {= $\square a$}, to be read "the rectangle xy is given, which means that it is equal to some unseen area, say the square on (the line segment) a". And if the text says "the ratio of $\sqsubset\sqsupset$DCB to $\square$AB is given", I transcribe it as "gvn.$\sqsubset\sqsupset w,y$:$\square x$ {= $c:d$}", to be read "the ratio of the rectangle wy to the square on x is the same as some unseen ratio $c:d$".

In the *protasis* certain areas are said to be given; for all we know they may be given as squares. The line segments x and y are to be proved 'given'.

If, because of this notation, you feel that you are reading algebra, the reason is that you were brought up nowadays. No Greek could possibly feel that (although I must confess that we cannot know what they felt), because algebra was not yet invented; he would know that he was manipulating ratios and proportions and applying areas; but then, probably no Greek would supply what I am doing in those braces; whatever we write in the braces is our own concoction, which may (or may not) help us to follow the operations more easily.

CHAPTER 13. INTERSECTING HYPERBOLAS. (86)

Four line segments occur, $AB = x$, $BC = y$, $BD = z$, and $CD = y-z = w$. Five rectangular figures (with their names from the text in parentheses),

(rectangle ABC)	⊏⊐AB,BC	= ⊏⊐x,y,
(square on CB)	□CB	= □y,
(rectangle CBD or DBC)	⊏⊐CB,BD	= ⊏⊐y,z,
(rectangle DCB or BCD)	⊏⊐DC,CB	= ⊏⊐w,y,
(square on AB)	□AB	= □x.

The manuscripts present only a poor figure (86 left), a rectangle with letters A, B, C, and D. In my supplemental diagrams (86.2–5) I introduce auxiliary letters B′, C′, D′. The rule is that, say, B′C′ is congruent with BC, etc.

Since the ratios of the sides in a right triangle are given if one acute angle is given [Dt 40], we may without loss of generality let the parallelogram AB,BC be a rectangle. Euclid did so, tacitly; or rather, the figure in the manuscripts does not represent the parallelogram ABC, but the rectangle contained by AB and BC, which causes some confusion. We invent another diagram (figure 86.1, inspired by Menge's lemma p. 224) showing both the parallelogram ABC with the given angle ABC [v], and the rectangle ABC, with AB = AB. We note that the ratio of the parallelogram to the rectangle is given {= sin v, anachronistically} because of the given angle [Dt 40]. So, since the parallelogram ABC is given, ⊏⊐ABC is given, too.

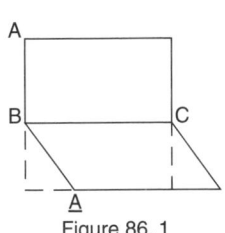
Figure 86.1

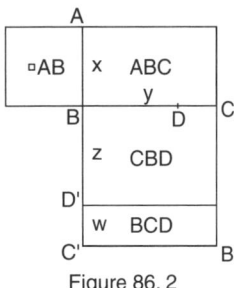
Figure 86.2

We may let the given area of the parallelogram (AC) be equal to an unseen square {□α}; the ratio of that square to ⊏⊐x,y (AC) is given {= sin v}, so ⊏⊐x,y equals another unseen square {□α}. Now we are ready to follow Euclid's arguments and introduce other 'latent coactors' as we need them, with an eye on figure 86.2:

Let the two straight lines AB, BC contain the given area (AC) in the given angle ABC, and let ⊓CB be by a given [area] greater than in [a given] ratio to ⊓BA. I say that each of AB,BC is also given.

Since ⊓CB is by a given area greater than in [a given] ratio to ⊓BA, let the given [area] have been subtracted as ⊏⊐CBD′; then the ratio of the remaining ⊏⊐DCB to ⊓AB is given [Def. 11].

And since ⊏⊐ABC is given, and ⊏⊐CBD′ is also given, therefore the ratio ⊏⊐ABC:⊏⊐CBD′ is given.

But ⊏⊐ABC:⊏⊐CBD′ : AB:BD′ [VI.1]; therefore also the ratio AB:BD′ is given [Def. 2]; so the ratio ⊓AB:⊓BD′ is also given [Dt 50] (figure 86.3).

gvn.parallelogram x,y {= ⊓a}   (1
gvn.(⊓y−⊓):⊓x   {∷ c:d}   (2
{⊓ is the given area, = ⊏⊐y,z in the following; c and d are (unseen) line segments, the usual way of representing ratio, even ratio of areas.}
Assertion: gvn.x and gvn.y.

gvn.⊓ = gvn.⊏⊐y,z   {= ⊓b}   (3
put $y-z = w$ (4
gvn.⊏⊐w,y:⊓x   {∷ c:d}   (2a

gvn.⊏⊐x,y {= ⊓a}   (1
gvn.⊏⊐y,z {= ⊓b}   (3a
gvn.⊏⊐x,y:⊏⊐y,z {∷ ⊓a:⊓b}   (5

⊏⊐x,y:⊏⊐y,z ∷ $x:z$ (5a
{ ∷ ⊓a:⊓b ∷ a:e, if $a:b :: b:e$ [VI.21 corr.]; we shall need a ratio of line segments in the next step, so we take the third proportional to a and b.}
gvn.$x:z$ {∷ a:e}   (6
gvn.⊓x:⊓z {∷ ⊓a:⊓e}   (7

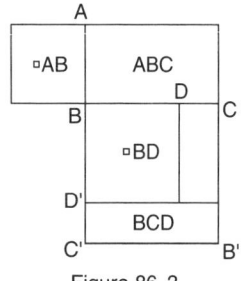

Figure 86.3

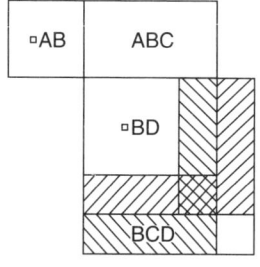

Figure 86.4

CHAPTER 13. INTERSECTING HYPERBOLAS. (86)

The ratio □AB:⊏⊐B′C′D′ is given [by hypothesis, Def. 11];

gvn.□x:⊏⊐w,y   (2b
{:: d:c :: □a:□g, if d:c :: a:f and a:g :: g:f; so f is the fourth proportional to d,c,a, and g is the mean proportional to a and f.}

therefore the ratio ⊏⊐B′C′D′:□DB is also given [compounding, Dt 8]; so the ratio of 4 ⊏⊐B′C′D′:□BD is also given [Dt 6] (figure 86.4).

gvn.⊏⊐w,y:□z   {:: □g:□e}   (8
{compounded from □g:□a and □a:□e, cf. 2b and 7.}
gvn.4⊏⊐w,y:□z   {:: 4□g:□e}   (9

Therefore the ratio of 4 ⊏⊐B′C′D′ + □BD′ to □BD′ is also given [Dt 6].

gvn.(4⊏⊐w,y+□z):□z   (10
{:: (4□g+□e):□e :: □h:□e if h is hypotenuse in a right triangle with sides 2g and e.}

But 4 ⊏⊐B′C′D′ + □BD′ is [equal to] □(BC+CD) [II.8] (figure 86.5); therefore the ratio □(BC+CD):□BD is given [Dt 6]; so the ratio of BC[+C]D to BD is given [Dt 54];

$4\sqsubset\!\!\!\sqsupset w,y + \square z = \square(y+w)$ (11
gvn.□(y+w):□z   {:: □h:□e}   (12
gvn.(y+w):z   {:: h:e}   (13

therefore, *synthenti*, the ratio 2CB:BD is given [Dt 6]; so the ratio CB:BD is given [Dt 8].

[y+w+z = 2y, since w = y−z]
gvn.2y:z   {:: (h+e):e :: 2k:e   (14
              if h+e = 2k}
gvn.y:z   {:: k:e}   (15

But CB:BD :: ⊏⊐ CBD′:□BD′.

y:z :: ⊏⊐y,z:□z   {:: k:e}

Therefore the ratio ⊏⊐ CBD′:□BD′ is given [Def. 2].

gvn.⊏⊐y,z:□z   (15a
{:: k:e :: □b:□m if k:e :: b:n and b:m :: m:n, cf. 2b, i.e. if n the fourth proportional to k, e, b, and m is the mean proportional to b and n.}

⊏⊐CB,BD′ is given; therefore □BD′ is also given [Dt 2]; therefore BD′ is given;

gvn.⊏⊐yz   {= □b}   (2
gvn.□z   {= □m}   (16
gvn.z   {= m}   (17

so BC, too, is given, because the ratio of CB	$y{:}z$ $\{::k{:}e :: p{:}m$, if p is the fourth proportional to $e, k, m\}$
	gvn.$y{:}z$ $\{::p{:}m\}$
	gvn.y $\{=p\}$
and the parallelogram (AC) is given, and the angle B is given; therefore AB is also given [Dt 57]. Therefore each of AB and BC is given.	gvn.(AC) $\{=\Box a$ hypot.$\}$
	gvn.$\sqsubset\!\!\!\!\supset\!x,y$ $\{=\Box a\}$ (1
	gvn.x $\{=q$, if the square $\Box a$ applied to the line segment p makes the width $q\}$.

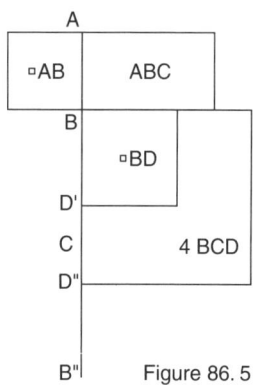

Figure 86. 5

Résumé: The magnitudes in { } disclose a kind of reasoning, a pseudocalculation, that *could* be made, but is not explicit in the text. The proposition proves what it promised to prove, namely that *the sides* of the parallelogram ABC *are given* if the conditions of the protasis are fulfilled. Nevertheless, a geometer *could* put geometrical values, viz. line segments, squares and rectangles into the several steps, as we have seen above, but there are no traces that Euclid did so. The text has only the repetitive "is (also) given".

In fact, it is no easy matter to formulate the 'output':

If the given area of the parallelogram (x,y) is $\Box a$ (unseen) and if the given area of $\sqsubset\!\!\!\!\supset$ (y,z) (to be subtracted from the square on y) is $\Box b$ (also unseen), and if c and d are line segments that represent the given ratio, so that c is to d as the 'remaining rectangle' (BCD, $\sqsubset\!\!\!\!\supset\!w,y$) to the square on AB (x),

then the sides of the parallelogram x and y are equal to line segments q and p respectively, where

q is the width of a rectangle equal to the given square $\Box a$ applied to a line segment,
p which is the fourth proportional to e, k, m,
e is the third proportional to a and b (the sides of the given squares),
m is the mean proportional to b and n,
n is the fourth proportional to k, e, b,

CHAPTER 13. INTERSECTING HYPERBOLAS. (86)

k is $(h+e)/2$,
h is hypotenuse in a right triangle h, e, $2g$,
g is the mean proportional to a and f,
f is the 4'th proportional to d, c, a.

If all this reminds you more of a Lewis Caroll puzzle than of Greek geometry, there is little need to underline the fact that none of it is visible in the *Data*'s text.

Why would Zeuthen insist that the use of geometry plays hardly any role in *Data* 86? I am prepared to swear that we have been doing geometry all the time. He might answer:

> Line segments and plane rectilinear figures are continuous magnitudes according to Aristotle; therefore we may, without trespassing, interpret them as positive real numbers, if we abstract lengths and size from the actual figures, and consider areas as products of lengths, sides of squares as square roots of areas, etc. Adding and subtracting geometric figures (as we learn in the *Elements*, Book I and II) can be interpreted as adding and subtracting real numbers. In short: it is possible to translate the geometrical world into the real number world and forget about the geometry.

But, why should one want to do that? In order to facilitate for the modern reader the opaque methods of the Ancient Greeks, the modern who is supposed to be *ageométrêtos*, not to be admitted to and initiated in The Wisdom of the Greeks. Or is it the other way round: The Greeks were not initiated?

> When God created the Greeks, he forgot to teach them algebra, so they had to make do with squares and rectangles. When he arrived at creating the Arabs, he took great pains, as he had already done with the Babylonians, to teach them algebra as well as geometry; but although they knew the difference quite well, they mixed it up, and so much so as to lead the moderns (when they came along) to believe that it was one and the same thing. The Greeks had, unconsciously, been doing algebra all the time.

This may be a quite tenable position; I do not know that Zeuthen entertained myths like that, but he insisted on translating Greek geometry into 19'th century algebra, as if he did. He was a bit careless, though: one must keep the two worlds strictly apart; otherwise, the result is a mixed interpretation, where one never knows which language is being used at any particular point of reasoning. But even if one endeavours to keep the two levels apart, something creative, or *heuristic*, is missing in this algebraic treatment of geometry; as if one reads a prose translation of a passage from Homer's

Iliad, never really representing the dactylic wrath of Achilles.

The algebraic interpretation of Greek geometry is philologically unwarranted, but in the case of Dt 86 it is not the worst sin. What more, then, is wrong with Zeuthen's interpretation? First and foremost: it is not History. It is quite legitimate to show that a Greek proposition has modern equivalents, even though that does not always (rather: hardly ever) lead to an understanding of the Greek way of thinking; but, when treating Dt 86, Zeuthen *invented* at least half of his documentation.

By the equation $xy = a$ Zeuthen renders the hypothesis: that two line segments contain a given area. By this interpretation Zeuthen deviates widely from Euclid's text: everything Zeuthen writes on the right side of the equation must by necessity be his own invention; therefore, to maintain that Euclid is solving equations, is misleading from the very beginning. *Zeuthen is solving equations*, right – but that does not prove his case.

Even if we admit that an equation conveys the meaning of 'given' (namely: that the rectangle $\sqsubset\!\sqsupset x,y$ is equal to another area which is 'given' – whatever the latter *given* may mean), we shall run into difficulties if we do not respect the fact that the letter a represents a magnitude (namely a geometrical area, a *chorion*) of a different kind from x and y (line segments); we must, so to speak, keep dimensions visible. That will be evident in a moment.

The equation $y^2 - mx^2 = b$ renders the hypothesis that the square of one side (y) of the given rectangle is too great by a given area (b) to be in a given ratio to the square on the other side (x). In fact, $y^2 - b = mx^2$ would be a better transcription by Zeuthen's own standard; the surprise, however, of Zeuthen's notation is the m, which undoubtedly must signify a given *ratio*.

Zeuthen was not the first to denote a ratio by one letter, but he ought to have been the last, as more bewilderment than understanding is caused by it: in any definition and in any use of ratio in Greek mathematics, at least two magnitudes or numbers are engaged in one ratio; in the *Data* it certainly is so. The notation which Zeuthen normally uses is the horizontal stroke, as of fractions, with the antecedent as numerator and the consequent as denominator. But here he treats m on the same level as a and b, being obviously quite at ease with rendering ratio as a real number, as if ratios were magnitudes.

Whichever interpretation of the Greek theory of *logos* one may prefer, one must admit that Zeuthen by his m takes leave of the Greeks: mx^2 does not fairly render the Greek text, which says that the square on x has a given ratio to the remaining rectangle.

By putting $b = yz$ he renders the subtraction of the given *chorion b* as a rectangle with side y. Zeuthen calls z an auxiliary magnitude, and it is, to be sure, a clever trick to make the given area have y for one side – especially if the situation which inspired the whole theorem was the hyperbola problem mentioned above, in which case the given area was a square.

Keeping in mind what I said about Zeuthen's representation of ratio, let us follow his steps, which he considers interpretations of Euclid's (I have put numbers to the steps, counting his equation marks):

Step (3) renders the *construction*: if from the square on y (☐BC) the given area ⊏⊐z,y (CBD) be subtracted, then the ratio of the remaining rectangle ⊏⊐w,y (BCD) to the square on x (☐AB) is given.

Skipping a couple of Euclid's statements he arrives at step (4), quite correctly. Now he gets into trouble, but remains tacit about it: if a and b are areas, *choría*, what is the geometrical interpretation of the ratio of a^2 to b^2 (in step 5)?

A Euclidean mathematician might transform the ratio of a to b into a ratio of squares, say ☐r to ☐s and then into r to t, with t the third proportional to the segments r and s, in order to have a ratio of line segments from which to get step (5) ☐x:☐z :: ☐r:☐t (cf. Dt 24 and *Elements* VI.20). Of course Zeuthen knew as much, so why he assigned the values a^2 and b^2 I do not understand; unless he wanted to expose (what he thought to be) the *illusion* of geometric representation.

The next step (6) is, as it stands in Zeuthen's text, rather unmotivated, nor does it get any explanation. Euclid, on the other hand, needed to reverse step (3) in order to use Dt 8 directly. Zeuthen's attitude is very significant, however, as it discloses that he meant m to be understood as the ratio of m to 1. Later in the same chapter Zeuthen admits that the given ratio must be a ratio between line segments. It seems to me that he has, tacitly, chosen a unit segment, which is quite legitimate from a modern point of view, though Euclid would never choose one. There are no units in Euclid's theory of magnitudes. – Further: if m is a line segment, mx^2 is a solid, which cannot be subtracted from y^2. Thus, by so doing, Zeuthen has definitively left the Greek context.

Zeuthen would not need (6) since he could combine (3) and (5) to get immediately to the step before (7) – let us call it (6b) – which he skips, multiplying by 4 at once. By skipping step (6b) he avoids touching the problem of compound ratio, which underlies Dt 8: from two ratios of line segments to form one ratio of rectangles. Euclid, of course, could not do without that; but Zeuthen had to, if 'geometry plays hardly any role'.

From now on, the right sides (apart from not being in the text) are chaos from a Greek point of view, while the left sides may still be interpreted as geometry, inspired as they are by Euclid's reasonings. – The right sides are, in fact, comprehensible expressions in real numbers, but they have no analogy in Greek geometry. They are no longer *ratios*, but *calculations* in Zeuthens mind; he now seems (as he said) to interpret the predicate 'is given' as 'can be determined step by step', when a, b, and m are given. What he means by 'determined' must be 'calculated', which is definitely wrong, as can easily be demonstrated if the phrase of the protasis 'in a given angle' be taken into consideration. No calculating involving any angle can be done without some table of chords; so calculation was some centuries away, waiting for Ptolemy. What is meant by 'given' must be 'constructible', *i.e.* obtainable by the methods of Book I of the *Elements*.

But then, apart from other objections one may have to Zeuthen's interpretations, we cannot admit that geometry plays hardly any role. Why insist that constructibility must be un-geometric? Further, while some of Zeuthen's expressions may still be recognized as *ratios*, that in (9), right side, is no longer so. Adding 1 to a doubtful *logos* does not make Greek sense (even though some modern commentators interpret the *synthenti* operation, *Elements* V.18, that way). Also, to Zeuthen, the right side is a kind of entity which has a square root, in step (10); a 'perfect' square, Zeuthen calls it; but the left side in (9) is a *ratio* of squares.

Let me confine myself to pointing to the right sides of his equations, without trying to read them aloud. If anyone of my readers can easily associate his concluding equation (17) with any word or statement in the Greek text, he should tell me. Perhaps at the same time he could disclose the value of x.

Is Zeuthen still worth reading, then? Of course he is; he was undisputedly one of the best *connoisseurs* of Greek mathematical results, if not always presenting Greek methods. One way of reading him is to retranslate him, with an eye in the Greek text or some reliable translation; read in that way he is an endless source of 'aha's!

Some twenty years ago I first read about Dt 86 in Zeuthen's book; I do not remember what I thought about it, only that I put it away, deciding to find out what the *Data* could be about. I was quite sure that this was not the *Data*. I have been walking in that enigmatic mathematical and philosophical quick-sand off and on ever since, sinking deeper into bewilderedness the more I try to understand, but I have until now managed to get out again; I am sure that I know more about the *Data* now than I did then; among other things, Zeuthen did that to me. Why he would do it that way, I shall probably never understand.

I suggested, with Zeuthen, that this might have something to do with the hyperbola, but what?[154] In Book I of the *Conics*, proposition 12, Apollonius proved (what was already known to Archimedes) the *symptoma* of a given hyperbola:

---

[154] Ken Saito, with no knowledge of Zeuthen's Danish paper, proposed the same conjecture in his penetrating analysis of Book II of the *Elements* (1985, p. 51 ff.) while looking for an application of II.8 in the Conics and opposing Tannery's algebraic interpretation of *Data* 86. So there are strong reasons for accepting the conjecture.

CHAPTER 13. INTERSECTING HYPERBOLAS. (86)

The square on the ordinate has a constant ratio to the rectangle contained by the abscissas.

Apollonius represents that ratio by the ratio of a certain line segment, ἡ ὀρθία, the 'upright side' or the *parameter*[155] to the so-called *transverse diameter*, the part of the diameter between the two branches. How he may have hit on that idea, can be seen from figure 86.6, which shows the rectangle contained by the abscissas AD (=DF) and BD. That the square on the ordinate DP has a given ratio to this rectangle may be illustrated by applying the square to the lesser abscissa as a rectangle ADE, producing a certain width DE.

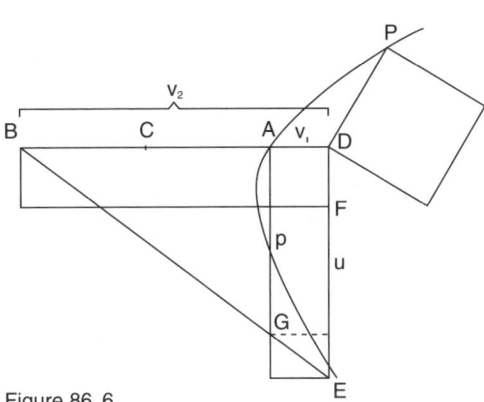

Figure 86. 6

Thus
□DP : ADB :: ADE : ADB
and by *Elements* VI.1,
□DP : ADB :: DE : BD;
by drawing the diagonal BE
□DP : ADB :: AQ : AB
that is
□DP : ADB :: p : $2d$,
if we rename AQ as p and give the name d to *half* of the transverse diameter AB, observing that d is half the difference between the abscissas.

Thus, for a given hyperbola, the ratio of p to $2d$ is given; and as the transverse diameter is given, p will also be given, and therefore the *eidos*, the shape or form of the rectangle ⊏⊐p,d (QAB) is characteristic of the hyperbola in question, very properly called its *symptoma*, "what befalls it".

Figure 86.6 also shows the square □DP applied to the segment p with a rectangular excess (ὑπερβολή) (QE) which is similar to the *eidos*. We understand why Apollonius was inspired to give the name *hyperbola* to this conic section.

To Greek mathematicians like Euclid or Apollonius, the rectangle contained by the two abscissas of the hyperbola invites gnomonic thinking (figure 86.7):

---

[155] The term *parameter* is not ancient, but was introduced as late as the 16'th century.

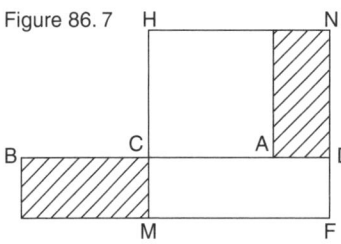

Figure 86.7

From *Elements* II.6 we know that the said rectangle (BF) equals the gnomon MFNAC, that is: the square (HF) minus the square (HA). The square (HA) is given, being the square on *d*, half the diameter. Therefore, the square on the ordinate, □DP, has a given ratio to the gnomon and vice versa; to use the idiom of Def. 11, the square (HF) on half the sum of the abscissas is greater by the given square (HA) than a given ratio to the square on the ordinate DP.

Now we learn from Dt 86 that *if the parallelogram contained* (in a given angle) *by the ordinate* DP (AB in figure 86.2) *and half the sum of the abscissas*, CD (BC in figure 86.2), *has a given area, those segments will be given*, and therefore particularly the ordinate point on the curve. But is that parallelogram given in magnitude?

Consider another given hyperbola (figure 86.8), having the conjugate diameters of this one for its asymptotes, and passing through the ordinate point; then the parallelogram in question is given in magnitude, as proved in the *Conics* II.12. Thus we may believe Zeuthen and Ken Saito (cf. note 154) and consider Dt 86 as a demonstration of the givenness of the point of section of two given hyperbolae, one of which has its asymptotes on the conjugate diameters of the other one. And, of course, Euclid did write four books on conics.

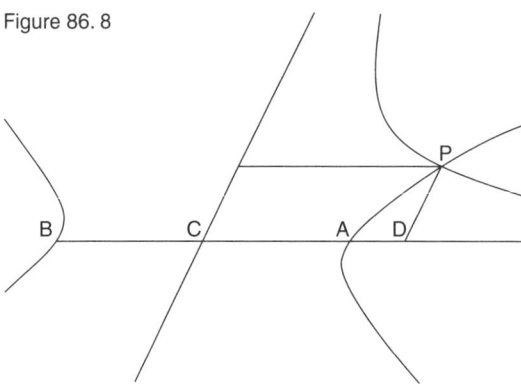

Figure 86.8

◆

Chapter 14. Circles
Dt 87–94

Preliminary remarks: The last eight propositions, Dt 87–94, consider given circles, whether given in magnitude or in position. Def. 5 defines that *a circle is given in magnitude if its radius is given in magnitude*. Def 6. defines that it is *given in position and magnitude if its centre is also given* (*viz.* in position). Of course, a circle given in magnitude must have a centre (which can even be found by III.1) but if it does not matter *where* it is, it is said to be given in magnitude only (cf. my remarks on Dt 2). One could expand those definitions by mentioning that position involves magnitude according to Dt 26: a circle is given in position, if its centre and any one point on the circumference (*viz.* the endpoint of a given radius) are given, or if three points on the circumference are given. The latter criterion is not mentioned in the Data.[156]

Θε 87. Ἐὰν εἰς κύκλον δεδομένον τῷ μεγέθει εὐθεῖα γραμμὴ ἀχθῇ ἀπολαμβάνουσα τμῆμα δεχόμενον γωνίαν δοθεῖσαν, δέδοται ἡ ἀχθεῖσα τῷ μεγέθει.

εἰς γὰρ κύκλον δεδομένον τῷ μεγέθει τὸν ΑΒΓ διήχθω ἡ ΑΓ ἀπολαμβάνουσα τμῆμα τὸ ΑΕΓ δεχόμενον γωνίαν δοθεῖσαν· λέγω, ὅτι ἡ ΑΓ δέδοται τῷ μεγέθει.

εἰλήφθω γὰρ τὸ κέντρον τοῦ κύκλου τὸ Δ, καὶ ἐπιζευχθεῖσα ἡ ΑΔ διήχθω ἐπὶ τὸ Ε, καὶ ἐπεζεύχθω ἡ ΓΕ· δοθεῖσα ἄρα ἐστὶν ἡ ὑπὸ τῶν ΑΓΕ· ὀρθὴ γάρ· ἔστι δὲ καὶ ἡ ὑπὸ ΑΕΓ δοθεῖσα· καὶ λοιπὴ ἄρα ἡ ὑπὸ τῶν ΓΑΕ δοθεῖσά ἐστιν· δέδοται ἄρα τὸ ΑΓΕ τρίγωνον τῷ εἴδει· λόγος ἄρα ἐστὶ τῆς ΑΕ πρὸς τὴν ΑΓ δοθείς. δοθεῖσα δὲ ἡ ΕΑ τῷ μεγέθει, ἐπεὶ καὶ ὁ κύκλος δέδοται τῷ μεγέθει· δοθεῖσα ἄρα ἐστὶν ἡ ΑΓ τῷ μεγέθει.

Dt 87. *If in a circle given in magnitude a straight line be drawn cutting off a segment admitting a given angle, the line drawn is given in magnitude.*

For, in the circle ABC given in magnitude let the [chord] AC have been drawn cutting off the segment AEC admitting a given angle. I say that AC is given in magnitude.

For, let the centre D of the circle have been taken [III.1], and let the joining line AD have been produced to E, and let CE have been joined; then ∡ACE is given, because it is a right angle [III.31, post. 4]; and ∡AEC is also given [by hypothesis];

---

[156] Although in the propositions circles are (almost) always denoted by three letters on the circumferemce and not by their centre and the endpoint of a radius. (In fact, we meet a couple of two-points-circles in theorem 90 and 91). Marinos gives the threepoint criterion as an example of something *tetagménon*, 'fixed', 'placed in a fixed position', Menge 238, 11. Cf. Apollonius' term 'drawn τεταγμένως' about an ordinate in a conic section.

therefore the remaining ∠CAE is given [I.32]; therefore △ACE is given in form [Dt 40]; therefore the ratio AE:AC is given [Def. 3].

And the line EA [the diameter] is given in magnitude because the circle is also given in magnitude [Def. 5]; therefore AC is given in magnitude [Dt 2]. ∎

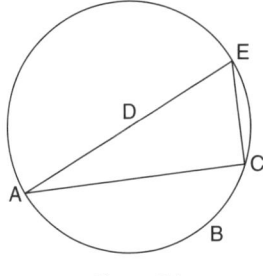

Figure 87

Remarks: Dt 87 and its converse Dt 88 enclose a more general theorem, which is the foundation of trigonometry; while in III.34 we learn how to cut off from a given circle a segment admitting an angle equal to a given rectilineal angle, we are now taught (to speak in modern terms) that the length of the cutting chord is a function of the circle's radius:

Dt 87* *In any circle a cord subtending* [157] *a given angle has a given ratio to the diameter.*

And its converse:

Dt 88* *In any circle a cord having a given ratio to the diameter subtends a given angle.*

This one-to-one correspondence was used by Ptolemy (and before him by Hipparchos) to tabulate lengths of chords (substantially equivalent to calculating tables of sines) expressed as a number of parts in a system where the diameter is divided in 120 parts (Ptolemy-Toomer 48).

◆

Θε 88. ʹΕὰν εἰς κύκλον δεδομένον τῷ μεγέθει εὐθεῖα γραμμὴ ἀχθῇ δεδομένη τῷ μεγέθει, ἀπολήψεται τμῆμα δεχόμενον γωνίαν δοθεῖσαν.

εἰς γὰρ κύκλον δεδομένον τῷ μεγέθει τὸν ΑΒΓ εὐθεῖα γραμμὴ ἤχθω ἡ ΑΓ δεδομένη τῷ μεγέθει· λέγω, ὅτι ἀπολήψεται τμῆμα δεχόμενον γωνίαν δοθεῖσαν.

εἰλήφθω γὰρ τὸ κέντρον τοῦ κύκλου τὸ Δ, καὶ ἐπιζευχθεῖσα ἡ ΑΔ διήχθω ἐπὶ τὸ Ε, καὶ ἐπεζεύχθω ἡ ΓΕ. ἐπεὶ δοθεῖσά ἐστιν ἑκατέρα τῶν ΕΑ, ΑΓ, λόγος ἄρα ἐστὶ τῆς ΕΑ πρὸς τὴν ΑΓ δοθείς. καί ἐστιν ὀρθὴ ἡ ὑπὸ τῶν ΑΓΕ γωνία· δέδοται ἄρα τὸ ΑΓΕ τρίγωνον τῷ εἴδει· δοθεῖσα ἄρα ἐστὶ καὶ ἡ ὑπὸ τῶν ΑΕΓ γωνία.

[157] The *Data* would have 'cutting off a segment admitting'; I prefer the shorter expression.

CHAPTER 14. CIRCLES (87-94)

Dt 88. *If in a circle given in magnitude a straight line be drawn given in magnitude, it will cut off a segment admitting a given angle.*

For, in the circle ABC given in magnitude let the straight line AC have been drawn given in magnitude; I say that it will cut off a segment admitting a given angle.

For, let the centre D of the circle have been taken [III.1] (figure 87), and let the joining AD have been produced to E, and let CE have been joined. Since both EA [Def. 5] and AC is given, therefore the ratio EA:AC is given [Dt 1]. And $\angle$ACE is right [III.31]; therefore $\triangle$AQE is given in form [Dt 43]; therefore $\angle$AEC is also given.

∎

Remarks: Instead of referring to Dt 44, this proposition uses Dt 43, which proves Dt 44 and is used nowhere else.

◆

Θε 89. ' Ἐὰν κύκλου δεδομένου τῇ θέσει ἐπὶ τῆς περιφερείας δοθὲν σημεῖον ληφθῇ, ἀπὸ δὲ τούτου πρὸς τὴν τοῦ κύκλου περιφέρειαν κλασθῇ τις εὐθεῖα δεδομένην γωνίαν ποιοῦσα, δέδοται τὸ ἕτερον πέρας τῆς κλασθείσης.

κύκλου γὰρ τῇ θέσει δεδομένου τοῦ ΑΒΓ εἰλήφθω ἐπὶ τῆς περιφερείας δοθὲν σημεῖον τὸ Β, ἀπὸ δὲ τοῦ Β κεκλάσθω εὐθεῖα ἡ ΒΑΓ δεδομένην ποιοῦσα γωνίαν τὴν ὑπὸ τῶν ΒΑΓ· λέγω, ὅτι δέδοται τὸ Γ σημεῖον.

εἰλήφθω γὰρ τὸ κέντρον τὸ Δ, καὶ ἐπεζεύχθωσαν αἱ ΒΔ, ΔΓ. ἐπεὶ δοθέν ἐστιν ἑκάτερον τῶν Β, Δ, θέσει ἄρα ἐστὶν ἡ ΒΔ. καὶ ἐπεὶ δοθεῖσά ἐστιν ἡ ὑπὸ τῶν ΒΑΓ γωνία, δοθεῖσα ἄρα ἐστὶν ἡ ὑπὸ ΒΔΓ. ἐπεὶ οὖν πρὸς θέσει εὐθείᾳ καὶ τῷ πρὸς αὐτῇ σημείῳ τῷ Δ εὐθεῖα ἦκται ἡ ΔΓ δεδομένην ποιοῦσα γωνίαν τὴν ὑπὸ τῶν ΒΔΓ, δοθεῖσα ἄρα ἐστὶν ἡ ΔΓ τῇ θέσει· θέσει δὲ δοθεὶς καὶ ὁ ΑΒΓ κύκλος· δοθὲν ἄρα ἐστὶ τὸ Γ σημεῖον.

Dt 89. *If in a circle given in position a given point be taken* [158] *on the circumference, and if from that point a straight line be inflected* [159] *at the circumference of the circle, making a given angle, the other endpoint of the inflected line will be given.*

---

[158] If we did not know it already, we learn that a point can be *taken* and called *given*.

[159] κλασθῇ: "The verb κλάω (to *break off*) was the regular technical term for drawing from a point a (broken) straight line which first meets another straight line or curve and is then *bent back* from it to another point on a curve or another straight line. κεκλάσθαι is one of the geometrical terms the definition of which must according to Aristotle be assumed (*Anal. Post.* I. 10, 76 b 9)" (Heath 1956, II .47).

For, of the circle ABC given in position let the given point B have been taken on the circumference, and from the point B let the straight line BAC have been inflected [at the circumference], making the given angle BAC; I say that the point C is given.

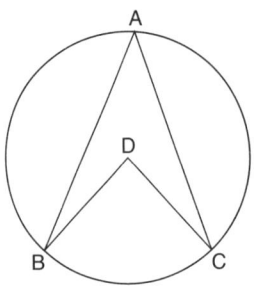

Figure 89

For, let the centre D have been taken, and let BD, DC have been joined. Since each of [the points] B,D is given [Def. 6], [the line] BD is given in position [Dt 26]. And since ∠BAC is given, ∠BDC is given [III.20, Dt 2].

Since then, at a straight line given in position, and at the given point D on it the straight line DC has been drawn making the given angle BDC, therefore DC is given in position [Dt 29]. And the circle ABC is given in position; therefore the point C is given [Dt 25]. ∎

Remarks: A simpler proof that does not involve the centre applies the theorems just proved, at the same time disclosing that there may be two solutions (as so often where givenness in position is proved):

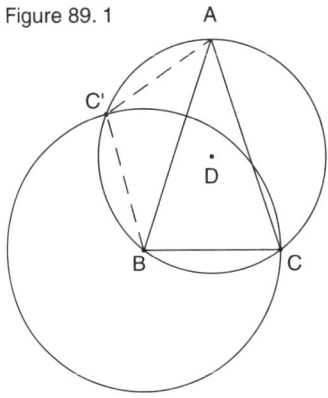

Figure 89. 1

Since the chord BC is given in magnitude [Dt 87], the circle with centre B and radius BC is given in position (and magnitude) [Def. 6], so that the point C is given [Dt 25]. If ∠BAC is less than ∠BCA, there will be two solutions, according as the reflection is to the left or to the right at A.

Dt 89 seems to belong in another stratum than the *Elements*, where the term *inflect* is found only once, whereas Apollonius uses it in his *Conica*, III.51 & 52 and elsewhere.

This feature and a couple of others (see remarks on Dt 94) in the propositions about circles suggest a connection between the *Data* and the *Conica* (or *some* conics); a scholium on Deff. 13–15 ascribes those to Apollonius.

◆

CHAPTER 14. CIRCLES (87–94)

Θε 90. Ἐὰν ἀπὸ δεδομένου σημείου θέσει δεδομένου κύκλου ἐφαπτομένη εὐθεῖα ἀχθῇ, δέδοται ἡ ἀχθεῖσα τῇ θέσει καὶ τῷ μεγέθει.

ἀπὸ γὰρ δεδομένου σημείου τοῦ Γ θέσει δεδομένου κύκλου τοῦ ΑΒ ἐφαπτομένη εὐθεῖα ἤχθω ἡ ΓΑ· λέγω, ὅτι ἡ ΓΑ εὐθεῖα δέδοται τῇ θέσει καὶ τῷ μεγέθει.
εἰλήφθω γὰρ τὸ κέντρον τοῦ κύκλου τὸ Δ, καὶ ἐπεζεύχθωσαν αἱ ΔΑ, ΔΓ. ἐπεὶ δοθέν ἐστιν ἑκάτερον τῶν Δ, Γ, δοθεῖσα ἄρα ἐστὶν ἡ ΔΓ. καί ἐστιν ὀρθὴ ἡ ὑπὸ τῶν ΔΑΓ γωνία· τὸ ἄρα ἐπὶ τῆς ΔΓ γραφόμενον ἡμικύκλιον ἥξει διὰ τοῦ Α. ἡκέτω καὶ ἔστω τὸ ΔΑΓ· θέσει ἄρα ἐστὶ τὸ ΔΑΓ. θέσει δὲ καὶ ὁ ΑΒ κύκλος· δοθέν ἐστιν ἄρα τὸ Α. ἀλλὰ καὶ τὸ Γ δοθέν ἐστιν· δοθεῖσα ἄρα ἐστὶν ἡ ΑΓ τῇ θέσει καὶ τῷ μεγέθει.

Dt 90. *If from a given point a straight line be drawn tangent to a circle given in position, the straight line drawn will be given in position and in magnitude.*

For, from the given point C let the straight line CA have been drawn tangent to the circle AB given in position [III.17]; I say that the straight line CA is given in position and in magnitude.

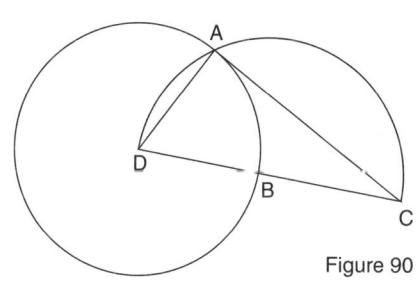

Figure 90

For, let the centre D of the circle have been taken, and let DA, DC have been joined. Since each of D, C is given, therefore [the line] DC is given [Dt 26]. And $\angle$DAC is right [III.18]; therefore the semicircle described on DC will pass through A [III.31].

Let it have passed, and let it be the [semicircle] DAC; then DAC is given in position [Def. 8].

And the circle AB is also given in position; therefore the [point] A is given [Dt 25]. But C is also given; therefore AC is given in position and in magnitude [Dt 26]. ■

Remarks: From Book III of the *Elements* we know that there are two equal tangents to a circle from a point outside it. Dt 90 can be seen as proving the uniqueness of either of them (cf. remarks on *symmetry*, p. 105). – If the circle is given in position, it is also given in magnitude, that is, its radius is given in magnitude. Since the point (outside the circle) from which the tangent is drawn is given, the line to the centre is given in magnitude, and therefore the length of the tangent and the size of its square are given by I.47.

Does Dt 90 prove that no more than two tangents can be drawn from a given point to a given circle? Certainly not; that could be proved by a *reductio ad absurdum*, but it also follows from III.10, *A circle does not cut a circle in more points than two*. Dt 90 proves that the given point and the given circle determines the tangent; this fact is used in Dt 91 and 92 about the 'power of a point with respect to a circle'.

◆

Θε 91. ' Ἐὰν κύκλου δεδομένου τῇ θέσει ληφθῇ τι σημεῖον ἐκτὸς δοθέν, ἀπὸ δὲ τοῦ σημείου εἰς τὸν κύκλον διαχθῇ τις εὐθεῖα, τὸ ὑπὸ τῆς ἀχθείσης καὶ τῆς μεταξὺ τοῦ σημείου καὶ τῆς κυρτῆς περιφερείας περιεχόμενον ὀρθογώνιον δοθέν ἐστιν.

κύκλου γὰρ δεδομένου τῇ θέσει τοῦ ΑΒΓ εἰλήφθω τι σημεῖον ἐκτὸς τὸ Δ, ἀπὸ δὲ τοῦ Δ σημείου διήχθω τις εὐθεῖα ἡ ΔΒ τέμνουσα τὸν κύκλον· λέγω, ὅτι δοθέν ἐστι τὸ ὑπὸ τῶν ΒΔ, ΔΓ.

ἤχθω ἀπὸ τοῦ Δ σημείου τοῦ ΑΒΓ κύκλου ἐφαπτομένη εὐθεῖα ἡ ΑΔ· δοθεῖσα ἄρα ἐστὶν ἡ ΑΔ τῇ θέσει καὶ τῷ μεγέθει. ἐπεὶ οὖν δοθεῖσά ἐστιν ἡ ΑΔ, δοθὲν ἄρα ἐστὶ καὶ τὸ ἀπὸ τῆς ΑΔ. καί ἐστιν ἴσον τῷ ὑπὸ τῶν ΒΔΓ· δοθὲν ἄρα ἐστὶ καὶ τὸ ὑπὸ τῶν ΒΔΓ.

Dt 91.   *If a given point be taken outside a circle given in position, and from the point a straight line be drawn through the circle, the rectangle contained by the straight line and the straight line between the point and the convex circumference is given.*

For, let a given point D have been taken outside the circle ABC given in position, and from the point D let some straight line BD have been drawn cutting the circle; I say that the rectangle BD, DC is given.

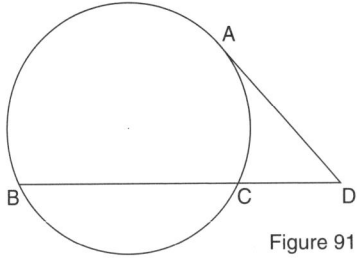

Figure 91

From the point D let the straight line AD have been drawn tangent to the circle ABC [III.17]; then AD is given in position and in magnitude [Dt 90]. Then, since AD is given, ▫AD is also given. [160] And it is equal to ⊏⊐BDC; therefore ⊏⊐BDC is also given.   ∎

Remarks: The equality ▫AD = ⊏⊐BD,DC (the "power of D with respect to the circle") is proved in III.36. That both areas are given follows from the givenness of the tangent, established in Dt 90.

---

[160] Menge (and McDowell/Sokolik) refer to Dt 52; but that one is proved by assuming the truth of 24*B, *a square is given* in magnitude *if its side is given.*

CHAPTER 14. CIRCLES (87–94)

Θε 92. ’Εὰν κύκλου δεδομένου τῇ θέσει ληφθῇ τι σημεῖον ἐντὸς δοθέν, διὰ δὲ τοῦ σημείου διαχθῇ τις εὐθεῖα εἰς τὸν κύκλον, τὸ ὑπὸ τῶν τῆς ἀχθείσης τμημάτων περιεχόμενον ὀρθογώνιον δοθέν ἐστιν.

κύκλου γὰρ δεδομένου τῇ θέσει τοῦ ΒΓ εἰλήφθω τι σημεῖον ἐντὸς τὸ Α δοθέν, διὰ δὲ τοῦ Α διήχθω τις εὐθεῖα ἡ ΓΒ· λέγω, ὅτι δεδομένον ἐστὶ τὸ ὑπὸ τῶν ΓΑΒ.

εἰλήφθω γὰρ τὸ κέντρον τοῦ κύκλου τὸ Δ, καὶ ἐπιζευχθεῖσα ἡ ΑΔ διήχθω ἐπὶ τὰ Ζ, Ε. ἐπεὶ οὖν δοθέν ἐστιν ἑκάτερον τῶν Δ, Α, θέσει ἄρα ἐστὶν ἡ ΔΑ. θέσει δὲ καὶ ὁ ΓΒΖ κύκλος· δοθὲν ἄρα ἐστὶν ἑκάτερον τῶν Ζ, Ε. ἔστι δὲ καὶ τὸ Α δοθέν· δοθεῖσα ἄρα ἐστὶν ἑκατέρα τῶν ΖΑ, ΑΕ· δοθὲν ἄρα ἐστὶ τὸ ὑπὸ τῶν ΖΑ, ΑΕ. καί ἐστιν ἴσον τῷ ὑπὸ ΒΑΓ· δοθὲν ἄρα ἐστὶ καὶ τὸ ὑπὸ τῶν ΓΑΒ.

Dt 92. *If a given point be taken inside a circle given in position, and through the point a straight line be drawn in the circle, the rectangle contained by the segments of the straight line is given.*

For, let the given point A be taken inside the circle BC given in position, and through A let some straight line CB have been drawn; I say that the rectangle CAB is given.

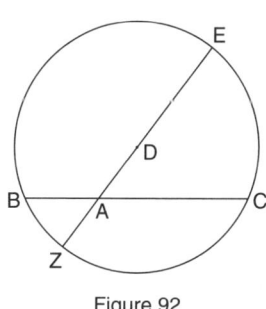

Figure 92

For, let the centre D of the circle have been taken, and let AD joined have been drawn through to the points Z, E. Since each of the points D,A is given, therefore [the line] DA is given in position [Dt 26].

And the circle CBZ is given in position; therefore each of Z, E is given [Dt 25]; and A is given; therefore each of ZA, AE is given [Dt 26]; therefore ⊏⊐ZA,AE is given [Dt 24*A]. And it is equal to ⊏⊐ BAC [III.35]; therefore ⊏⊐ BAC is also given. ∎

Remarks: The points Z and E are given (by the axiomatic Dt 25) because the straight line DA is given in position, that is, its *direction* is fixed and determines the points of section with the given circle. The equality of the rectangles is proved in III.35, so that the givenness of one of them entails the givenness of the other.

Θε 93. ’Εὰν εἰς κύκλον δεδομένον τῷ μεγέθει εὐθεῖα γραμμὴ ἀχθῇ ἀπολαμβάνουσα τμῆμα δεχόμενον γωνίαν δοθεῖσαν, καὶ ἡ ἐν τῷ τμήματι γωνία δίχα τμηθῇ, συναμφότεροι αἱ τὴν δεδομένην γωνίαν περιέχουσαι πρὸς τὴν δίχα τέμνουσαν τὴν

γωνίαν λόγον ἕξουσι δεδομένον, καὶ τὸ ὑπὸ συναμφοτέρου τῶν τὴν δεδομένην γωνίαν περιεχουσῶν εὐθειῶν καὶ τῆς κάτω ἀπολαμβανομένης ἀπὸ τῆς δίχα τεμνούσης τὴν γωνίαν πρὸς τῇ περιφερείᾳ δοθὲν ἔσται.

εἰς γὰρ κύκλον δεδομένον τῷ μεγέθει τὸν ΑΒΓ εὐθεῖα ἤχθω ἡ ΒΓ ἀπολαμβάνουσα τμῆμα δεχόμενον γωνίαν δοθεῖσαν τὴν ὑπὸ ΒΑΓ, καὶ τετμήσθω ἡ ὑπὸ ΒΑΓ γωνία δίχα τῇ ΑΔ εὐθείᾳ· λέγω, ὅτι λόγος ἐστὶ συναμφοτέρου τῆς ΒΑΓ πρὸς τὴν ΑΔ δοθείς, καὶ ὅτι δοθέν ἐστι τὸ ὑπὸ συναμφοτέρου τῆς ΒΑΓ καὶ τῆς ΕΔ.

ἐπεζεύχθω ἡ ΒΔ. καὶ ἐπεὶ εἰς κύκλον δεδομένον τῷ μεγέθει τὸν ΔΑΓ διῆκται εὐθεῖα ἡ ΒΓ ἀπολαμβάνουσα τμῆμα τὸ ΒΑΓ δεχόμενον γωνίαν δοθεῖσαν τὴν ὑπὸ τῶν ΒΑΓ, δοθεῖσα ἄρα ἐστὶν ἡ ΒΓ τῷ μεγέθει. διὰ τὰ αὐτὰ δὴ καὶ ἡ ΒΔ δοθεῖσά ἐστι τῷ μεγέθει· λόγος ἄρα ἐστὶ τῆς ΒΓ πρὸς τὴν ΒΔ δοθείς. καὶ ἐπεὶ ἡ ὑπὸ τῶν ΒΑΓ γωνία δίχα τέτμηται τῇ ΑΔ εὐθείᾳ, ἔστιν ἄρα ὡς ἡ ΒΑ πρὸς τὴν ΑΓ, οὕτως ἡ ΒΕ πρὸς τὴν ΕΓ· ἐναλλὰξ ἄρα ὡς ἡ ΑΒ πρὸς ΒΕ, οὕτως ἡ ΑΓ πρὸς τὴν ΓΕ· καὶ ὡς ἄρα συναμφότερος ἡ ΒΑΓ πρὸς τὴν ΒΓ, οὕτως ἡ ΑΓ πρὸς τὴν ΓΕ. καὶ ἐπεί ἐστιν ἴση ἡ ὑπὸ τῶν ΒΑΕ γωνία τῇ ὑπὸ τῶν ΕΑΓ, ἔστι δὲ καὶ ἡ ὑπὸ τῶν ΑΓΕ τῇ ὑπὸ τῶν ΒΔΕ ἴση, λοιπὴ ἄρα ἡ ὑπὸ τῶν ΑΕΓ λοιπῇ τῇ ὑπὸ τῶν ΑΒΔ ἐστιν ἴση. Ἰσογώνιον ἄρα ἐστὶ τὸ ΑΕΓ τρίγωνον τῷ ΑΒΔ τριγώνῳ· ἔστιν ἄρα ὡς ἡ ΑΓ πρὸς τὴν ΓΕ, οὕτως ἡ ΑΔ πρὸς τὴν ΒΔ. ἀλλ' ὡς ἡ ΑΓ πρὸς τὴν ΓΕ, οὕτως συναμφότερος ἡ ΒΑ, ΑΓ πρὸς τὴν ΒΓ· ἔστιν ἄρα ὡς συναμφότερος ἡ ΒΑ, ΑΓ πρὸς τὴν ΒΓ, οὕτως ἡ ΑΔ πρὸς τὴν ΔΒ· ἐναλλὰξ ὡς συναμφότερος ἡ ΒΑΓ πρὸς τὴν ΑΔ, οὕτως ἡ ΒΓ πρὸς τὴν ΒΔ· λόγος δὲ τῆς ΒΓ πρὸς τὴν ΒΔ δοθείς· λόγος ἄρα καὶ συναμφοτέρου τῆς ΒΑΓ πρὸς τὴν ΑΔ δοθείς.

λέγω, ὅτι καὶ τὸ ὑπὸ συναμφοτέρου τῆς ΒΑΓ καὶ τῆς ΕΔ δοθέν ἐστιν.
ἐπεὶ γὰρ ἰσογώνιόν ἐστι τὸ ΑΕΓ τρίγωνον τῷ ΔΕΒ τριγώνῳ, ἔστιν ἄρα ὡς ἡ ΒΔ πρὸς τὴν ΔΕ, οὕτως ἡ ΑΓ πρὸς τὴν ΓΕ. ὡς δὲ ἡ ΑΓ πρὸς τὴν ΓΕ, οὕτως ἐστὶ συναμφότερος ἡ ΒΑΓ πρὸς τὴν ΒΓ· καὶ ὡς συναμφότερος ἄρα ἡ ΒΑΓ πρὸς τὴν ΓΒ, οὕτως ἐστὶν ἡ ΒΔ πρὸς τὴν ΔΕ· τὸ ἄρα ὑπὸ συναμφοτέρου τῆς ΒΑΓ καὶ τῆς ΕΔ ἐστιν ἴσον τῷ ὑπὸ τῶν ΓΒ, ΒΔ. δοθὲν δὲ τὸ ὑπὸ τῶν ΓΒ, ΒΔ· δοθὲν ἄρα καὶ τὸ ὑπὸ συναμφοτέρου τῆς ΒΑΓ καὶ τῆς ΕΔ.

Dt 93. *If in a circle, given in magnitude, a straight line be drawn cutting off a segment admitting a given angle, and if the angle in the segment be bisected, the sum of the straight lines containing the given angle will have a given ratio to the bisector, and the rectangle contained by the sum of the lines containing the given angle and* [by] *that part of the bisector which is cut off below towards the circumference will be given.*

For, in the circle ABC, given in magnitude, let the straight line BC have been drawn cutting off a segment admitting the given angle BAC, and let ∠BAC have been bisected by the straight line AD; I say that the ratio of the sum BAC to AD is given, and that the rectangle [contained by] the sum BAC and [by] ED is given.

Let BD have been joined. And since in the circle DAC, given in magnitude, the straight line BC have been drawn cutting off the segment BAC admitting the given angle BAC, therefore BC is given in magnitude [Dt 87]. For the same reason BD is also given in magnitude; therefore the ratio BC:BD is given.

And since △BAC have been bisected by the straight line AD, therefore BA:AC :: BE:EC [VI.3]. *Enallax* AB:BE :: AC:CE [V.16].; and therefore as the sum BAC is to [the sum] BC, so is AC to CE [V.12].

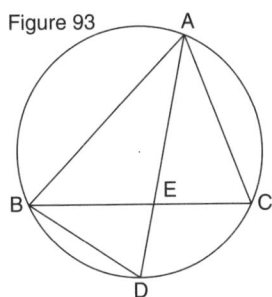

Figure 93

And since ∡BAE = ∡EAC, and also ∡ACE = ∡BDE [III.21], therefore also the remaining ∡AEC is equal to the remaining ∡ABD. Therefore △AEC is equiangular with △ABD; therefore AC:CE :: AD:BD [VI.4].

But AC:CE :: (BA+AC):BC; therefore (BA+AC):BC :: AD:DB; *enallax* BAC:AD :: BC:BD; and the ratio BC:BD is given; therefore the ratio BAC:AD is also given.

I say that ⊏⊐BAC, ED is also given.

For, since △AEC is equiangular with △DEB [III.21], therefore BD:DE :: AC:CE [VI.4]. But AC:CE :: BAC:BC; and BAC: BC :: BD:DE [V.11]. therefore ⊏⊐BAC, ED = ⊏⊐CB,BD [VI.16]. And ⊏⊐CB,BD is given [Dt 24*B]; therefore ⊏⊐BAC,ED is also given. ∎

Remarks: Dt 93 is easily seen to be a special case of Ptolemy's theorem (Almagest I.10, Heiberg 36)[161] if DC is joined to make an inscribed quadrilateral. (BAC means the sum of the sides BA,AC):

Dt 93 proves BAC:AD = BC:BD
    ∴ ⊏⊐BAC,BD = ⊏⊐AD,BC [VI.16], which [II.1] can be resolved into
    ⊏⊐BA,BD + ⊏⊐AC,BD = ⊏⊐AD,BC,
that is ⊏⊐BA,CD + ⊏⊐AC,BD = ⊏⊐AD,BC, since BD = CD.

Therefore, I think, one should modify Toomer's footnote 59 (Ptolemy-Toomer, 50) to say that Ptolemy's *general* theorem is not attested before him. Dt 93 may have served an early table-maker, say Hipparchus?

Is the second part of the theorem interesting? I do not see it.

◆

---

[161] As also Heath seems to realise (1956, II, 227) in his long comment on VI.16, since he gives an alternative proof of Dt 93 by means of Ptolemy.

Θε 94. Ἐὰν κύκλου δεδομένου τῇ θέσει ἐπὶ τῆς διαμέτρου δοθὲν σημεῖον ληφθῇ, ἀπὸ δὲ τοῦ σημείου πρὸς τὸν κύκλον προσβληθῇ τις εὐθεῖα καὶ ἀπὸ τῆς τομῆς πρὸς ὀρθὰς ἀχθῇ τῇ διαχθείσῃ, διὰ δὲ τοῦ σημείου, καθ' ὃ συμβάλλει ἡ πρὸς ὀρθὰς τῇ περιφερείᾳ, παράλληλος ἀχθῇ τῇ διαχθείσῃ, δοθέν ἐστι τὸ σημεῖον, καθ' ὃ συμβάλλει ἡ παράλληλος τῇ διαμέτρῳ, καὶ τὸ ὑπὸ τῶν παραλλήλων περιεχόμενον ὀρθογώνιον δοθὲν ἔσται.

κύκλου γὰρ τῇ θέσει δεδομένου τοῦ ΑΒΓ ἐπὶ διαμέτρου τῆς ΒΓ εἰλήφθω δοθὲν σημεῖον τὸ Δ, διὰ δὲ τοῦ Δ πρὸς τὸν κύκλον προσβεβλήσθω τις τυχοῦσα ἡ ΔΑ, ἀπὸ δὲ τοῦ Α τῇ ΔΑ πρὸς ὀρθὰς γωνίας εὐθεῖα ἤχθω ἡ ΑΕ, διὰ δὲ τοῦ Ε τῇ ΑΔ παράλληλος ἤχθω ἡ ΕΖ· λέγω, ὅτι δοθέν ἐστι τὸ Ζ, καὶ ὅτι τὸ ὑπὸ τῶν ΑΔ, ΕΖ χωρίον δοθέν ἐστιν.

διήχθω ἡ ΕΖ ἐπὶ τὸ Θ, καὶ ἐπεζεύχθω ἡ ΑΘ. ἐπεὶ ὀρθή ἐστιν ἡ ὑπὸ τῶν ΘΕΑ γωνία, ἡ ΘΑ διάμετρός ἐστι τοῦ ΑΒΓ κύκλου· ἔστι δὲ καὶ ἡ ΒΓ· τὸ Η ἄρα κέντρον ἐστὶ τοῦ ΑΒΓ κύκλου· δοθὲν ἄρα ἐστὶ τὸ Η. ἔστι δὲ καὶ τὸ Δ δοθέν· δοθεῖσα ἄρα ἐστὶν ἡ ΔΗ τῷ μεγέθει. καὶ ἐπεὶ παράλληλός ἐστιν ἡ ΑΔ τῇ ΕΘ, καί ἐστιν ἴση ἡ ΘΗ τῇ ΗΑ, ἴση ἄρα ἐστὶ καὶ ἡ μὲν ΔΗ τῇ ΗΖ, ἡ δὲ ΑΔ τῇ ΖΘ· δοθεῖσα δὲ ἡ ΔΗ· δοθεῖσα ἄρα καὶ ἡ ΖΗ· ἀλλὰ καὶ τῇ θέσει· ἑκατέρα ἄρα τῶν ΗΖ, ΗΔ δοθεῖσά ἐστιν. καί ἐστι δοθὲν τὸ Η· δοθὲν ἄρα καὶ τὸ Ζ ἐστιν. καὶ ἐπεὶ κύκλου δεδομένου τῇ θέσει τοῦ ΑΒΓ εἴληπται σημεῖον τὸ Ζ δοθέν, καὶ διῆκται ἡ ΕΖΘ, δοθὲν ἄρα ἐστὶ τὸ ὑπὸ τῶν ΕΖΘ· ἴση δὲ ἡ ΘΖ τῇ ΔΑ· δοθὲν ἄρα ἐστὶ τὸ ὑπὸ τῶν ΑΔ, ΕΖ· ὅπερ ἔδει δεῖξαι.

Dt 94. *If on the diameter of a circle given in position a given point be taken, and from that point some straight line be drawn to the circle, and* [if] *from the point of intersection a perpendicular be dropped to the line so drawn, and* [if] *through the point at which the perpendicular meets the circumference a parallel be drawn to the line so drawn,* then *the point at which that parallel meets the diameter is given, and the rectangle contained be the parallel lines will be given.*

For, on the diameter BG of the circle ABG given in position let the given point D have been taken, and through D let an arbitrary straight line DA have been drawn to the circle, and from A let the straight line AE have been drawn at right angles to the straight line AD, and through E let the straight line EZ have been drawn parallel to AD; I say that the point Z is given, and that the rectangle [162] AD, EZ is given.

Let EZ have been produced to Q, and let AQ have been joined.

Since $\angle$QEA is right [I.29], QA is a diameter of the circle ABG [III.31]; and so is BG. Therefore H is the centre of the circle ABG; therefore H is given [Def. 6]. And D is given; therefore DH is given in magnitude [Dt 26].

And since AD is parallel to EQ, and QH is equal to HA, therefore DH is equal to HZ and AD is equal to ZQ [I.29, I.15, I.26]. And DH is given; therefore ZH is also given; but also in position; therefore each of HZ, HD is given.

---

[162] χωρίον in the sense of rectangle.

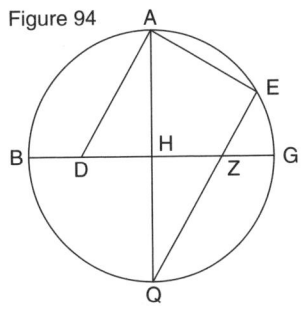

Figure 94

And H is given; therefore Z is given.

And since in the circle ABG given in position the given point Z has been taken, and the straight line EZQ has been drawn through, therefore ⊏⊐EZQ is given [Dt 92]; and QZ is equal to DA; therefore ⊏⊐AD,EZ is given.

Being what it was required to prove. ∎

Remarks:

III.35 ensures that ⊏⊐EZ,ZQ = ⊏⊐GZ,ZB; since ZQ = DA (by congruent triangles), also ⊏⊐EZ,AD = ⊏⊐GZ,ZB, which is a given rectangle according to Dt 92. Note that the position of the point Z (symmetrical to D about the centre H) depends only on the choice of D and is independent to the direction of DA.

A connection with the foci propositions in Apollonius' *Conica*, (III.42, 45-52) seems pretty clear. The following arguments are due to Michael N. Fried:

AE is tangent to an ellipse with foci D and Z, and ⊏⊐BD,DC is equal to the fourth part of "the *eidos* on BC", that is the square on half the conjugate diameter of the said ellipse.

First, consider the following proposition (figure 94.1):

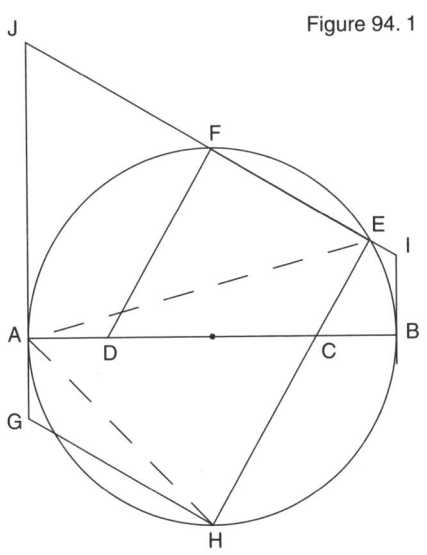

Figure 94.1

Let AFEB be a circle with diameter AB; let DF be drawn to the circle from a point D on AB, let EF be perpendicular to FD and CE perpendicular to EF (therefore, AC = DB, and ⊏⊐FD,EC = ⊏⊐AC,CB or ⊏⊐BD,DA [Dt 94]). Let BI and AJ be tangent to the circle, and let them meet EF at I and J respectively. Then, ⊏⊐FD,EC = ⊏⊐BI,JA = ⊏⊐BD,DA.

Extend EC and JA, and draw HG parallel to EF.

Plainly, then [because CHGA is congruent with DFIB], CH = FD and AG = IB, so it will be sufficient to show that ⊏⊐HC,CE = ⊏⊐GA,AJ = ⊏⊐BD,DA.

Join AH and AE. Then, △JAE = CHA and △GAH = CEA. And, since △CAG, △JEC, △GHC are right angles, △HAC = AEJ and △GHA = CAE. Therefore, △HAC is similar to △AEJ and △AGH is similar to △ECA. Therefore, HC:AH::AJ:AE and AH:GA::AE:CE, so that, ex aequali, HC:GA::AJ:CE, from which the proposition follows. ∎

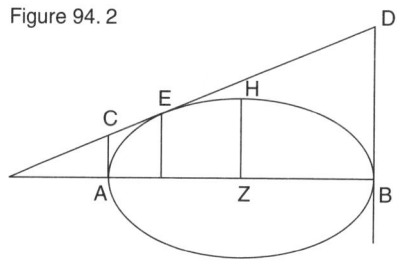

Figure 94. 2

Now, in Conica III.42, Apollonius proves that if AB is the diameter of an ellipse and AC and BD are drawn parallel to an ordinate and CD touches the ellipse at E, then ⊏⊐AC,BD = fourth part of the figure on AB, that is ⊔ZH (figure 94.2). ∎

Is the line AE in figure 94 tangent to an ellipse with diameter BC? Yes. The construction for the ellipse called the pedal curve construction runs as follows:

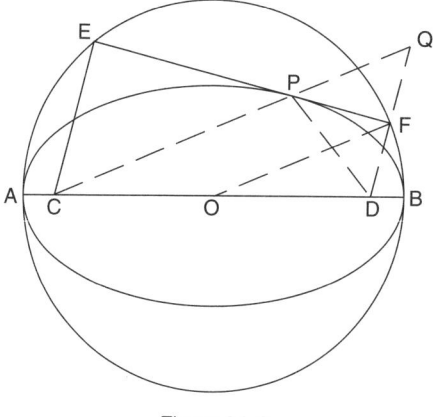

Figure 94. 3

Let a circle with center O and diameter AB be given (figure 94.3); let D be a point on AB (for the corresponding construction of the hyperbola, D is taken on the extension of AB), and let DF be drawn from D to the circle; let EF be perpendicular to DF, and draw EC perpendicular to EF. Then EF is tangent to an ellipse having foci C and D.

Let DF = FQ. Join CQ, and let CQ intersect EF at P; join DP. Then DP = PQ, and CQ = CP+PQ = CP+PD.

But since O bisects CD and F bisects DQ, CQ = 2OF = 2(radius of circle O). So, P is on an ellipse with foci C and D and axis AB; that EF is tangent is clear since △EPC = △QPF = △FPD. ∎

The converse to this, in effect, is III.49. There Apollonius shows that if there is an ellipse with axis AB and focus H, and CD is drawn touching the ellipse and HQ is drawn perpendicular to CD, then AQB will be a right angle (so that Q lies on a circle with diameter AB)

Of course, when Apollonius introduces the foci in III.45, he defines them so that if D (or C) is the focus then ⊏⊐BD,DA (or ⊏⊐AC,CB) = fourth part of the figure on axis AB. The key step in the proof of III.45 is that ⊏⊐BI,JA = ⊏⊐BD,DA (using the lettering of figure 94.1).

Whether this points to a knowledge of propositions III.42, 45-52 predating Apollonius or whether Apollonius had something like Dt 94 in mind when he defined the foci is hard to say. If the latter, then I (M F) think it is likely that Apollonius' path would have been via the first proposition above, to III.42 to III.45 and the rest.

◆

Dt 94 is the last theorem in the *Data*. Why the booklet should end here is not evident, for there is in fact no obvious end to the chain of statements about Givens, (and in this respect the *Data* resembles *Elements* Book II.). The last sentence, the Q.E.D. formula so well known from the *Elements*, (the one and only occurrence of it in the *Data*) intimates that the author considered his task done. But then, the formula may be a later addition. How much else of the work falls in the same category, who knows? Even the name of the author should be suspected, although I know from inside how different can be one's writings at different times. So why should not Euclid of Alexandria have changed his mode of writing on his way?

◆

Deductive Structure of Dt 84–94

From the *Data*

		Dt 84	85	86	87	88	89	90	91	92	93	94
Data Def.	1	.	.	.	+	.	.	.	+	+	+	+
	2*	.	.	+	.	.	.	.	.	.	+	.
	3	+	+	.	+	+	.	.	.	.	.	.
	5	.	.	.	+	.	.	.	.	.	.	.
	6	.	.	.	.	.	+	.	.	.	.	+
	8	.	.	.	.	.	.	+	.	.	.	.
	11	.	.	+	.	.	.	.	.	.	.	.
Dt	1	+	.	+	.	+	.	.	.	.	+	.
	2	.	.	+	+	.	+	.	.	.	.	.
	3	+	+	.	+	+	.	.	.	.	.	.
	4	.	+	.	+	.	.	.	.	.	.	.
	6	.	.	+	.	.	.	.	.	.	.	.
	8	.	.	+	.	.	.	.	.	.	.	.
	24*B	.	.	.	.	.	.	.	.	+	.	.
	25	.	.	.	.	+	+	.	+	.	.	.
	26	.	.	.	.	+	+	.	+	.	.	+
	27	.	.	.	.	.	.	.	.	.	.	+
	29	.	.	.	.	.	+	.	.	.	.	.
	40	.	.	.	+	.	.	.	.	.	.	.
	43	.	.	.	.	+	.	.	.	.	.	.
	50	.	.	+	.	.	.	.	.	.	.	.
	52	.	.	.	.	.	.	+	+	+	.	.
	54	.	+	.	.	.	.	.	.	.	.	.
	55	.	+	.	.	.	.	.	.	.	.	.
	57	.	+	.	.	.	.	.	.	.	.	.
	58	.	+	.	.	.	.	.	.	.	.	.
	59	+	.	.	.	.	.	.	.	.	.	.
	87	.	.	.	.	+	.	.	.	.	+	.
	90	.	.	.	.	.	.	+	.	.	.	.
	92	.	.	.	.	.	.	.	.	.	.	+
		Dt 84	85	86	87	88	89	90	91	92	93	94

CHAPTER 14. CIRCLES (87–94)

Deductive Structure of Dt 84–94

From the *Elements*

Elem		Dt	84	85	86	87	88	89	90	91	92	93	94
I.	post 4		.	.	.	+	.	.	.	.	.	.	.
	Def. 17		.	.	.	.	.	.	.	.	.	.	+
	15		.	.	.	.	.	.	.	.	.	+	+
	26		.	.	.	.	.	.	.	.	.	.	+
	29		.	.	.	.	.	.	.	.	.	.	+
	32		.	.	.	+	.	.	.	.	.	+	.
	34		+	+	.	.	.	.	.	.	.	.	.
II.	2		.	.	+	.	.	.	.	.	.	.	.
	8		.	.	+	.	.	.	.	.	.	.	.
III.	1		.	.	.	+	+	+	.	.	+	.	.
	17		.	.	.	.	.	.	+	+	.	.	.
	18		.	.	.	.	.	.	+	.	.	.	.
	20		.	.	.	.	.	+	.	.	.	.	.
	21		.	.	.	.	.	.	.	.	.	+	.
	31		.	.	.	+	+	.	+	.	.	.	+
	34		.	.	.	+	.	.	.	.	.	.	.
	35		.	.	.	.	.	.	.	.	+	.	.
	36		.	.	.	.	.	.	.	+	.	.	.
V.	7		.	.	+	.	.	.	.	.	.	.	.
	11		.	.	.	.	.	.	.	.	.	+	.
	12		.	.	.	.	.	.	.	.	.	+	.
	16		.	.	.	.	.	.	.	.	.	+	.
VI.	1		.	.	+	.	.	.	.	.	.	.	.
	3		.	.	.	.	.	.	.	.	.	+	.
	4		.	.	.	.	.	.	.	.	.	+	+
	16		.	.	.	.	.	.	.	.	.	.	+
		Dt	84	85	86	87	88	89	90	91	92	93	94

Appendix A
Marinus' Commentary [163]

Preliminary remarks: Many keywords and specific terms have been left untranslated (printed in *italics* and in a transliteration from Greek, mostly adjectives in neuter nominative); they are listed below and given explanations that we hope will make sense in the places where they occur.

agnoston "unknown", "not knowable"; the opposite of *gnorimon*.

alogon "irrational", the opposite of *rheton*. Used of line segments and (rectilineal) figures. The Greeks did not know irrational 'numbers'.

aporon "not available", "not to be provided"; the opposite of *porimon*.

atakton the opposite of *tetagménon*, whose meaning is best understood from Apollonius' *Conica*. In most places the translation "ambiguous" will render the meaning intended.

ek duo onomaton "binomi(n)al line", a line segment "with two names", is defined in the *Elements* X.36: If two *rhetai* (see below) straight lines commensurable in square only are added together, the whole is irrational, *alogos*; and let it be called "[line] with two names". (The terms 'binomial' and 'binomium' were probably misreadings for 'binominal' and 'binominum' in Latin manuscripts with only a stroke to denote the nasal.)

gnorimon "known" or "knowable". Cognate with *gnosis*, "cognition", "knowledge".

onomastikos (adv.) <defined> by a special name, "One-word definition".

---

[163] For the following translation I am in debt to Richard McKirahan of Santa Barbara, who tore some precious hours out of his spare time to help me understand and interpret this peculiar and in many ways difficult text. The responsibility for any errors and misunderstandings that may still be found, is my own – either because I did not follow Richard or because I did.

porimon	"available", cognate with the verb *porizesthai*, "to provide", and the noun *porisma*, a true statement or proposition (often) found as a corollary to another proposition. The term "provide" is discussed on p. 26 ff.
rheton	"rational", defined in Euclid's *Elements* Book X as follows: A line segment *r* is set out and called *rhete*, whose literal meaning is "expressible", "that can be said", *i.e.* that can be associated with a number (a positive integer). The square on *r* and any area that is commensurable with that square is *rheton*, and any line segment whose square is commensurable with the square on *r* is *rhete*. A anglolatin adjective "quadrational" would probably give the right associations. – The definition gives rise to a partition of *rhetai* line segments into such that are commensurable in length, and such that are commensurable in square only.
tetagmenon	"orderly", organized according to some *taxis*, "order". We may quote from Apollonius' *Conica*, I, def. 4, where he defines the ordinates of a conic section: "In any curved line I call that straight line 'diameter' which, drawn from the curved line, bisects all the straight lines drawn in the curved line parallel to some straight line. And I say that each of the parallels is drawn 'ordinatewise', *tetagmenos*, to the diameter. – In Marinus' commentary *tetagmenon* may sometimes be "unique", sometimes "unambiguous". A point or line that is "given in position" is *tetagmenon*, cf. the discussion on p. 93 ff.

Commentary on Euclid's *Givens*, dictated by Marinus the Philosopher.[164]

[Menge 234] It is necessary first to posit what *the given* is; next, to state what is the utility of the treatise about it [*i.e.*, the *Dedomena* itself]; and third <to state> under which science it falls.

Now people define the given in many ways and differently in recent times than before. This is why the true explanation of it turns out to be difficult. In fact, some

---

[164] Marinus from Sikem (in Samaria, also called Neapolis, in modern Arabic Nablus, whence 'Marinus Neapolitanus'), in 485 succeeded Proclus as head of the Neoplatonic Academy. According to the mss this is not written ny Marinus himself but 'dictated' (ἀπὸ φωνῆς), that is, taken down by some student in his audience.

have not even offered any definition if it, but have attempted to discover some property of the given. Others have attempted to define it by putting together the <properties stated> by the first group, but they do not agree with one another either. But it appears that they all begin from one and the same idea and notion in anything about it: they supposed the given to be something comprehensible. For this reason, some of those who proposed to describe the given by a single differentia, such as Apollonius in his book "On Vergings" and in his "Universal Treatise", [165] defined it as *tetagmenon*, while others, such as Diodorus, describe it as *gnorimon*. (He says that radii and angles are given in this way, and also everything accessible to any kind of knowledge, even if it is not *rheton*.) Some declared it to be *rheton*, as Ptolemy seems to [166] when he calls given those magnitudes whose measure is *gnorimon* either precisely or approximately.

[236] Further, some have supposed what is set out [167] in the hypothesis by a person proposing <a mathematical problem> to be given. In the first elementary expositions [168] they also speak of the given <point> and the given <line> in another way, *i.e.* a straight line that someone determines <by an endpoint> and gives the length of. All these <properties> aim to signify some comprehension. And this is why the most highly regarded definitions are those that best reveal what is comprehensible; that will be evident to us when we go on <studying the several definitions>.

Let us now set out the differences among those who do not characterize the nature of the given simply by some single <property>, but produce a kind of definition of it. When summed up the ways <of defining the given> used by these people are easy to count, too. For some determined the given as that which is both *tetagmenon* and *porimon*, others as that which is both *tetagmenon* and *gnorimon*, and still others as that which is both *gnorimon* and *porimon*. All of these too seem to define in the way stated above, namely, with a view to the comprehension or grasping and discovery of the given. But in order to establish this idea of theirs securely and further to choose the true definition of the matter in question from among the many definitions that have been handed down, we must first consider the significance of each of the simple <terms> and of their opposites: *atakton*, *agnoston*, *aporon*, and *alogon* as they apply

---

[165] Those two works are lost. See (Toomer, 1990, xii).

[166] Almagest I.10, Heiberg 32.

[167] ἐκτιθέμενον, set out in that part of a Euclidean proposition that is called the *ekthesis*: "The setting-out marks off what is given, by itself, and adapts it beforehand for use in the investigation" (Proclus, Friedlein 203, 7–9).

[168] He may refer to *Elements* I.1–3, 10, 22, 46. Or simply 'elementary textbooks'.

to the subject matter of geometry, which is being discussed at present (for such <terms> have application also to the objects studied by natural philosophy and to the other mathematical sciences).

[238] They outline the *tetagmenon* as that which always remains the same in the respect in which it is *tetagmenon* – in size, shape or some other such respect.[169] Alternatively <they outline it> as that which does not admit becoming different at different times, but <occurs> in only one way in some determinate place. For example, to speak roughly, the straight line drawn through two fixed points is said to be *tetagmenon* because it cannot be drawn otherwise and variably. By contrast, a circumference through two points is *atakton*, since it can be drawn in many ways and variably – since both circles that are larger and smaller ad infinitum can be drawn through the two points. On the other hand, the circumference through three points is *tetagmenon*.

Also the following sorts of things are *tetagmena*: constructing an equilateral triangle on a given straight line – for even though it occurs in two ways, still it happens in only one way and unchangeably on either side of the straight line; and cutting a given straight line in a given ratio; for this too occurs in only one way on either side of the midpoint.

Things opposite to these are *atakta*, for example constructing a scalene triangle and cutting a straight line indeterminately. "In the respect in which it is *tetagmenon*" is added to the definition because one and the same thing can be *tetagmenon* in one respect and *atakton* in another – for example the equilateral triangle is *tetakton* in that it is equilateral, but is not wholly determined in respect of size.

Gnorimon is that which is known and grasped as evident to us, and *agnoston* is that which is not known or grasped by us. For example, the length of a road is said to be *gnorimon* [240] in that I have grasped how many stades long it is, and the facts that the interior angles of a triangle are equal to two right angles and that the line *ek duo onomaton* is *alonos*. Further the following things are called *gnorimon*: the fact that there is <exactly >one tangent to a spiral on each side from a given external point. For if there were a second, two straight lines would enclose an area, which is impossible. *Aloga* are not <in general> *agnosta* – rather, those *aloga* that are not known or grasped by us.

That which we are now able to make and construct – *i.e.*, to bring to our thought -- is *porimon*. The *porimon* is also defined otherwise, either as that which is furnished (*porizomenon*) by proof or when something is obvious even without demonstration. For example, drawing a circle given the center and distance [*i.e.*, radius], constructing a

---

[169] E.g. 'given in position', which (together with 'given in magnitude' and 'given in form') plays a dominant role in the *Data*.

triangle (not only an isosceles but even a scalene triangle), finding the line *ek duo onomaton*, and finding three *rhetai* straight lines that are commensurable in square only.[170] Even things that occur in an indefinite number of ways are *porimon*, for example, drawing a circle through two points.

The opposite is *aporon*. For example, squaring a circle, for this is not yet in our power,[171] even if it can be attained and falls under some science, for the scientific knowledge of it has not yet been grasped. At present we are discussing what is already in our power, which is what we name *porimon* in the strict sense. For what is not yet in our power even though it can be attained they label attainable (*poriston*) in a special sense. But as has been stated, that which is opposite to the *porimon* is *aporon* – that whose investigation is undecided.[172]

A thing is *rheton* when we can state its size, shape or position. [242] This definition is more general. What is specifically and per se *rheton* is that which we know by some number in relation to an arbitrary measure, for example a palm or a finger.

Now that these <terms> have been provisionally distinguished in these ways, it will next be easier to spot their common nature and their differences, and first, how the *tetagmenon* is related to the *gnorimon* and how their opposites are related to one another. Now things like these are not convertible nor are they such that one term extends more widely than the other. For even though they have much in common, e.g. to draw a straight line through two points and a circle through three, and to construct an equilateral <triangle on a line>, still, to square the circle is *tetagmenon* but is *agnoston*. And that there is <exactly> one tangent to a spiral from a single point is *tetagmenon* and necessary, but its demonstration and construction are not known. Further, division ad infinitum and the construction of a scalene <triangle> are known, but not for all that *tetagmenon*. So it is obvious that some things that are *tetagmenon* will be *gnorimon* and others will be *agnoston*, and contrariwise, some things that are *gnorimon* will be *tetagmenon* and others *atakton*. They are related to one another in the same way as eloquence is to prose: such things are not co-exensive nor does either

---

[170] e.g. the sides of a right triangle a, b, and c where $a^2 = 5$, $b^2 = 8$, and $c^2 = 13$. Such problems are treated in *Elements* X. 30 ff.

[171] *I.e.* among the possible constructions. Marinus presents the case of squaring the circle as if it were still an open question.

[172] What Marinus should be saying here is that things are *aporon* when there is not a satisfactory result; this can be either when it is not known whether there is a solution or when it is known that there is no solution; but *adiakritos*, 'undecided', seems to cover only the first of these cases.

extend more widely than the other.

The same holds for the relation between *tetagmenon* and *atakton* on the one hand and *porimon* and *aporon* on the other. They have a very great deal in common, and they differ from one another in the way stated above. [244] For the spiral is *tetagmenon* but was not *porimon* for mathematicians before Archimedes, and things that occur in an indefinite number of ways and *ataktos* are *porima* if someone devises their construction and formation, but they are not *tetagmena* for all that. For example, to devise a scalene triangle and to bring one's thought from the construction of an equilateral [triangle] to the construction of a scalene one, is not difficult but easy to achieve, even though it is *atakton* and indefinite.

The *tetagmenon* and the *atakton* are related in this same way to the *rheton* and the *alogon*: they have much in common and they differ in the way that has been stated. For they too are not coextensive nor does either include the other. For the line *ek duo onomaton* and the *alogoi* lines that are grasped in that way are *tetagmena* but not *rheta*, and likewise for the ratio of the diagonal to the side. Further, many things that are *rheta* are *atakta*, such as things that occur in many ways and in an indefinite manner. For a scalene triangle can be measured by a proposed definite *rheton* measure, even though it is *atakton*.

The similarity between the *gnorimon* and the *porimon* is easy for anyone to distinguish, but the difference is hard to grasp. For by nature they are close to one another, so that they seem to be coextensive. But even here a certain inherent difference will be seen if we observe them accurately. For it is evident and *gnorimon* that there is [precisely] one tangent to a spiral from a single point, but for all that the problem is not now *porimon* as well, since it has not yet been grasped. [246] And so not everything that is *gnorimon* is *porimon*. However everything that is *porimon* is also *gnorimon*. Therefore the *gnorimon* extends more widely than the *porimon*.

Also, the *gnorimon* and the *rheton* in one way have a commonality and in another differ from one another in the way stated above. For the *alogoi* lines we have mentioned are *gnorimoi*, but not *rhetai*, whereas every number is *rheton*, but not every one is *gnorimon* as well.[173] Also, what is *rheton* is similarly *rheton* for those who have the same custom, and a length will not be *rheton* for one and not for another, for they will refer it to the same measure. But the same length proves to be *gnorimon* to one person but not to another, even if they share the same custom. Perhaps here too it is hard to find something that is both *rheton* and *agnoston*, for the *gnorimon* seems to extend more widely than the *rheton*.

It is obvious from the following considerations that the *porimon* and the *aporon* differ from the *rheton* and the *alogon*. For even some things that are *aloga* can be

---

[173] Because there are infinitely many numbers?

porima, but nothing *rheton* is *alogon*. The kinship of these very things, just like that of the others, is obvious to everyone. However, these too are related to one another in such a way that the *porimon* seems to be wider than the *rheton*.

It is possible to view the difference between the things mentioned above in the following way too. The *rheton* and *alogon* are defined with reference to the measure, without being referred to our knowledge. For it is possible that it is not *gnorimon* to us how something that is *rheton* is *rheton*, and it can fail as yet to be grasped that it is *rheton*. But the *tetagmenon* and [248] the *atakton* are considered <to be attributes that hold> per se and in virtue of a thing's own nature, even if they are not yet grasped by us. In any case Archimedes proved many things that are *tetagmenon* by nature which had not been considered to be *tetagmenon* by his predecessors. However, *gnorimon* and *agnoston* are defined by reference to us. And so the features that have been mentioned differ from one another if indeed one case contains a reference to us, the second to the nature <of the things involved>, and the third to the measure.

Now that the commonality and the difference among the terms proposed have been determined, the next task is, then, to consider precisely what the given is. Now all those who think that the given is that which is being given by hypothesis by the person who sets a problem completely miss the object of investigation. For all the *Elements of Givens* are not composed about the kind of thing that is given by hypothesis – as can be seen if you refer to the treatises on this topic. For this reason we must abandon this notion and examine the accounts given by those who define it differently; what is given by hypothesis will be what is considered as following from the principles.

Some define <the given> making use of "one-word definitions" and characterizing it by one or another of the aforementioned terms, as was stated at the beginning. Practically everyone seems to have a common notion about the given; they have supposed that it is something grasped, as the very name "given" makes plain. This is particularly true for those who outline it as what is given by hypothesis. Some others have focused on the fact that it is granted. We too will use what is said as a standard and criterion, and thus will be able to [250] find the complete definition of the given. Clearly it will have to be co-extensive or convertible with the definiendum: this is a requirement for properly rendered definitions. In the present case, among the more simply stated definitions, the one that defines <the given> as the *porimon* is of this sort, as is the complex definition that defines it as what is both *gnorimon* and *porimon*. All the others are incomplete. The one that defines it as the *tetagmenon* is insufficient as regards the extension of the given, since neither all nor only that which is *tetagmenon* is comprehended, but some things that are *atakta* <are comprehended>, as has been shown. Nor is the definition sufficient which defines it as the *gnorimon*, since again not everything of this sort is comprehended, even if this is the only kind of thing that is comprehended (for what is *agnoston* could not be comprehended). Nor is the one that declares it to be *rheton* a complete definition, for this is not the only thing that is comprehended, since some *aloga* are comprehended. Perhaps not even everything that is *rheton* is comprehended either as it was defined earlier. Among the definitions rendered with one word (*onomastikos*), there remains <the definition of it as> the *porimon*, which seems best to display the comprehension. For all and only that which is *porimon*, is comprehended. In fact Euclid too made use of this kind of definition when he was outlining all the kinds of the given. The only complex definition that is complete is the one that defines the given as both *gnorimon* and *porimon*, since it treats the *gnorimon* analogously with the *genus* and the *porimon* analogously with the *differentia*. The definition that calls it both *tetagmenon* and *porimon* is incomplete, since such things are not the only things that are given. Also the one <defining it as> *tetagmenon* and *rheton* likewise has a narrower extension than the [252] given, whereas that which defines it as both *gnorimon* and *tetagmenon* is unsound because it extends beyond the subject, since not everything of that kind is given. The only people, then, that seem to reach the notion of the given are those who declare it to be both *gnorimon* and *porimon*. For all and only things of this kind are comprehended, and both these properties [*i.e.*, "all and only"] must hold of scientifically rendered definitions. Close by to these are those who frame <the definition> as follows: given is what we can provide (*porisasthai*) on the basis of what is posited in the first hypotheses and principles. Euclid too belongs to the previously mentioned group, since he everywhere uses the word "furnish," even if he fails to mention the *gnorimon* as accompanying the *porimon*.[174] Someone might reasonably accuse him of not first defining the given in general but immediately <defining> each of its kinds, even though in his Elements of Geometry he clearly defined line without qualification before the kinds of line, and similarly for the other things.

Now that the given has been analysed more generally and for present purposes, the next task is to explain the utility of the treatise dedicated to it. This too belongs to the class of things that have reference to something else: for the knowledge of this subject is absolutely necessary with reference to the technique called analysis. The power of analysis in the mathematical sciences and in related sciences of optics and canonics has been determined elsewhere, as well as the fact that analysis [254] is the discovery of demonstration, and how it contributes to our discovery of the demonstration of similar cases, and that possessing the ability to perform analysis is more important than knowing many demonstrations of particular results.

Now since the study of the given is useful for such sciences and particularly since it also makes a great contribution to analysis, it can reasonably be said that it has to do not with any one science, but to with what is called universal mathematics. This is

---

[174] See the discussion about "known" on p. 18.

the <mathematical science> concerned with multitudes, magnitudes, times, speeds, and all such things, just like the <mathematical science> that deals with ratios, proportions and means of all kinds. For the purpose of this scientific comprehension of the givens, which is so useful, the book of Givens was brought to perfection by Euclid, whom they also named the composer of Elements par excellence. He arranged elements as a sort of introduction for practically every mathematical science, handing down to posterity elements of the whole of geometry in 13 books, the elements of astronomy in the Phaenomena, and likewise of music and optics. And in particular, in the present work he made an analytical arrangement of the elements of the entire treatment of the given. But since he was above all a geometer, he applied the general statements of the given to magnitudes in particular, the same way as he dealt with ratios in general, treating them in the fifth book of the plane geometry as if holding of magnitudes in particular.

Now we have said generally what the given is and to [256] which science it is referred. We have also said that the study of it is extremely useful. But let us add to what has been said a description of the science concerned with it. It will consist, as is obvious from what has been said, in the comprehension of things that are given in every way and of their attributes. In particular and with a view to the present book, let it be said to be an investigation that contains the elements of the entire science of givens. Consequently both it and the remaining matters as well will possess utility in virtue of their relation to the given. The book is divided with reference to the kinds of the given. Its first section contains things given in ratio, the second those given in position. Next are those given in form. Material on things given in magnitude is a single topic, but is scattered in fragments in the other sections, especially in the <section on> things given in form. He began with things given in ratio and position since things given in form are composed of these. There is a different division of as well: into magnitudes in general and into lines, planes and propositions about circles. He employed the same arrangement for the definitions or hypotheses of the book. The manner of exposition that he followed here is not that of synthesis but of analysis, as Pappus demonstrated sufficiently in his commentary on the book.

Appendix B. Synopsis of Dt 39 and I.22

Mutatis mutandis the text of Dt 39 is almost *verbatim* the same as Elements I.22 (cf. p. 123). The author of the Data took the text of I.22 and inserted the 'given'-idiom where needed, which explains the syntactical syncopes that are evident in this synopsis:

Dt 39
ἐκκείσθω γὰρ εὐθεῖα τῇ θέσει δεδομένη ἡ ΔΜ, πεπερατωμένη μὲν κατὰ τὸ Δ, ἄπειρος δὲ κατὰ τὸ λοιπόν, καὶ κείσθω τῇ μὲν ΑΒ ἴση ἡ ΔΕ: δοθεῖσα δὲ ἡ ΑΒ: δοθεῖσα ἄρα καὶ ἡ ΔΕ: ἀλλὰ καὶ τῇ θέσει: καί ἐστι δοθὲν τὸ Δ: δοθὲν ἄρα καὶ τὸ Ε:
τῇ δὲ ΒΓ ἴση ἡ ΕΖ: δοθεῖσα δὲ ἡ ΒΓ: δοθεῖσα ἄρα καὶ ἡ ΕΖ: ἀλλὰ καὶ τῇ θέσει: καί ἐστι δοθὲν τὸ Ε: δοθὲν ἄρα καὶ τὸ Ζ:

τῇ δὲ ΑΓ ἴση ἡ ΖΗ. δοθεῖσα δὲ ἡ ΑΓ: δοθεῖσα ἄρα καὶ ἡ ΖΗ. ἀλλὰ καὶ τῇ θέσει. καί ἐστι δοθὲν τὸ Ζ: δοθὲν ἄρα καὶ τὸ Η.

καὶ κέντρῳ μὲν τῷ Ε, διαστήματι δὲ τῷ ΕΔ κύκλος γεγράφθω ὁ ΔΚΘ: θέσει ἄρα ἐστὶν ὁ ΔΚΘ.
πάλιν κέντρῳ μὲν τῷ Ζ, διαστήματι δὲ τῷ ΖΗ κύκλος γεγράφθω ὁ ΗΚΛ: θέσει ἄρα ἐστὶν ὁ ΗΚΛ: θέσει δὲ καὶ ὁ ΔΘΚ κύκλος:
δοθὲν ἄρα ἐστὶ καὶ τὸ Κ σημεῖον. ἔστι δὲ καὶ ἑκάτερον τῶν Ε, Ζ δοθέν: δοθεῖσα ἄρα ἐστὶν ἑκάστη τῶν ΚΕ, ΕΖ, ΖΚ τῇ θέσει καὶ τῷ μεγέθει: δέδοται ἄρα τὸ ΚΕΖ τρίγωνον τῷ εἴδει.
καί ἐστιν ἴσον τε καὶ ὅμοιον τῷ ΑΒΓ: δέδοται ἄρα τὸ ΑΒΓ τρίγωνον τῷ εἴδει.

Elements I.22
ἐκκείσθω τὶς εὐθεῖα ἡ ΔΕ, πεπερασμένη μὲν κατὰ τὸ Δ, ἄπειρος δὲ κατὰ τὸ Ε, καὶ κείσθω τῇ μὲν Α ἴση ἡ ΔΖ,

τῇ δὲ Β ἴση ἡ ΖΗ,

τῇ δὲ Γ ἴση ἡ ΗΘ.

καὶ κέντρῳ μὲν τῷ Ζ, διαστήματι δὲ τῷ ΖΔ κύκλος γεγράφθω ὁ ΔΚΛ
πάλιν κέντρῳ μὲν τῷ Η, διαστήματι δὲ τῷ ΗΘ κύκλος γεγράφθω ὁ ΚΛΘ:

καὶ ἐπεζεύχθωσαν αἱ ΚΖ, ΚΗ. λέγω ὅτι ἐκ τριῶν εὐθειῶν τῶν ἴσων ταῖς Α, Β, Γ τρίγωνον συνέσταται τὸ ΚΖΗ.

(I.22 ends with a proof that the sides are respectively equal to the given lines.)

Dt 39

For, let a straight line given in position DM have been laid, terminated at D, infinite in the other direction, and let DE lie equal to AB;

and AB is given; therefore DE is also given, but in position, too; and D is given, therefore E is also given.

And [let] EZ [lie] equal to BC; and BC is given, therefore EZ is also given, but in position, too; and E is given, therefore Z is also given.

And [let] ZH [lie] equal to AC; and AC is given, therefore ZH is also given, but in position too; and Z is given, therefore H is also given.

And with centre E and radius ED let the circle DKG have been described; then the [circle] DKG is given in position.

Again with centre Z and radius ZH let the circle HKL have been described; then the [circle] HKL is given in position. And the circle DGK is given in position, too; therefore the point K is given.

And each of the points E and Z is also given; therefore each of KE, EZ, ZK is given in position and in magnitude; therefore the triangle KEZ is given in form. And it is equal and similar to the triangle ABC; therefore the triangle ABC is given in form.

Elements I.22 (Heath)

Let there be set out a straight line DE, terminated at D but of infinite length in the direction of E, and let DF be made equal to A,

FG equal to B,

and GH equal to C.

With centre F and distance (*i.e.* radius) FD let the circle DKL be described;

Again, with centre G and distance GH let the circle KLH be described;

And let KF, KG be joined. I say that the triangle KFG has been constructed out of three straight lines equal to A, B, C.

***

I.22 ends with a proof that the sides are respectively equal to the given lines.

List of Definitions and Enunciations

A survey of the 15 definitions and the enunciations (προτάσεις) of the 94 propositions in the *Data*, indicating the pages where the definition or proposition is translated and discussed. You should also consult the list of supplementary definitions and lemmas after this one.

Chapter 1. Definitions

Def 1   Given in magnitude is said of figures and lines and angles for which we can provide equals.   17, 23, 26

Def 2   A ratio is said to be given for which we can provide the same.   17, 24, 31

Def 3   Rectilineal figures are said to be given in form if each angle is given and the ratios of the sides to one another are given.   17, 24, 33, 115

Def 4   Given in position is said of points and lines and angles which always hold the same place.   17, 33

Def 5   A circle is said to be given in magnitude if its radius is given in magnitude.
                                                                                                       34

Def 6   And a circle is said to be given in position and in magnitude if its center is given in position and its radius in magnitude.   34

Def 7   Segments of circles are said to be given in magnitude if the angles in them and the bases of the segments are given in magnitude.   34

Def 8   And segments of circles are said to be given in position and in magnitude if the angles in them are given in magnitude and the bases of the segments are given in position and in magnitude.   34

Def 9   A magnitude [M] is by a given [G] greater than a magnitude [N] if, when the given magnitude [G] be subtracted [from M], the remainder is equal to the same[N].   34

Def 10 A magnitude [M] is by a given less than a magnitude [N] if, when the given magnitude be added [to M], the sum is equal to the same [N].   34

Def 11   A magnitude is by a given greater than in ratio to a magnitude if, when the given magnitude be subtracted, the remainder has a given ratio to the same.
35, 57

Def 12   A magnitude is by a given less than i n ratio to a magnitude if, when the given magnitude be added, the sum has a given ratio to the same.   35

Def 13   A straight line is dropped if it is drawn in a given angle from a given point to a straight line given in position.   35

Def 14   A straight line is raised if it is drawn in a given angle from a given point on a straight line given in position.   35

Def 15   A straight line is drawn parallel in position if it is drawn through a given point parallel to a straight line given in position.   35

◆

Chapter 2. Magnitudes and Ratio I

Dt 1   The ratio of given magnitudes to one another is given.   37

Dt 2   If a given magnitude have a given ratio to some other magnitude, the other is also given in magnitude.   39

Dt 3   If any number of given magnitudes be added together, the magnitude composed of them will also be given.   42

Dt 4   If a given magnitude be subtracted from a given magnitude, the remainder will be given.   43

Dt 5   If a magnitude have a given ratio to some part of itself, it will also have a given ratio to the remainder.   46

Dt 6   If two magnitudes having a given ratio to one another be added together, the whole will also have a given ratio to each of them.   49

Dt 7   If a given magnitude be divided in a given ratio, each of the parts is given. 50

Dt 8   [Magnitudes] which have a given ratio to the same, will also have a given ratio to one another. 52

Dt 9   If two or more magnitudes have a given ratio to one another, and if the same magnitudes have given ratios to some other magnitudes, even if not the same ratios, then those magnitudes will have given ratios to one another. 54

Chapter 3. By a Given Greater than in Ratio

Dt 10   If a magnitude [M] be by a given [magnitude] greater than in [a given] ratio to a magnitude [N], the sum [M+N] will be by a given [magnitude] greater than in [a given] ratio to the same [N];
and if the sum [M+N] be by a given[magnitude] greater than in [a given] ratio to the same [N], the remainder [M] is either by a given [magnitude] greater than in [a given] *ratio to the same* [N], or the remainder together with the adjacent to which the other one [N] has a given ratio, is given. 61

Dt 11   If a magnitude be by a given greater than in ratio to a magnitude, the same will be by a given greater than in ratio to the sum;
and if the same be by a given greater than in ratio to the sum, the same will be by a given greater than in ratio to the remainder. 65

Dt 12   If there be three magnitudes, and the first plus the second be given in magnitude, and the second plus the third be given in magnitude, the first is either equal to the third, or the one is greater than the other by a given magnitude.
68

Dt 13   If there be three magnitudes, and the first have to the second a given ratio, while the second be greater than in ratio to the third by a given [magnitude], the first will be greater than in ratio to the third by a given [magnitude]. 70

Dt 14   If two magnitudes have a given ratio to one another, and a given magnitude be added to each of them, the wholes will either have a given ratio to one another, or one of them is by a given magnitude greater than in ratio to the other. 72

Dt 15   If two magnitudes have a given ratio to one another, and a given magnitude be subtracted from each of them, the remainders either will have a given ratio to one another, or one of them is by a given greater than in ratio to the other. 74

Dt 16 If two magnitudes have a given ratio to one another, and from the one a given magnitude be subtracted, while to the other a given magnitude be added, the whole will be greater than in ratio to the remainder by a given magnitude. 75

Dt 17 If there be three magnitudes, and the first be greater than in ratio to the second by a given magnitude, while the third be greater than in ratio to the same by a given magnitude, then the first will either have a given ratio to the third, or one of them will be greater than in ratio to the other by a given magnitude. 77

Dt 18 If there be three magnitudes, and one of them be greater than in ratio to each of the remaining ones by a given magnitude, the remaining ones either will have a given ratio to one another, or one of them is greater than in ratio to the other by a given magnitude. 78

Dt 19 If there be three magnitudes, and the first be greater than in ratio to the second by a given magnitude, and the second be greater than in ratio to the third by a given magnitude, the first will be by a given greater than in ratio to the third by a given magnitude. 79

Dt 20 If there be two given magnitudes, and magnitudes having a given ratio to one another be subtracted from them, the remainders either will have a given ratio to one another, or one of them is by a given greater than in ratio to the other. 81

Dt 21 If there be two given magnitudes, and magnitudes having a given ratio to one another be added to them, the wholes will either have a given ratio to one another, or one of them is by a given greater than in ratio to the other. 82

Chapter 4

Dt 22 If two magnitudes have a given ratio to a third, their sum will have a given ratio to the same. 85

Dt 23 If a whole have to a whole a given ratio, and the parts have to the parts given ratios, but not the same, all [the magnitudes] will have given ratios to all. 87

Dt 24 If three straight lines be proportional, and the first have to the third a given ratio, it will have to the second a given ratio. 89

Chapter 5. Position. Distance, Direction, Parallels

Dt 25   If two lines given in position cut one another, their point of section is given in position.   93

Dt 26   If the extremities of a straight line be given in position, the line is given in position and in magnitude.   99

Dt 27   If the one extremity of a straight line given in position and in magnitude be given, the other will also be given.   100

Dt 28   If through a given point a straight line be drawn parallel to a straight line given in position, the straight line drawn is given in position.   100

Dt 29   If at a straight line given in position and at a given point on it a straight line be drawn making a given angle, the straight line drawn is given in position.   102

Dt 30   If from a given point to a straight line given in position a straight line be drawn making a given angle, the line drawn is given in position.   103

Dt 31   If from a given point a straight line given in magnitude be drawn to meet a straight line given in position, the line drawn is also given in position.   104, 106

Dt 32   If on parallel straight lines given in position a straight line be drawn making given angles, the straight line drawn is given in magnitude.   106

Dt 33   If on parallel straight lines given in position a straight line given in magnitude be drawn, it will make given angles.   107

Dt 34   If a straight line be drawn from a given point to parallel straight lines given in position, it will be cut in a given ratio.   108

Dt 35   If from a given point to a straight line given in position, a straight line be drawn and cut in a given ratio, and [if] a straight line be drawn through the point of section parallel to the line given in position, the latter straight line drawn will be given in position.   110

Dt 36   If from a given point to a straight line given in position a straight line be drawn and a straight line having a given ratio to it be added to it, and [if] a straight line be drawn through the extremity of the added line parallel to the line given in position, the [latter] straight line drawn is given in position. 111

Dt 37   If on parallel lines given in position a straight line be drawn and cut in a given ratio, and through the point of section a straight line be drawn parallel to the lines given in position, the [latter] line drawn is given in position. 112

Dt 38   If on parallels given in position a straight line be drawn and a straight line having a given ratio to it be added to it, and if through the extremity a parallel straight line be drawn parallel to the lines given in position, the [latter] line drawn is given in position. 113

Chapter 6. Form. Triangles and Polygons

Dt 39   If each of the sides of a triangle be given in magnitude, the triangle is given in form. 119

Dt 40   If each of the angles of a triangle be given in magnitude, the triangle is given in form. 124

Dt 41   If a triangle have one given angle, and the sides about the given angle have a given ratio to one another, the triangle is given in form. 126

Dt 42   If the sides of a triangle have a given ratio to one another, the triangle is given in form. 121

Dt 43   If in a right-angled triangle the sides about one of the acute angles have a given ratio to one another, the triangle is given in form. 127

Dt 44   If a triangle have one given angle, and the sides about another angle have a given ratio to one another, the triangle is given in form. 129

Dt 45   If a triangle have one angle given, and the sum of the sides about the given angle has a given ratio to the remaining side, the triangle is given in form. 130

Dt 46   If a triangle have one angle given, and the sum of the sides about another angle have a given ratio to the remaining side, the triangle is given in form. 132

LIST OF DEFINITIONS AND ENUNCIATIONS

Dt 47 Rectilineal figures given in form are divisible into triangles given in form.
133

Dt 48 If on the same straight line two triangles be described, given in form, they will have a given ratio to one another.
134

Dt 49 If on the same straight line two arbitrary rectilineal figures given in form be described, they will have a given ratio to one another.
136

Dt 50 If two straight lines have a given ratio to one another, the similar and similarly described rectilineal figures on them will have a given ratio to one another.
137

Dt 51 If two straight lines have a given ratio to one another, and arbitrary rectilineal figures given in form be described on them, they will have a given ratio to one another.
138

Dt 52 If on a straight line given in magnitude a form given in form be described, the [form] described is given in magnitude.
139

Dt 53 If two forms be given in form, and one side of one form have a given ratio to one side of the other, the other sides will also have a given ratio to the other sides.
140

Dt 54 If two forms given in form have a given ratio to one another, their sides will have a given ratio to one another.
141

Dt 55 If an [rectilineal] figure be given in form and in magnitude, its sides will be given in magnitude.
116, 142

Chapter 7. Equiangular parallelograms I

Dt 56 If two equiangular parallelograms have a given ratio to one another, as the side of the first parallelogram is to the side of the second, so the other side of the second will be to that [line] to which the other side of the first has the given ratio which the first parallelogram has to the second.
147

Chapter 8. Application of areas I

Dt 57 If a given [area] be applied to a given [straight line] in a given angle, the width of the applied [area] is given.
151

Dt 58 If a given [area] be applied to a given [straight line] deficient by a form given in form, the dimensions of the defect are given.   154

Dt 59 If a given [area] be applied to a given straight exceeding by a figure given in form, the dimensions of the excess are given.   157

Dt 60 If a parallelogram given in form and in magnitude be augmented or diminished by a given gnomon, the dimensions of the gnomon are given.   159

Dt 61 If to one side of a form given in form a parallelogrammic area be applied in a given angle, and [if] the form have a given ratio to the parallelogram, the parallelogram is given in form.   161

Chapter 9. Ratio and Angles

Dt 62 If two straight lines have a given ratio to one another, and on one of them a form, given in form, be described, on the other a parallelogrammic area in a given angle, and [if] the form have a given ratio to the parallelogram, the parallelogram is given in form.   163

Dt 63 If a triangle be given in form, the squares on each of its sides will have a given ratio to the triangle.   165

Dt 64 If a triangle have a given obtuse angle, the area by which the square on the side subtending the obtuse angle is greater than [the sum of] the squares on the sides containing the given angle, [that area] will have a given ratio to the triangle.   167

Dt 65 If a triangle have a given acute angle, the area by which the square on the side subtending the acute angle is less than [the sum of] the squares on the sides containing the given angle, [that area] will have a given ratio to the triangle.   169

Dt 66 If a triangle have a given angle, the rectangle contained by the lines that contain the given angle has a given ratio to the triangle.   170

Dt 67 If a triangle have a given angle, a certain area will have a given ratio to the triangle, namely that area by which the square on the sum of the sides containing the given angle is greater than the square on the third side.   171

Chapter 10. Equiangular parallelograms II

Dt 68 If two equiangular parallelogras have a given ratio to one another, and one side have a given ratio to one side, the other side will have a given ratio to the other side. 178

Dt 69 If two parallelograms have given angles and a given ratio to one another, and one side have a given ratio to one side, the other side will have a given ratio to the other side. 179

Dt 70 If in two parallelograms the sides about equal angles, or about unequal but given angles have a given ratio to one another, the parallelograms themselves will have a given ratio to one another. 181

Dt 71 If in two triangles the sides about equal angles, or about unequal but given angles, have a given ratio to one another, the triangles themselves have a given ratio to one another. 182

Dt 72 If in two triangles the bases be in a given ratio, and the straight lines drawn to them from the [subtending] angles making either equal angles, or unequal but given angles, at the bases, be in a given ratio, the triangles themselves will have a given ratio to one another. 183

Dt 73 If in two parallelograms the sides about equal angles, or about unequal but given angles, be such that, as one side of the first parallelogram is to one side of the second, so is the other side of the second parallelogram to some straight line, and [if] the other side of the first parallelogram have a given ratio to that line, the parallelograms themselves will have a given ratio to one another. 185

Dt 74 If two parallelograms have a given ratio, either in equal angles, or unequal but given angles, then as the side of the first parallelogram is to the side of the second, so the other side of the second will be to that straight line to which the other side of the first parallelogram has the given ratio. 187

Dt 75 If two triangles have a given ratio to one another, either in equal angles, or unequal but given angles, then as the side of the first triangle is to the side of the second, so the other side of the second will be to that straight line to which the other side of the first triangle has the given ratio. 189

Chapter 11. Duplicates and Outsiders

Dt 76 If in a triangle given in form a perpendicular be drawn from the vertex to the base, the line drawn has a given ratio to the base.   191

Dt 77 If two forms, given in form, have a given ratio to one another, any one side of one of the forms will also have a given ratio to any one side of the other form.   192

Dt 78 If a given form have a given ratio to some rectangle, and one side [of the form] have a given ratio to one side [of the rectangle], the rectangle is given in form.   193

Dt 79 If two triangles have one angle equal to one angle, and from the equal angles to the bases straight lines be drawn perpendicularly, and if as the base of the first triangle be to the perpendicular, so is the base of the second triangle to the perpendicular, then the triangles will be equiangular.   195

Dt 80 If a triangle have one angle given, and the rectangle contained by the lines about the given angle have a given ratio to the square on the third side, the triangle is given in form.   198

Dt 81 If three proportional straight lines have to three proportional straight lines the extremes in a given ratio, the means will also be in a given ratio. And if the extreme have a given ratio to the extreme and also the mean to the mean, then the other extreme will have a given ratio to the other extreme.   200

Dt 82 If four lines be proportional, as the first is to that to which the second has a given ratio, so the third will be to that to which the fourth has a given ratio.   202

Dt 83 If four straight lines be to each other in such a way that, any three of them taken together with a fourth to which the last of the original four has a given ratio, the four straight lines be proportional, then as the fourth is to the third, so the second will be to that to which the first has the given ratio.   203

Chapter 12. Application of areas II

Dt 84 If two straight lines contain a given area in a given angle, and one of them be greater than the other by a given [straight line], each of them will be given.   207

LIST OF DEFINITIONS AND ENUNCIATIONS

Dt 85 If two straights contain a given area in a given angle, and their sum be given, each of them will be given.   209

Chapter 13. Intersecting Hyperbolas. Zeuthen's conjecture

Dt 86 If two straight lines contain a given area in a given angle and if the square on one of them be by a given [area] greater than to be in a given ratio to the square on the other, then each of them will also be given.   211

Chapter 14. Circles

Dt 87 If in a circle, given in magnitude, a straight line be drawn cutting off a segment admitting a given angle, the line drawn is given in magnitude.   225

Dt 88 If in a circle given in magnitude, a straight line be drawn given in magnitude, it will cut off a segment admitting a given angle.   227

Dt 89 If in a circle given in position a given point be taken on the circumference, and if from that point a straight line be inflected at the circumference, making a given angle, the other endpoint of the inflected line will be given.   227

Dt 90 If from a given point a straight line be drawn touching a circle given in position, the straight line drawn will be given in position and in magnitude.   229

Dt 91 If a given point be taken outside a circle given in position, and from the point a straight line be drawn into the circle, the rectangle contained by the straight line and the straight line between the point and the convex circumference is given.   230

Dt 92 If a given point be taken inside a circle given in position, and through the point a straight line be drawn in the circle, the rectangle contained by the segments of the straight line is given.   231

Dt 93 If in a circle, given in magnitude, a straight line be drawn cutting off a segment admitting a given angle, and if the angle in the segment be bisected, the sum of the straight lines containing the given angle will have a given ratio to the bisector, – and the rectangle contained by the sum of the lines containing the given angle and [by] that part of the bisector which is cut off below towards the circumference will be given.   232

Dt 94 If on the diameter of a circle given in position a given point be taken, and from that point some straight line be drawn to the circle, and [if] from the point of section of the line a perpendicular be dropped to the line so drawn, and [if] through the point at which the perpendicular meets the circumference a parallel be drawn to the line so drawn,
then the point at which that parallel meets the diameter is given, and the rectangle contained be the parallel lines will be given. 234

◆

Supplementary definitions and lemmas.

Ax. 0* Any point or line segment or angle may be (taken and) appointed given.
25, 30, 94

Def. 1* A figure or a line or an angle is given in magnitude if and only if it is equal to one that is given. 29

Def. 1*A An angle is given in position and in magnitude if its vertex and one point on each side is given in position. 116

Def. 2* A ratio is given if it is the same as a given ratio. 33

Def. 3*A' A triangle is given in form if its vertices are given in position. 33

Def. 3*A A rectilineal figure is given in form if its vertices are given in position. 115

Def. 3*B' A triangle is given in form if it is similar to one that is given in form. 33

Def. 3*B A rectilineal figure is given in form if it is similar to one that is given in form. 115

Def. 4*A Given in position is said of points and lines and angles which do not *metapipt*. 93

Def. 4*B A point P is said to be given in position if and only if it is *assigned* to a point in the Plane so that no other point in the Plane can claim the name P. A line L is said to be given in position if and only if it is *assigned* to a line

LIST OF DEFINITIONS AND ENUNCIATIONS 265

in the Plane so that no other line in the Plane can claim the name L or do the same job.

An angle is given in position if its sides are given in position.   95

Dt 8*   Given ratios compound into a given ratio.   53

Dt 10*A   If a sum (AC) is greater than a part (BC) than in ratio, then the remainder (AB) will either be by a given greater than the part (BC) than in ratio, or the remainder (AB) will be less than a given by a magnitude which has to the part (BC) a given ratio.   64

Dt 24*A   If two squares have a given ratio to one another, their sides will also have a given ratio to one another.   89

Dt 24*B   If two line segments are given in magnitude, their rectangle is also given in magnitude.   90

Dt 24*C   If a square is given in magnitude, its side is also given in magnitude.   90

Dt 24*D   If $\square(A):\square(B) = \square(D):\square(E)$ then $A:B = D:E$.   90

Ax. 25*   If two lines given in position cut one another, their point of section is given in position.   97

Ax. 26*A If the extremities of a straight line be given in position, the line is given in position and in magnitude.   97

Ax. 26*B If two points are given, [the infinite straight line that passes through them is given in position; and] the segment whose endpoints are the two points is given in position and in magnitude; [and the circle with one of the points as centre and the segment as radius is given in position and in magnitude.]   99

Ax. 27*   If the one extremity of a straight line given in position and in magnitude be given, the other will also be given.   98

Ax. 29*   If a given angle has its vertex and its one side given in position, the other side is also given in position.   103

Dt 56*A Let P and Q be parallelograms with sides a, b and c, d respectively.
If   P is equiangular with Q
and   if e is the fourth proportional to a, c, and d (so that $a:c :: d:e$)
then $b:e :: P:Q$. 148

Dt 80*   If a triangle have a given angle, and the ratio of the perpendicular from the given angle to the base is given, the triangle will be given in form. 196

Dt 87*   In any circle a cord subtending a given angle has a given ratio to the diameter. 226

Dt 88*   In any circle a cord having a given ratio to the diameter subtends a given angle. 226

◆

Select Bibliography

Berggren, J. L. and Glen Van Brummelen. 2000. "The Role and Development of Geometric Analysis and Synthesis in Ancient Greece and Medieval Islam", in *Ancient & Medieval Traditions in the Exact Sciences: Essays in Memory of Wilbur Knorr*. Center for the Study of Language and Information. Stanford, California.

Fowler, David H. [1987] 2000. *The Mathematics of Plato's Academy*, 2nd ed. Oxford, Clarendon.

Fowler, David H. & C.M. Taisbak. 1999. "Did Euclid's Circles Have Two Kinds of Radius?" *Historia Mathematica* 26.

Heath, Thomas L. 1921. *A History of Greek Mathematics*, 2 vols., Oxford, Clarendon.

Heath, Thomas L. [1926] 1956. *The Thirteen Books of Euclid's Elements*, 2nd ed., 3 vols. Dover publ. (unabridged and unaltered from the 1926 edition, Cambridge, England).

Hartshorne, Robin. 2000. *Geometry: Euclid and Beyond*. Springer, New York etc..

Hertz-Fischler, Roger. 1984. "What are propositions 84 and 85 of *Euclid's Data* all about?" *Historia Mathematica* 11, 86–91.

Hogendijk, Jan P. 1987. "On Euclid's Lost *Porisms* and Its Arabic Traces". *Bolletino di Storia delle Scienze Matematiche* vol. VII, fasc. 1, 93–115.

Jones, Alexander. 1986. *Pappus of Alexandria, Book 7 of the Collection*. 2 vols. Springer, New York.

Knorr, Wilbur Richard. 1986. *The Ancient Tradition of Geometric Problems*. Birkhäuser, Boston.

Manitius, K. 1963. *Ptolemaeus, Handbuch der Astronomie*. Teubner, Leipzig (revised by O. Neugebauer).

McDowell, George L., Merle A. Sokolik. 1993. *The Data of Euclid* translated from the text of Menge; with Introduction by Richard Delahide Ferrier. Union Square Press, Baltimore.

Menge, H. 1896. *Euclids Data*, in *Euclidis Opera omnia* vol. VI, ed. (J. L. Heiberg et) H. Menge. Teubner, Leibzig.

Mueller, Ian. 1981. *Philosophy of Mathematics and Deductive Structure in Euclid's Elements*. MIT Press, Cambridge, Mass.

Netz, Reviel. 1999. *The Shaping of Deduction in Greek Mathematics.* Cambridge University Press.

Petersen, Julius. 1866. *Metoder og Theorier til Loesning af geometriske Konstruktionsopgaver*. Copenhagen, Schoenbergs. English translation *'Methods and Theories for the Solution of Geometrical Constructions*. London, Samson Low et al.

Ptolemy-Heiberg, J. L. Heiberg, *Claudii Ptolemaei Opera quae exstant omnia*, Vol. I. *Syntaxis Mathematica*. 2 vols. Teubner, Leipzig 1898-1903.

Ptolemy-Toomer, G. J. Toomer. 1984. *Ptolemy's Almagest*. London: Duckworth.

Saito, Ken. 1986. "Compounded Ratio in Euclid and Apollonius." *Historia Scientiarum* no. 31.

Taisbak, Chr. Marinus. 1982. *Coloured Quadrangles. A Guide to the Tenth Book of Euclid's Elements*. Museum Tu.sculanum Press, Copenhagen 1982.

Taisbak, Chr. Marinus. 1986. "Zeuthen and Euclid's Data 86. Algebra – or a Lemma about intersecting Hyperbolas?" *Centaurus* vol. 38.2-3.

Taisbak, Chr. Marinus. 1991. "Elements of Euclid's Data." In *Peri ton Mathematon*, ed. by Ian Mueller (= Apeiron vol. XXIV no. 4). Alberta, Canada.

Thaer, Clemens. 1962. *Die Data von Euklid* nach Menges Text aus dem Griechischen uebersetzt. Springer, Berlin-Goettingen.

Toomer, G. J. 1990. *Apollonius, Conics, Books V–VII*. Springer.

Vitrac, Bernard. 1990–2001. *Euclide d'Alexandrie, Les Eléments*. Traduction et commentaires. 4 vols. Presses Universitaire de France, Paris.

Index

addition of ratios 86
analysis 8, 15, 26, 95, 127, 174, 248, 249, 267
anapalin 45
anastrepsanti 45, 46, 48
angles 30, 102, 116
Apollonius' circle 59
appointment 25, 31
Archimedes' helix 94
assign 31, 95, 96
choría 30
clone 38
common denominator 86
complements 156
compound 45, 53, 149
compound ratio 53, 149, 166, 179, 201
coordinate 20, 25, 97
cosine law 168
dedoménon 95
di' isou 45, 53
diagonal notation 10
dielonti 45, 51
enallax 37, 38
endless loop 25
epidemic 45
equal 26
equality 23-26
Eudoxus 58
ex aequali = di'isou 53
existence 19, 95, 152, 153, 208
fallacy 102
formula 26
fourth proportional 40
Fried, Michael N. 235
Gardies, Jean Louis 41
gnomonic situations 155
Hartshorne, Robin 26, 267

Heath, Thomas L. 16, 17, 26, 90, 91, 94, 137, 152, 166, 171-173, 208, 227, 233, 267
height 135
Helping Hand 28
Heron's formula 122
Herz-Fischler, Roger 208
isosceles adjunct 131, 173
Knorr, Wilbur R. 7, 15, 16, 126, 267
known 18, 24, 98, 214, 241
latent co-actor 24, 26, 40, 48, 51, 62
locomotion 94
magnitude 20, 30
magnitudinified 32
Manitius 13
Marinus 7, 18, 27, 103, 241
mathematicals 96
McDowell, George L. & Merle A. Sokolik 7
Menge 7, 51, 63, 122, 125
metapipt 93, 94, 97, 98, 100, 101, 104
missing proofs 90
movement 98
Mueller, Ian 19, 27, 32, 40, 53, 94, 95, 98, 102, 136, 137, 166, 268
multiplication of ratios 86
Netz, Reviel 11, 109, 151, 153, 268
order relation 58, 77
Pappus 27, 39, 165, 192, 196, 249, 267
parallel results 16, 126
petitio principii 42, 44
Plane 19, 25, 27, 30, 47, 94
Plato 27
Proclus 20, 91, 96, 101, 242, 243
provide 27
Ptolemy 13, 29, 103, 122, 128, 152, 222, 226, 233, 243, 268
Ptolemy's theorem 233
Q.E.D. 8, 237
ratio 31
Realm of Intelligence 28
repeated construction 105
representative 31
ruler and compasses 27

Saito, Ken	15, 41, 45, 53, 222, 224, 268
same	23-25, 31
schema	156
Schmidt, Olaf	7
similarity	20
symmetrical positions	105, 229
symmetry	52, 130
synthenti	45, 49
tacit assumption	14, 15, 18, 25, 151
Thaer, Clemens	7, 26, 45, 51, 58, 67, 68, 76, 80, 83, 88, 122, 154, 158, 160, 165, 168, 192, 196. 268
Tool Box	15, 38
Toomer, G. J.	13, 243, 268
transitivity	52
trigonometry	128, 153, 159, 160, 166, 168, 170, 226
trisect	13, 29, 40, 102
uniqueness	95
Vitrac, Bernard	23, 41, 43, 44, 48, 86, 109, 137, 268
Zeuthen, H. G.	19, 211, 212, 214, 219